# FRONTLINE AND FACTORY

# *Archimedes*

## NEW STUDIES IN THE HISTORY AND PHILOSOPHY OF SCIENCE AND TECHNOLOGY

VOLUME 16

### EDITOR

JED Z. BUCHWALD, *Dreyfuss Professor of History, California Institute of Technology, Pasadena, CA, USA.*

### ADVISORY BOARD

HENK BOS, *University of Utrecht*
MORDECHAI FEINGOLD, *Virginia Polytechnic Institute*
ALLAN D. FRANKLIN, *University of Colorado at Boulder*
KOSTAS GAVROGLU, *National Technical University of Athens*
ANTHONY GRAFTON, *Princeton University*
PAUL HOYNINGEN-HUENE, *University of Hannover*
EVELYN FOX KELLER, *MIT*
TREVOR LEVERE, *University of Toronto*
JESPER LÜTZEN, *Copenhagen University*
WILLIAM NEWMAN, *Harvard University*
JÜRGEN RENN, *Max-Planck-Institut für Wissenschaftsgeschichte*
ALEX ROLAND, *Duke University*
ALAN SHAPIRO, *University of Minnesota*
NANCY SIRAISI, *Hunter College of the City University of New York*
NOEL SWERDLOW, *University of Chicago*

*Archimedes* has three fundamental goals; to further the integration of the histories of science and technology with one another: to investigate the technical, social and practical histories of specific developments in science and technology; and finally, where possible and desirable, to bring the histories of science and technology into closer contact with the philosophy of science. To these ends, each volume will have its own theme and title and will be planned by one or more members of the Advisory Board in consultation with the editor. Although the volumes have specific themes, the series itself will not be limited to one or even to a few particular areas. Its subjects include any of the sciences, ranging from biology through physics, all aspects of technology, broadly construed, as well as historically-engaged philosophy of science or technology. Taken as a whole, *Archimedes* will be of interest to historians, philosophers, and scientists, as well as to those in business and industry who seek to understand how science and industry have come to be so strongly linked.

# Frontline and Factory: Comparative Perspectives on the Chemical Industry at War, 1914–1924

*Edited by*

ROY MACLEOD
*University of Sydney*

and

JEFFREY ALLAN JOHNSON
*Villanova University*

A C.I.P. Catalogue record for this book is available from the Library of Congress.

TP
20
.F765
2006

ISBN-10 1-4020-5489-0 (HB)
ISBN-13 978-1-4020-5489-1 (HB)
ISBN-10 1-4020-5490-4 (e-book)
ISBN-13 978-1-4020-5490-7 (e-book)

Published by Springer,
P.O. Box 17, 3300 AA Dordrecht, The Netherlands.

*www.springer.com*

*Printed on acid-free paper*

All Rights Reserved
© 2006 Springer
No part of this work may be reproduced, stored in a retrieval system, or transmitted
in any form or by any means, electronic, mechanical, photocopying, microfilming,
recording or otherwise, without written permission from the Publisher, with the exception
of any material supplied specifically for the purpose of being entered and executed
on a computer system, for exclusive use by the purchaser of the work.

'Above all, we still owe a word to those who work in the powder and explosives factory, who in the war economy stand, so to speak, right at the frontline – and did so long before 1939.'

'Ein Wort müssen wir nun vor allem noch den Menschen widmen, die hier in der Pulver- und Sprengstoff-Fabrik arbeiten und die im kriegswirtschaftlichen Sinne sozusagen direkt an der Front stehen – und zwar schon lange vor 1939.'

V. Muthesius (Berlin, 1941)

# CONTENTS

Preface ix

Illustrations xi

Introduction xiii
*Jeffrey Allan Johnson and Roy MacLeod*

## I. Mobilization and Industrial Policy

1. Technological Mobilization and Munitions Production: Comparative Perspectives on Germany and Austria 1
   *Jeffrey Allan Johnson*

2. Mobilization and Industrial Policy: Chemicals and Pharmaceuticals in the French War Effort 21
   *Sophie Chauveau*

3. First World War Explosives Manufacture: The British Experience 31
   *Wayne D. Cocroft*

4. Transforming a Village into an Industrial Town: The Royal Prussian Powder Plant in Kirchmöser (Brandenburg) 47
   *Sebastian Kinder*

## II. Systems of Innovation: Allies and Neutrals

5. Wartime Chemistry in Italy: Industry, the Military, and the Professors 61
   *Giuliano Pancaldi*

6. Munitions, the Military, and Chemistry in Russia 75
   *Nathan M. Brooks*

7. Technical Expertise and U.S. Mobilization, 1917–18: High Explosives and War Gases 103
   *Kathryn Steen*

8. Operating on Several Fronts: The Trans-National Activities of Royal
   Dutch/Shell, 1914–1918    123
   *Ernst Homburg*

### III. Science and Industry at War: Contrasting Styles

9. Kuhlmann at War, 1914–1924    145
   *Erik Langlinay*

10. Organizing for Total War: DuPont and Smokeless Powder
    in World War I    167
    *John Kenly Smith, Jr.*

11. Science and the Military: The Kaiser Wilhelm Foundation
    for Military-Technical Science    179
    *Manfred Rasch*

12. Managing Chemical Expertise: The Laboratories of the French
    Artillery and the Service des Poudres    203
    *Patrice Bret*

13. The War the Victors Lost: The Dilemmas of Chemical Disarmament,
    1919–1926    221
    *Jeffrey Allan Johnson and Roy MacLeod*

Epilogue    247
*Seymour Mauskopf*

Bibliography    257

Notes on Contributors    269

Index    271

# PREFACE

It has been said that history is a debate between the present and the past about the future. Nowhere are these lines drawn more significantly than in the study of science and war. And nowhere is the discourse more relevant, than in the study of science and technology as foundations and multipliers of military power. This book is concerned with one particularly seminal aspect of this development — the history of chemical munitions during and immediately after the First World War. The Great War, as it came to be known, was not the first industrial war, but it was the first to involve all the major industrial nations of the world. Within four years, the world witnessed unprecedented feats of industrial development, many of which drew upon and extended pre-war reservoirs of scientific and technological knowledge. The experience comes down to us as a *conjuncture* of scientific, economic, political and, ultimately, military departures, which by their nature involved new ways of meeting crises, and eventually new forms of critical thinking. That these new forms emerged only gradually and unexpectedly is not to underestimate their capacity to endure, or to minimize their relevance. From the Great War came patterns, assumptions, and practices which were to make an indelible mark on science and technology for the rest of the twentieth century and beyond.

To understand how different nations, with different industrial and technical traditions, came to contrive very similar weapons, for very similar purposes — producing, in the process, very different outcomes — requires us to take a comparative view of developments from laboratory to field, and across the 'front', during the period 1914–1920. What for historians of science has often been glibly accepted (or dismissed) as a 'chemists' war', was in fact a complex history of industrial systems and economic networks, of scientific disciplines and government departments. Taken as a whole, their history formed part of the collective experience that produced the first modern 'academic-military-industrial complex'. They reflected, in particular, the changing role of munitions production in the structure of modern industry. Together, they informed a new discourse – of technological mobilisation, industrial efficiency, technology transfer, and 'dual use' – that began during the war, and gathered impetus in its wake.

To improve our grasp of these issues, we convened a workshop at the Institute of Advanced Studies of the University of Bologna in June 2002. To this we invited colleagues from Italy, France, Britain, Germany, The Netherlands, and the United States. We would like to express our warm thanks to the Director of the Institute, and especially to Professor Roberto Seazziere. Complementing facilities contributed by the Institute came generous assistance for American participants from the National

Science Foundation, and for French colleagues, from the CNRS. We wish to express our thanks to these and other organisations, including English Heritage, for their timely attempts to retrieve the architectural heritage of munitions manufacture.

We wish also to express our thanks to all who joined in our discussions, not least those who are not with us at the end of the day. At Villanova University, we are grateful for the assistance of the Research Office and the Department of History, and at the University of Sydney, we are grateful for the superb assistance of our interlibrary loan librarians, and for the help of Christopher Hewitt, Alanta Colley and Penelope Grist.

*Roy MacLeod*
*Jeffrey Johnson*

ILLUSTRATIONS

**Chapter 2.   (Chauveau)**

Table 1.   French Pharmaceuticals Production, 1913–1917        26

**Chapter 3.   (Cocroft)**

Figure 1.   The Distribution of Explosives Factories during the First
            World War                                           32
Figure 2.   Royal Naval Cordite Factory, Holton Heath, Dorset   34
Figure 3.   Chilwell, Nottinghamshire, National Filling Factory No.6   37
Figure 4.   Cunard's Shellworks, Liverpool                      40
Figure 5.   Quedgeley, Gloucestershire, National Filling Factory No.5   41
Figure 6.   Royal Gunpowder Factory Waltham Abbey, Essex        42
Figure 7.   His Majesty's Factory, Avonmouth, near Bristol      43

**Chapter 4.   (Kinder)**

Figure 1.   German State-owned Powder Plants, 1914–1918         49
Figure 2.   Layout of the Royal Prussian Powder Plant in Kirchmöser   53
Figure 3.   Output development at the Royal Prussian Powder Plant
            in Kirchmöser, 1916–1918                            55

**Chapter 8.   (Homburg)**

Figure 1.   The Process of Extracting Aromatic Intermediates from Borneo
            Oil for the Dyestuffs and Explosives Industries     129
Figure 2.   The Royal Dutch Plant for the Production of TNT     133

**Chapter 9.   (Langlinay)**

Table 1.   French Heavy Industry, 1913–18 (in millions of tons)   148
Table 2.   French Imports of Sulfuric Acid (metric tons)        149
Table 3.   Compagnie National des Matières Colorantes, 1917 Main
           share-holders and members of the board               153
Table 4.   Établissements Kuhlmann-CNMC, 1925                   158
Table 5.   Établissements Kuhlmann Finances, 1900 to 1925       161
Figure 1.   Éstablissements Kuhlmann:. Annual Production, 1873 to 1924   156
Figure 2.   Éstablissements Kuhlmann:. Dyestuffs Production, 1919 to 1924   157

xi

## Chapter 12. (Bret)

| | | |
|---|---|---|
| Figure 1. | Chemical Explosives Production Sites in France, 1914–1918 | 204 |
| Figure 2. | Studies of German Artillery Material, 1916–1918 | 212 |
| Figure 3. | Chemical Establishments in France, 1914–1918 | 213 |
| Table 1. | Studies and Analyses of the Chemical Laboratory of the Artillery Technical Section LSTA | 215 |

# INTRODUCTION

The First World War was a significant turning point in modern history, in many ways marking, to use the well-worn phrase, the coming of 'nowadays'. With the end of three empires and the re-making of a fourth, and the rise of the United States to the status of a great power, came impressive transformations in military technique and technology that rendered many traditional ideas of warfare obsolete. The war embraced collective sacrifice on a vast scale. Ten million dead stood as a reproach to humanity. At the same time, the world brought a new set of relationships between knowledge and power.[1] Among those most affected were the institutions of science, and the industries concerned with the manufacture, supply and use of what later generations would call weapons of mass destruction. The consequences of this new relationship were profound. For some, it registered the idea that modern war was becoming simply too terrible to contemplate.[2] For others, it marked merely a fearful instalment in the history of the 'warfare state'.[3] It is within the context of this emerging dilemma and debate that this book has been written.

Historians have viewed the Great War in many ways. Yet, no one fresh to the subject now begins without reference to the fact that this was, in unprecedented ways, an industrial war, or as Hew Strachan puts it, a '*Materialschlacht*', a 'battle of materials'.[4] Armaments, produced in massive quantities, dominated Europe. Mobilisation led the conversion of peacetime producers into a war economy, which found both purpose and profit in working raw materials into weapons. For the first time, nation states were unable to look to the past for military solutions. As van Creveld has put it, warfare became completely permeated by technology, and governed by it. Thirty years of industrial progress preceding 1914 had put all nations on what van Creveld calls a technological 'treadmill', which required constant attention to remain in the same place.[5] The result came unevenly, but ineluctably. The machine gun, with its crew of two or three, produced a level of firepower equal to a Napoleonic battalion. Similar developments brought a vast increase in logistic requirements, with major implications for training, equipment, and lines of communication and supply. In its wake, came the development and repositioning of massive industrial infrastructures, throughout Europe, and across the Atlantic.

Unprecedented destruction was an inevitable result of this new 'war of *matériel*'. By the end of November 1914, the war had seen the end of pre-war strategic assumptions based on men and mobility. The Western Front had become a warren of trenches and obstacles, with armies reduced to deadlock and stalemate. Innovations in weapons were driven by tactical necessities. New weapons systems dominated the terrain. Above all, there was the artillery, as Strachan has put it, 'In the First World War, the artillery was the agent of industrial and social mobilization'.[6] While the tragedies of gas warfare remain indelibly printed on war art and literature, it was shrapnel and high explosive that were responsible for 80% of the casualties on the Western Front.[7] At the core of this contest was the 'shell', singing in its different calibres, a chorus of destruction.[8] The sounds of shell-fire – what Sassoon called the 'din of war' – brought to all sides a fear and loathing of modern technologies, and a contempt for those who produced them. At the same time, they were talismanic objects, the quintessence of modernism, yet the embodiment of tradition. Shells were the servants of the military, but the system that produced them was the work of science and commerce. Among its outcomes, were the ruins of Verdun and the carnage of the Somme. Their effects made a lasting impression on modern memory. By 1916, the stalemate seemed as if it would last forever. Then, the sciences came into their own.[9] The 'protection' ironically afforded by the trenches gave time needed to launch a second mobilisation. And among the industries most affected, those of chemistry led the way.

For most observers of the Great War, the chemical industry was the 'ghost in the machine'. And although poison gas gave the war its chemical sobriquet, and its chemists everlasting notoriety, it was the explosives industry that claimed most attention, most resources, and most lives. In both cases, the chemical industry classically exemplified the nature of 'dual-use' technologies, well expressed by Brigadier (Sir) Harold Hartley, when overseeing the perfectly preserved Rhineland factories of defeated Germany in 1919: 'In the future ... every chemical factory must be regarded as a potential arsenal ...., ' he said.[10] Chemicals good for agriculture were equally useful in making explosives. After the war, no one could turn back the clock, although the victors at Versailles tried to do so. A new form of warfare had begun.

Between 1914–1919, the chemical industries fought as intensely as the military. But the history of their struggle is not well known and, beyond post-war apologetics, has never been thoroughly assessed. To do so requires fresh research, and at least three critical elements—new sources, a comparative perspective, and a set of fresh questions.

Within the last twenty years, and certainly since the end of the Cold War, it has become possible to consult records to which earlier generations of historians had no access. These give us a chance to re-examine the emerging relationship between science-based industry and modern industrial warfare, as it changed in ways which were all the more remarkable for being unplanned and unprecedented in speed, scale and impact. However, many of these sources remain geographically and commercially difficult to consult, and their successful use requires a degree of familiarity born of long association.

For this reason, the subject is best approached with a comparative perspective. Indeed, to neglect this dimension would risk missing points that, in their different ways, French, British, American and German historians consider essential. We agree with Niall Ferguson, writing in a different context, that 'The literature on war ... perfectly illustrates the dangers of writing national history without adequate comparative perspective'.[11] Accordingly, it is essential to cast our exercise in terms of the advancing front of scholarship on what French scholars call 'l'histoire totale de guerre'.[12] As the war became increasingly 'total' – which is to say, as opposing nations sought to mobilize resources of all kinds – solutions to unplanned problems provided at least part of the key to victory or defeat.[13] In this changing picture, the resources of science came to play an increasingly important role, as did the leadership of industrial science and engineering, which came to enjoy a high degree of integration with government and the military. Its history enables us to assess the extent to which the war accelerated the emergence of a 'triple helix' between science, industry and government, with its many consequences for an emerging 'academic-military-industrial' complex.[14]

To explore these relationships, with all the fresh materials and perspectives at our disposal, we have found it useful to cast our questions in a comprehensive context, and to look at the collective experience of the belligerent powers – and also the leading neutrals—in terms of what we can identify and describe as 'national systems' of applied science and production. In appropriating the idea of 'national systems', we have drawn on language made popular by Thomas Hughes, Richard Nelson, Nathan Rosenberg, and Giovanni Dosi, who have in different ways explored more recent 'national systems of innovation'.[15] The 'national systems' that dominated the 1914–18 War were products of a century of imperial and international development, which brought in train a world economy driven in large part by free trade.

At first, the coming of war disturbed few illusions – except perhaps the idea that any such war was absurd, and would be over quickly – but by early 1915, it was clear that many critical chemicals, from nitrates and sulphates to dyes and pharmaceuticals, were in short supply on all sides. All belligerents proceeded to make major adjustments, creating new organizations, rationalizing resources, replacing essential materials with import-substitutes, and expanding the production of war *materiel*. On both sides, new systems were grafted onto old, and new systems were created, shaped by and shaping developments in the war. Month by month, year by year, these competing systems were brought to the test. Efficient mobilization, followed by rapid deployment, could make long-term deficiencies irrelevant; equally, inefficient mobilization, or inefficient lines of production and supply, could nullify the advantages of the richest nation in the world.

A critical factor in mobilizing chemical munitions was the industry's relative level of what Hughes has called 'technological momentum'.[16] From late 1914, the chemical industry slowly adjusted itself— reluctantly, as we shall see, in Germany; and with difficulty in France and Britain – to wartime demands. The 'technological systems' of industrial warfare required not only plant and equipment, but also a supporting 'technological culture'. In each belligerent country, a national 'system' emerged, acquiring 'momentum' through expansion and innovation to overcome obstacles to growth ('reverse salients', in Hughes' terminology). In the broadest sense, the characteristics of a country's 'technological system' were reflected in rates of innovation and use of scientific expertise, resources and organization. They were also reflected in the changing relationship between military and civil production, and in the conflicting priorities of the public and the private sector. Looking at these factors across national frontiers can tell us much about relative 'success' and 'failure'—and the relative values accorded research, development, testing, safety and speed. These were the 'winning weapons' in the new business of 'machine warfare'. These were also to become vital to the success of the chemical industry for at least the next fifty years.

+++++

In examining different national responses to the challenges of war, it would be ideal to take comparative cases that consider the same questions, similar institutions, and at similar moments in time. This ideal we have not achieved. Instead, as a first step, we have assembled a series of national studies that look – in some cases, for the first time – at emerging systems of innovation, and at their immediate consequences, during the war and the *après-guerre*. Using a European perspective, each author has looked through a slightly different national lens to explore comparative questions of mobilization and production. Together, they offer a range of insights into the many ways in which competing national systems deployed their differing traditions, where possible to advantage, in attempting to meet similar objectives.[17] That they did so at different rates, and with differing levels of success, forms an important part of our story. We have focused on the Western Front, although many of the features of industrial warfare were equally evident in other theatres, and outside Europe. Given the particular circumstances of static warfare in these areas, the industrial effort of the Allies and the Central Powers concentrated on an expanding variety and quantity of guns, shells and fuses.

The war was not, of course, an entirely 'chemical war', and no serious historian questions the relevance of other material resources – notably, steel and brass for shell manufacture, and the coal, iron and copper on which they were based – nor the wide range of other wartime technical innovations, including aircraft, tanks, and submarines, let alone the contributions made by agriculture, medicine, and the social sciences.[18] However, in ways that parallel the interests of artists, anthropologists, and cultural historians, it is important to make a start, and in this case, by examining how the great 'chemical munitions feat' was conceived, contrived, and remembered. In its overwhelming military success, it became a foretaste of the future use of science as a weapon against humanity, and in its widest expression, a tragic consequence of the Enlightenment.[19]

For convenience, the book is organized in three parts. In Part I, we examine the principal systems that went to war in 1914. Comparing the Central Powers, the German and the Austrian systems, Jeffrey Johnson follows two phases of the war – the first in response to the emergencies of 1914, the second, in response to the dramatic intensification of war in the West in 1916. A critical factor in saving Germany's war effort from collapse in 1915 was the capacity of its dye industry to adapt quickly, with relatively little change, to wartime needs. It was difficult to alter the 'technological momentum' of an industry dependent upon international markets, in which the Entente Powers were major customers. Yet, as Johnson shows, 'dual-use' technologies were the key factor in enabling peacetime industry to meet wartime needs. After the summer of 1916, both the German and Austrian systems were 'overtaxed' by the 'second mobilization' – the Hindenburg programme and its Austrian equivalent – with consequences that were to prove disastrous for both countries.

Comparable causes of relative success and 'failure' are recounted in the essays by Sophie Chauveau (who studies French mobilization in the pharmaceutical industry), by Wayne Cocroft (who surveys British chemical munitions, from the perspective of industrial archaeology), and by Sebastian Kinder, who has carefully revisited a Prussian powder factory near Berlin. The British followed the French in creating a virtually new industrial system, using a mixture of public and private strategies, and mobilizing technical expertise from the Empire, which was almost completely dissipated in the years following the war. In Germany, wartime mobilization was followed by successful re-conversion, a process which, as Kinder hints, is continuing to this day.

In Part II, four studies pose broadly similar questions in parallel, rather than in competing, contexts – Tsarist Russia, Italy and the United States; and The Netherlands which, as a neutral, remained precariously poised between the Allies and the Central Powers. Giuliano Pancaldi shows how the war brought a vital stimulus to Italian chemistry, which quickly disappeared when the conflict ended. Nathan M. Brooks traces the role of central government in managing munitions science in Tsarist Russia, while Kathryn Steen shows how American industry adapted private sector methods to meet Allied, then American military demands. Ernst Homburg shows how Dutch industry prospered during the war, delicately balancing profits and principle in its commitment to global trade.

Finally, the five studies in Part III explore contrasting national systems and their intersection with military and political objectives. Eric Langlinay shows how, with mixed success, one of France's largest chemical firms sought competitive advantage from wartime circumstances. John Kenly Smith, Jr., shows, through the history of DuPont – the largest American explosives manufacturer – how American industry decisively shifted the balance of power between the competing systems. In this light, the lesser-known history of Germany's scientific mobilization, as reconstructed by Manfred Rasch, comes in some ways as an unexpected story of institutional delay and poor communication between science and the military, with far-reaching consequences for Germany. Against this, Patrice Bret emphasizes the close relationship between science and the military in France, particularly in the context of the remarkable *Service des Poudres* – not alone, but in large part responsible for the 'miracle' that was the French response to the loss of its major industrial centres and natural resources to Germany at the start of the war. In the Epilogue, we have invited Seymour Mauskopf to reflect on wider European comparisons illuminated by the papers, not least in their relation to the *longue durée* and earlier visions of 'total war'. We are grateful to him for this exercise in stressing points that are particularly relevant to an understanding of the history of chemistry and chemical munitions.

In the concluding chapter, we return to some of these considerations in the light of the 'chemical occupation' of Germany in 1919 and afterwards. It is well known that, after Versailles, disarmament became a hopeless cause.[20] It is less well appreciated that it was doomed at birth. Chemical disarmament was, in turn, an illusion born of diplomacy and politics, rather than an idea nurtured by science or industry. In seeking a lasting peace, the victors sowed the seeds of a future war. That war, and its aftermath, would raise in a new form the same questions concerning the proliferation of 'dual use' technologies, and the consequences of an increasingly close relationship between science, government, and the military throughout the industrialised world.

## NOTES

[1] A fact fully recognised at the time. See A.N. Whitehead, *Science and the Modern World* (Cambridge: Cambridge University Press, 1926), 298 *et seq.*; see also Lewis Mumford, *Technics and Civilization* (New York: Harcourt Brace, 1934), 84–87, for its meaning in reverse: 'modern industrialism may equally well be termed a large scale military operation' (84).

[2] For a powerful expression of the apocalyptic view, see Robert L. O'Connell, *Ride of the Second Horseman: The Birth and Death of War* (New York: Oxford University Press, 1995).

[3] David Edgerton, *Warfare State: Britain, 1920–1970* (Cambridge: Cambridge University Press, 2006).

[4] The literature is extensive. By way of introduction, see Roger Chickering and Stig Förster (eds.), *Great War, Total War: Combat and Mobilization on the Western Front, 1914–1918* (Cambridge: Cambridge University Press, 2000; Hew Strachan, *The First World War, Vol. I: To Arms* (Oxford: Oxford University Press, 2001); Trevor Wilson, *The Myriad Faces of War* (Cambridge: Polity Press, 1986 ); Robin Prior and Trevor Wilson, *The First World War* (London: Cassell, 2001); David Stevenson, *Armaments and the Coming of War: Europe, 1904–1914* (Oxford: Clarendon Press, 1996).

[5] Martin van Creveld, *Technology and War, from 2000 BC to the Present* (New York: Free Press, 1989), 224.

[6] Hew Strachan, 'La Notion de Puissance de Feu: La Revolution de l'Artillerie', delivered to the Conference on 'La Mobilisation de la Nation', Centre des Hautes Études de l'Armement (CHEAr), Paris, 26–28 October 2004.

[7] On the artillery war, see Tim Travers, *The Killing Ground: The British Army, the Western Front, and the Emergence of Modern Warfare, 1900–1918* (London: Unwin Hyman, 1987); David Zabecki, *Steel Wind: Colonel Georg Bruchmüller and the Birth of Modern Artillery* (Westport: Praeger, 1994); Erich Schoen, *Geschichte des Deutschen Feuerwerkswesen der Armee und Marine mit Einschluss des Zeugwesens* (Berlin: Paul, 1936); J. B.A. Bailey, *Field Artillery and Firepower* (Annapolis.: Naval Institute Press, 2nd ed., 2004); Frank E. Comparato, *Age of Great Guns: Cannon Kings and Cannoneers who Forged the Firepower of Artillery* (Harrisburg, PA: The Stackpole Co, 1965).

[8] Nicholas Saunders, 'The Ironic 'Culture of Shells' in the Great War and Beyond', in John Schofield, William G. Johnson, and Colleen Beck (eds.), *Matériel Culture: The Archaeology of 20th Century Conflict* (London: Routledge, 2002), 22–40; and N.J. Saunders (ed.), *Materialities of Conflict: Anthropology and the Great War, 1914–2002*. London: Routledge, 2004).

[9] Nowhere better encapsulated than in a work published for the US National Academy of Sciences, R[obert] M. Yerkes (ed.), *The New World of Science: Its Development during the War* (New York: The Century Co., 1920).

[10] National Archives (Kew) MUN 4/6737, H. Hartley, 'Report of the British Mission Appointed to Visit Enemy Chemical Factories in the Occupied Zone Engaged in the Production of Munitions of War' (typescript, 26 February 1919), 13. A longer, printed version is in MUN 4/7056; the first part, containing this quotation, was published in 1921.

[11] Niall Ferguson, *The Pity of War: Explaining World War I* (New York: Basic Books, 1999), 256–257.

[12] Jay Winter, Geoffrey Parker, and Mary R. Habeck, *The Great War in the Twentieth Century* (New Haven: Yale University Press, 2000), 9.

[13] Hew Strachan, 'On Total War and Modern War', *International History Review*, XX (2000), 341–370.

[14] For the re-invention of the 'Triple Helix' model in our time, see Loet Leydesdorff and Henry Etzkowitz, 'Triple Helix of Innovation: Introduction', *Science and Public Policy*, XXV (6), (1998), 358-364; Henry Etzkowitz and Loet Leydesdorff, 'The Endless Transition: A 'Triple Helix' of University-Industry-Government Relations', *Minerva*, XXVI (3), (1998), 203–218; Loet Leydesdorff and Henry Etzkowitz, 'The Triple Helix as a Model for Innovation Studies', *Science and Public Policy*, XXV (3), (1998),195–203; Loet Leydesdorff and Henry Etzkowitz (eds.), *A Triple Helix of University-Industry-Government Relations: The Future Location of Research?* (New York: Science Policy Institute, State University of New York, 1998); Henry Etzkowitzand Loet Leydesdorff (eds.), *Universities and the Global Knowledge Economy: A Triple Helix of University-Industry-Government Relations* (London: Cassell, 1997).

[15] Thomas Parke Hughes, *Networks of Power: Electrification in Western Society, 1880–1930* (Baltimore: Johns Hopkins University Press, 1983); and 'Technological Momentum in History: Hydrogenation in Germany, 1898–1933', *Past & Present*, No. 44 (August 1969), 106–132; Richard R. Nelson, *National Innovation Systems: A Comparative Analysis* (New York: Oxford University Press, 1993); Giovanni Dosi, *Technical Change and Industrial Transformation* (New York: St. Martin's Press, 1984); Giovanni Dosi, Keith Pavitt, and Luc Soete (eds.), *The Economics of Technical Change and International Trade* (New York: New York University Press, 1990); Giovanni Dosi, Renato Giannetti, and Pierangelo Maria Toninelli (eds.), *Technology and Enterprise in a Historical Perspective* (Oxford: Oxford University Press, 1992); Giovanni Dosi, *Innovation, Organization and Economic Dynamics: Selected Essays* (Cheltenham: Edward Elgar, 2000).

[16] Hughes, 'Technological Momentum', *op. cit.* note 15.

[17] For a comparison of wartime 'mentalities', see Jeffrey A. Johnson and Roy MacLeod, 'War Work and Scientific Self-Image: Pursuing Comparative Perspectives on German and Allied Scientists in the Great War', in Rüdiger vom Bruch and Brigitte Kaderas (eds.), *Wissenschaftssystem und Wissenschaftspolitik: Bestandsaufnahmen zu Formationen, Brüchen und Kontinuitäten im Deutschland des 20. Jahrhunderts* (Stuttgart : F. Steiner, 2002),169–179.

[18] See also J.P. Harris, *Men, Ideas and Tanks: British Military Thought and Armoured Forces, 1903–1939* (Manchester: Manchester University Press, 1995).

[19] For instructive cultural comparisons, see Douglas Mackaman and Michael Mays (eds.), *World War I and the Cultures of Modernity* (Jackson, Miss: University Press of Mississippi, 2000).

[20] For a recent treatment of these issues, see Richard Shuster, *German Disarmament after World War I: The Diplomacy of International Arms Inspection, 1920–1931* (London: Routledge, 2006).

JEFFREY ALLAN JOHNSON

# TECHNOLOGICAL MOBILIZATION AND MUNITIONS PRODUCTION: COMPARATIVE PERSPECTIVES ON GERMANY AND AUSTRIA

## INTRODUCTION

The First World War is rightly considered a milestone in twentieth-century history, marking a fundamental change in the system of production of modern industrial nations. This paper examines the relative successes and failures of the German and Austro-Hungarian systems in mobilizing their chemical industries for the development and production of explosives and propellants for munitions. The key to 'chemical mobilization' lay in part in the speed and efficiency with which industrial systems could grasp the inherent possibilities of their peacetime capacities. Once cut off from international trade, and involved in an extended war of attrition, these systems faced serious disadvantages in resources, both human and material. But superiority in technological mobilization was critical to victory. This required new measures to increase the productive capacity of existing private and government-owned plants, while finding the resources to supply them, and while converting peacetime chemical industries to wartime use.

The most important example of this conversion took place in the dyestuffs industry, one of Germany's most profitable and strongest near-monopolies. Following Thomas Parke Hughes, the successful conversion of an industry depended in part upon its 'technological momentum', a measure of the pace and direction of its development.[1] In the case of the German dye industry, despite the high prewar profitability of dyes, and the extensive factories and staff devoted to their production, the industry had only a relatively low technological momentum, as most of its facilities were inherently flexible. Hence its initial conversion to wartime production presented only minor problems, and the industry adapted itself smoothly to the unfamiliar needs of a war economy.

This paper will focus upon Germany's – and to a lesser extent, Austria's – wartime experience.[2] The story is a familiar one of military unreadiness and belated administrative leadership, coupled with a less familiar history of industrial flexibility in plant design and apparatus, a capacity to survive losses of raw materials and intermediates, and an ability to improvise and innovate in a domain of chemicals needed for explosives production. The war was also to expose stark contrasts between Germany's privately-owned research-intensive dyestuffs industry, and the less sophisticated Austrian chemical industry, in which state-owned or controlled factories played a dominant but less efficient role.

Wartime chemical mobilization in Germany and Austria took place in two stages, reflecting the intensification of the war. The first began as early as August 1914, in response to the unexpectedly high demands for munitions resulting from the ill-fated Schlieffen Plan. This 'makeshift mobilization,' which gathered speed by the spring of 1915, involved the impromptu completed expansion of temporary facilities. Anticipating an early return to peace, Germany's major dyestuffs firms refused for several months to manufacture explosives, until excess capacity led them reluctantly to respond to an explicitly temporary objective. Then, for nearly fifteen months, the Central Powers developed their munitions industries incrementally. However, in the autumn of 1916, following the losses at Verdun and on the Somme, came a second major 'mobilization, driven by the grandiose demands of the Hindenburg Programme, which called for dramatic increases in munitions production. As Hew Strachan has pointed out, this effort arose from the concept of *Materialschlacht*, the 'battle of materials'.[3] By the end of the war, Germany was beginning to build permanent, efficient 'readiness plants', which were to be converted to peacetime use after the war, but which were to be kept ready for quick re-conversion. In this way, Germany committed itself to something close to 'total' economic planning, and pressured the Austrians to follow suit.

In retrospect, these efforts came too late, and went too far, overtaxing the infrastructures and resources of their weakened systems and ultimately contributing to the collapse and defeat that so many pre-war observers had feared. Nonetheless, in some respects the Hindenburg Program substantially aided the German chemical industry's recovery in the postwar era, in particular by expanding its capacity to produce nitrates that could also be used for fertilizer. At the same time, however, the development of such dual-use technologies also helped to direct the postwar technological momentum of the chemical industry in ways that would keep it at the centre of German military planning for a future war.

## GERMANY AND AUSTRIA ON THE EVE OF WAR

Germany entered the war with a mixed system of explosives supply. Each of the major German states (Prussia, Bavaria, and Saxony) had state munitions complexes, including factories for explosives production. Prussia had a munitions complex in Berlin-Spandau, and supplementary facilities elsewhere, including a powder plant in Hanau. Bavaria had a powder plant in Ingolstadt, as did Saxony in Gnaschwitz. Most of these produced only propellant, as the principal shell in use by the army was shrapnel, rather than explosive.

Indeed, in the decades preceding the war, there was limited state interest in producing explosives. In 1904, Spandau had begun the production of high explosive, but in 1914 private companies still supplied half or more of Germany's military and naval munitions.[4] Most private firms belonged to a profit-sharing alliance of the British-German Nobel-Dynamite Trust [which was dissolved in 1915] and the German powder group. The principal German member of the Nobel trust was the Deutsche Sprengstoff AG, directed by Alfred Nobel's former associate, Gustav Aufschläger; the three main firms in the powder group were the

Vereinigte Köln-Rottweiler Pulverfabriken; the Rheinisch-Westfälische Sprengstoff AG in Troisdorf; and the Westfälisch-Anhaltinische Sprengstoff AG (WASAG) in Reinsdorf.[5] As the Kaiserreich had never established an Imperial Ministry of War or of Munitions, it was the Prussian War Minister who drew up the military budget and presented it to the Reichstag, and the Prussian War Ministry's *Feldzeugmeisterei* (Ordnance Department) was supposed to coordinate overall production and supply in time of war. Together with Prussia, the principal smaller German states, such as Bavaria and Saxony, also had their own War Ministries, which relied upon liaison officers to maintain communications.

Austria, by contrast, had a state monopoly over military propellants and explosives, which was administered from 1908 by the Powder Department (7P) of the Austro-Hungarian Imperial War Ministry. This was the one area of artillery munitions not placed under the competence of the War Ministry's Inspector of Technical Artillery (ITA) in 1912, in the wake of the first Balkan war, which led to legislation enabling the War Ministry to assume control over the economy during a war emergency. While the ITA attempted to modernize the artillery and set up a network of civilian contractors to supplement the state arsenals, the production of military propellants remained concentrated in a few locations – principally the Blumau state powder factory near Wiener Neustadt, which was not known for its efficiency. But until the outbreak of war in 1914, the Powder Department apparently saw no need to utilize civilian firms such as the Dynamit-Nobel AG in Poszony (today Bratislava).[6]

The private and state-owned manufacturers of explosives and propellants in both countries obtained critical raw materials and intermediates from commercial sources, which in turn were partially dependent upon foreign imports, and thus remained vulnerable to a naval blockade. Glycerine, required for nitroglycerine (NG) propellant, came mainly from soap manufacturers, which had to import some of their fats; cellulose, required for nitrocellulose (NC) propellant, came mainly from imported cotton. The two standard high explosives were trinitrotoluene (TNT), and picric acid (or trinitrophenol). Although toluene and phenol (once imported from Britain) came from German sources (coal tar distillation or as by-products of coking and urban gas plants), the lack of a process to synthesize toluene set limits on the production of TNT. The critical component common to all military explosives and propellants was, of course, nitric acid – in 1914, made primarily from imported Chilean nitrates. Concentrated sulphuric acid and oleum (concentrated sulphuric acid with anhydrous $SO_3$), needed for the production of nitric acid, was made largely from Spanish pyrites. Given a British blockade, the main hope for overcoming wartime shortages lay with the chemical industry – once it had converted from peacetime to wartime production.

## THE PRE-WAR GERMAN DYE INDUSTRY AND EXPLOSIVES PRODUCTION

In July 1914, the Germany dye industry seemed little interested in military production, indeed, recognising that a major war could destroy its 85% dominance of the global market. Yet, there were already signs of changes in the industry's technological momentum. Germany's most profitable pre-war producer of dyes and pharmaceuticals was the Rhineland company Bayer (more properly, the Farbenfabriken vorm. Friedr. Bayer & Co.), which in 1904 formed a profit-sharing alliance– the so-called 'little IG' – with Agfa (Aktiengesellschaft für Anilinfabrikation) in Berlin and BASF (Badische Anilin & Sodafabrik) in Ludwigshafen on the upper Rhine. Neither they nor their principal competitors – the Frankfurt/Main regional alliance (also formed in 1904) of Höchst (the Farbwerke vorm. Meister, Lucius & Brüning in Höchst) and the firms of Cassella, and Kalle – produced explosives. Yet, they were not unfamiliar with key explosives technologies, as in producing dyes and intermediates they commonly nitrated various aromatic compounds, including phenol and toluene – albeit normally only to the non-explosive mono- and di- (or bi-) nitro stages (though they produced picric acid in rather limited quantities as a dyestuff or intermediate).[7] Hence they could (and did) supply intermediates to the explosives industry, as well as to the state arsenals.

The third and largest company, BASF, had a longstanding interest in producing nitrates, which culminated shortly before the war in the achievement of the Haber-Bosch process for synthetic ammonia. However, BASF's relatively small output of 30,000 metric tons of Haber-Bosch ammonia in 1914 was used chiefly to make ammonium sulphate for agricultural fertilizer; there was no plant for oxidizing synthetic ammonia to nitric acid, an essential step before it could be used to produce explosives. Although the chemist Emil Fischer brought the process to the Kaiser's attention in January 1911, shortly before Fritz Haber's appointment as director of the Kaiser Wilhelm Institute for Physical Chemistry, the Prussian War Ministry rebuffed Haber's subsequent efforts to establish a connection between his Institute and the Prussian military. Hence, there is little evidence to support the postwar Allied myth that Haber-Bosch ammonia played a crucial role in pre-war German military planning.[8]

Military attitudes were slow to change. But commerce marched to a different rhythm. Whatever their views of military contracts, Germany's dye firms entered 1914 uncertain of their future and interested in prospects for diversification. BASF's move into nitrates arose in part from a perception, widespread from the turn of the century, that there would be few new developments in the field of synthetic dyestuffs to compare with the huge success of synthetic indigo. Despite the substantial sums spent in research and development, the most promising new dye group after indigo – the vat dyes – presented many technical problems to producers and dyers alike. Dyestuffs also seemed in many ways technologically unattractive. With the exception of alizarin and indigo, they generally entailed relatively small-scale, batch production, and were traditionally tied to the changing tastes of the textile industry, so required constant innovation to remain profitable. Dyes were also labour-intensive – on the factory floor, in sales, and in the laboratory. Of course, the production of synthetic indigo, together with its raw materials and intermediates, had demonstrated the profitable possibilities of large-scale, continuous processes. This, BASF successfully followed up with the Haber-Bosch process.[9] Turning plant to military use would require not new plant, but new policy.

## MOBILIZATION PLANS

Historians agree that German (and Austrian) pre-war military planning showed little regard for industrial mobilization, or the need to stockpile strategic materials. The Schlieffen Plan of 1905 was designed to effect a quick victory, which as a matter of principle made pointless any long-term considerations. Schlieffen intended to overwhelm Germany's opponents in the West with a quick strike with overwhelming force, while holding the line in the East against the slower-moving Russians. Artillery, especially heavy artillery for breaking down enemy fortifications, would play a crucial role. To this end, by 1914 Germany amassed more than seven thousand artillery pieces (including two thousand field howitzers and heavier pieces) with a reserve of some twenty million shells. This was more than the French and Russians had together, and was thought sufficient to last until the state-owned and private factories could begin meeting quotas set by mobilization plans. These plans anticipated that monthly production would not need to exceed 200,000 shells (requiring about 150–200 metric tons of propellants and perhaps 250 tons of high explosive) until the third month of the conflict.[10] Had the war lasted only six months, this would not have severely strained the German economy.

But what if it lasted longer? Considering this possibility, Schlieffen's successor, the younger Moltke, modified the Schlieffen Plan in 1911 to avoid marching through The Netherlands. The idea was to leave open a supply line through which to obtain critical imports in case the plan failed.[11] Ironically, this modification might have been sufficient in itself to cause the plan to fail, by narrowing the front through which the German army could march. Moreover, the likely imposition of a British blockade and British influence would reduce Germany's chances of obtaining material through neutral Holland.[12] As late as August 1914, the German military had failed to assemble a strategic reserve of critical nitrates and toluene. It is all the more remarkable that the government had begun discussing civilian mobilization with businessmen in 1912 – although the principal concern was food, not raw materials.[13]

## THE MUNITIONS CRISIS AND 'MAKESHIFT MOBILIZATION', 1914–1915

In the initial campaigns, the demand for munitions proved far greater than Germany had anticipated. As early as 13 August, when the modified Schlieffen Plan seemed likely to succeed in the West (while in the East, the Germans still awaited the advancing Russians), the War Ministry's Allgemeines Kriegs-Departement (General War Department) sent an anxious memorandum to the Ordnance Department: 'The consumption of munitions in the initial battles has been so great that we must employ all means possible to draw upon and make use of every potential source for munitions production'. The Ordnance Department was to provide a list of 'private companies, machines, raw materials, etc.' for this purpose.[14] One may ask why contingency plans had not been developed before the war; to be sure, in early July, the War Minister had ordered one – due on 1 January 1915![15]

The problem certainly worried thoughtful civilians. Walther Rathenau of the AEG (German General Electric Corporation) and a subordinate, Wichard von Moellendorf, were quick to recognize the failure of military planners to establish a system of rationing, or even a means of coordinating the confiscation and distribution of captured supplies. Rathenau seized the initiative, and communicated with the War Minister who, on 13 August – the same day as the memorandum to the Ordnance Department – authorised the creation of the *Kriegsrohstoffabteilung* (KRA, or War Raw Materials Department), under Rathenau's direction. The KRA coordinated the industrial mobilization through a series of 'war corporations', established as consortia for the production of various types of war materials, with the Imperial government as co-investor. The second of these (established on 30 September following a corporation for non-ferrous metals set up on 2 September), was the Kriegschemikalien AG (KCA, War Chemicals Corporation). This included twenty-six producers of explosives and intermediates, under the presidency of Gustav Aufschläger, of Deutsche Sprengstoff AG. Through confiscation, rationing, and controlled distribution, the KCA was to ensure reliable supplies of war chemicals.[16]

The KCA appeared after the death of the Schlieffen Plan on the Marne put an end to dreams of quick victory in the West, even as outnumbered Germans in East Prussia used superior artillery and mobility to drive back the Russians. The defensive now predominated over the offensive. Having used large-calibre guns to reduce the great fortresses of Liège in the first days of August, the German army (like their Austrian allies) expected to rely on shrapnel – or the newly developed *Einheitsgeschoß* (combined shell), joining shrapnel with high explosive – in the field.[17] In the East, where lines remained more fluid, this approach was effective, especially in the first year of the war. However, with the advent of trench warfare in the West, attention returned to intensive bombardments, using heavy artillery and high explosive. This required vast quantities of nitrates. Despite the military's failure to amass reserves, the KRA and KCA located supplies in domestic industry, in agriculture, and in importers' warehouses, as well as in captured Belgian ports, and purchased more in adjacent neutral countries. Without these stocks, equivalent to about 35,000 of the 55,000 tons of pure nitrogen equivalent used for explosives in 1914–1915, Germany could not have carried on the war. As it was, the Germans had time to find alternative sources.[18]

## DUAL-USE TECHNOLOGIES AND MOBILIZATION

The advantages of having 'dual-use' capacities soon became apparent, as Troisdorf and other firms that produced NC propellant as well as commercial products increased their military production by shutting down their celluloid factories immediately after the outbreak of war.[19] By contrast, the dye industry rejected the government's urgent appeal to begin production of TNT or picric acid or both.[20] Carl Duisberg of Bayer believed that the Prussian Ordnance Department was exaggerating the shortage, and he dreaded the idea of producing high explosives on a large scale. By the standards of the *Berufsgenossenschaft*, the chemical industry's insurance company, Bayer's employees were neither properly trained nor were its buildings properly equipped with safety features to handle explosives production (with protective earthen walls in areas where TNT was made).[21] The directors of Höchst put it more simply: 'we are not an explosives factory and, moreover, cannot create facilities to produce TNT and picric acid'.[22] In any case, why risk converting existing plant when the world dye market might reopen in a few weeks? In the meantime, they reasoned, the dye firms could simply increase their production of intermediates (e.g. dinitrotoluene) for the explosives firms.

Similarly, BASF's management rebuffed Haber's inquiries on behalf of the War Ministry, requesting the use of its synthetic ammonia to produce nitrates for explosives. The company had a process, but not yet a pilot plant for the oxidation of ammonia to nitric acid; the directors saw no way to achieve full-scale production in time to influence what they expected to be a short war. In the first few weeks of August, it had lost half its workers to the army, and been forced to shut down production lines, even its synthetic ammonia plant.

However, the existence of that plant was to initiate a decisive shift in the weeks following the Marne. Along with Haber, Emil Fischer now emerged as a mediator between the KRA and the branches of industry that could produce nitrogen products.[23] Although Haber could boast ties to BASF, and especially to Carl Bosch, the director of its nitrogen division, Fischer enjoyed much wider respect in the dye industry generally (reflected in his becoming the sole academic on the supervisory board of the KCA). His pre-war efforts to establish a *Kaiser Wilhelm Institut für Kohlenforschung* (Institute for Coal Research) had also brought him close to the Ruhr coal producers. To this group, assembled in Essen on 22 September, Fischer outlined the looming crisis, with an urgent plea to increase their production of by-product ammonia and coal-tar products, including toluene. He then asked his friend Duisberg to urge Bayer's partner firm, BASF, to develop its ammonia oxidation process, and to share this and related technologies with other firms. This BASF agreed to do, albeit reluctantly.[24]

By the end of September 1914, as BASF and other dye firms (again reluctantly) joined the explosives firms as co-founders of the KCA, BASF agreed to produce 5,000 tons of sodium nitrate from ammonia per month within six months – i.e., by April 1915.[25] This so-called 'saltpeter promise' represented a decisive shift toward war production, BASF's directors insisted that their first plant was an expedient, necessarily using imperfect technology that would have no commercial value in peacetime. Strongly supported by both Fischer and Haber, they requested subsidies for its construction. The Prussian War Ministry agreed to pay six million marks, and the *Landwirtschaftsministerium* [Ministry of Agriculture] added more for fertilizer production. Fischer also applied pressure on the Höchst Farbwerke, which had a process for producing nitrates from dilute by-product ammonia. In October 1914, while Bosch's team was feverishly transferring BASF's ammonia oxidation process from plan to full-scale production (bypassing a pilot plant), Höchst and Bayer (using an older BASF process) began building nitrate plants of a obsolescent type, which they expected to dismantle after the war. Similar plants were rushed to completion elsewhere in Germany and at the Blumau and Magyarovar state powder factories in Austria-Hungary.[26]

In the long run,, BASF and its Haber-Bosch ammonia would play a critical role in explosives production, if ammonia could be efficiently oxidized to nitric acid on a sufficiently large scale. As yet, only one small company, Zeche Lothringen, had a working plant for ammonia oxidation, but its scale (and that of similar processes) was limited by the need for an expensive platinum catalyst. BASF had, however, already identified a cheaper catalyst for its own plants, and the company was able to fulfil its promise. By mid-October, BASF's stockholders already perceived the long-term dual value of 'our ammonia manufacture', as likely to be of 'great significance...not only in peace but also in war', as well as a major source of future profits in either case.[27] Bosch's team rushed several new plants to completion, all without adequate testing, solving problems that arose as new plants were built. Hence, it was possible for BASF's partner, Agfa, to benefit from the improved technology in mid-to late 1915, when it finally agreed to produce nitrates.[28]

In November 1914, private companies were ordered to use chlorate-based explosives as 'substitutes' in non-military applications such as mining, in order to save nitrates for military use. By January 1915, the companies had complied.[29] However, nitrate mining explosives proved unsatisfactory for artillery shells, in part because they could not be poured but only packed in.[30] There were also frequent barrel detonations, especially when the Germans used unstable mixtures of ammonium nitrate and impure TNT with NG in cheap, easily-produced cast-iron shell-casings which the military arsenals tried to substitute for steel casings. By the time enough steel was available (summer 1915), Germany had millions of filled iron shells, whose explosives the chemical industry had then to recover. By this time, however, the Prussian *Militärversuchsamt* (Military Testing Office) at Spandau had also devised a '60/40' filling, a mix of TNT with ammonium nitrate in this ratio, which conserved toluene and could be poured, resulting in greater stability and reliability.[31]

THE AUSTRIAN CASE: COMPARATIVE ASPECTS

According to a postwar account prepared by a former official of the Austro-Hungarian War Ministry, the Austrian army entered the conflict far from prepared. The modernization of its artillery, begun in 1912, was incomplete; pending the adoption of new types and calibers, the army had minimal reserves for the older models. In mid-July 1914, the General Staff called for doubling the reserve of infantry propellants, and the Powder Department contracted with civilian firms (Poszony and Troisdorf), but the much smaller artillery reserve was thought to be adequate. Once the fighting began, Austria faced 'serious munitions crises' as neither Blumau nor civilian firms could cover even minimum demands for shells. Faced with its own crisis, Germany in August-September sent only thirty-two tons of TNT against the Austrian request for 1,000. By mid-October, with extensive losses of supplies as a result of early reverses on the Russian front, Austria was forced not only to halve its production of infantry ammunition, but also to strip supplies from the Italian border, leaving fortifications vulnerable when their erstwhile ally attacked in May 1915.[32]

Austria hoped to obtain resources from Germany, but the export of machine tools was banned, forcing the Austrians to seek these in neutral Switzerland. Like the Germans, they subsidised new and expanded plants in both state and civilian firms, including TNT works in Blumau and Poszony. Unlike the Germans, the Austrian War Ministry began by putting civilian plants under direct military supervision. Still, they could not catch up; as the generals' demands grew, so did the deficit. By the spring of 1915, their domestic explosives plants, including private companies, could still produce only 3.5 tons of TNT per day. By summer 1915, when they had reached a planned daily capacity of 16 tons (the amount needed in September 1914), the daily demand for high explosives had already reached 20 tons and rose to 30–35 by the end of the year. Only part of the deficit could be made up by using mixtures such as ammonal (ammonium nitrate, aluminium, and approximately 30% TNT), or by producing picric acid, for which a process was licensed from Bayer in October 1915.[33]

Much the same was true with propellant. In Germany, the private NC producers, with their 'dual-use' technologies, accrued advantages in increased output and overall efficiency. Having to produce for a commercial market evidently helped them fare better than the state factories, especially those in Austria. By subsidizing Nobel's expansion in Poszony and building new plants in Blumau and Magyarovar, the Austrian Powder Department had nearly doubled its output of NC by June 1915.[34] Even so, it was insufficient. In both Austria and Germany, the lack of propellants remained the chief limiting factor in artillery shell production during the first three years of the war.

In Austria, as in Germany, few senior officers or bureaucrats clearly understood the production capabilities of their chemical empires. But lacking a Rathenau, Austria created no KRA, and was slow to follow the German model of war corporations. In September 1914, Germany persuaded Austria to adopt this model of mobilization, leading in November to the first Austrian *Zentrale* (Centrals) to coordinate the production and use of strategic materials such as metals. Only the chemical industry remained 'decentralized'. The Austrian Powder Department followed the precedent of the German KCA, however, beginning confiscations of by-product ammonia, carbide products, and nitrate fertilizers in March 1915. These stocks proved insufficient, so by the end of the year, the Austrians contracted to receive regular shipments of nitrates as well as explosives from BASF, Bayer, and other sources.[35] Ultimately, both Central Powers endorsed a 'mixed' approach of state factories and private companies, with the Austrians putting more emphasis upon the former, despite their relative inefficiency. But for both, it was going to be a long, expensive war.

## EXPANDING EXPLOSIVES PRODUCTION, 1914–15: PLANTS AND MATERIALS

If the events of August and September 1914 had not demonstrated clearly enough the need to multiply Germany's capacity for propellants and explosives, by late October there was unmistakeable evidence of shortages. The Prussian General Staff estimated that the artillery would need 1.25 million shells in November, but would fall short by 150,000 shells, even after reducing supplies for the navy and substituting NG for NC. This forced the arsenals to shift propellant from infantry bullets to artillery shells, until additional propellant plant could come on-line in February 1915.[36]

With time the critical factor, the Reich began pouring funds into construction subsidies. Of more than 113 million marks that went to state-owned factories during the first year of the war, more than half (60,840,350 marks) went to powder factories (mainly for propellants). An additional 78 million marks in outright subsidies went to privately-owned factories, of which powder factories such as Köln-Rottwell received more than a third (27 million marks), with most of the rest going to nitrate or nitric acid plants.[37] Of course, it was not possible simply to expand every factory; the Prussian state factories, in particular, were either too close to the Western Front (like Hanau, near Frankfurt on the Main) or, like Spandau, too close to the expanding city of Berlin. In November 1914, the War Ministry decided to build a huge new powder plant, to be located in a sparsely inhabited, agricultural district on the Plauener See in the Province of Brandenburg. Despite the necessary haste of construction, the plant buildings were clearly built to last; the Prussian War Ministry seized the opportunity offered by the war for a large, permanent increase in its productive capacities. Within six months, the factory began large-scale production of NC propellant, later adding NG as well.[38]

The vulnerability of its plant to air attack led Köln-Rottweil to look eastwards. The plant begun in July 1915 in Premnitz, Brandenburg, ultimately added 50% to their NC capacity, making them Germany's leading manufacturer of propellants. The chemical stabilizer they had developed, Centralit II, made it possible to dispense with imported camphor. But their most significant contribution came from the Düneberg plant (near Hamburg), which produced NG propellant for heavy artillery. Just before the war, Köln-Rottweil had developed a new solventless NG technology that shortened the drying process from months to days, thus greatly accelerating production and saving acetone.[39]

Germany's expanded capacities were still in the future on 18 December 1914, when the Prussian Ordnance Department assumed responsibility for coordinating the distribution of all German military and naval explosives (having taken responsibility for regulating their production on 30 August). Stocks of 'high quality explosives' (mainly TNT) were rationed. Substitutes were sought in chlorate-based mining explosives, but these proved unsuitable.[40]

However. there was another way – through the production of toxic gases. This was an area in which the dye companies could easily apply their technical expertise, and both Bayer and Höchst tested chemical shells in the autumn and winter of

1914–1915.[41] Bayer and Höchst's early work on chemical shells, together with their production of nitrates, thus became an industrial bridge to the military. Invited to join a commission appointed by the High Command in October 1914, Duisberg developed ties to the munitions industry and with Major (later Colonel) Max Bauer, one of the few technically-savvy members of the Prussian General Staff. In November 1914, ten days after the first large-scale (and ineffective) test of Bayer's chemical shells on the Western Front, Duisberg agreed that Bayer would produce TNT, albeit 'solely from patriotic interest'.[42] In December, Höchst, which was developing a competing irritant shell, and which had begun producing ammonium nitrate, agreed to do the same.[43]

This change of heart can be partly explained by the fact that, by the close of 1914, it was clear that the war would not end quickly. It followed that international dye markets would not recover in the near future, and that firms would have excess capacities on their hands, which they might as well turn to both military and financial advantage. Thus Höchst, whose dye production in 1915 was less than a quarter of its output in 1913, began producing explosives and irritants in converted dye works.[44] In any case, by the autumn of 1915, if the dye companies were to produce anything in quantity, it had to be war *materiel*, as the KRA had confiscated, or the KCA was rationing, critical raw materials, leaving little or nothing for the commercial economy.[45]

The process of transition entailed plant conversions and expansions, and the establishment of on-site shell-filling stations. The Government subsidized some, but not enough to satisfy the firms – which complained that the Ordnance Department set prices that were too low. Procurement contracts were irregular, and it was difficult to avoid bottlenecks and maintain production.[46]

The experience of Bayer and Höchst are worth looking at more closely. During 1915, Bayer converted many of its facilities to explosives production; built a few new but explicitly temporary plants for nitric acid and picric acid production; and added shell-filling stations in Flittard, some distance to the south of its main Leverkusen plant. During the next two years, Bayer's production of TNT (using an improved process for the use of sulphuric and nitric acid) increased from 100 tons to 1,300 tons per month.[47] Bayer thereby became Germany's largest single producer of TNT – or at least until, as Duisberg had feared, an explosion destroyed the Flittard facility in January 1917.[48] By then, however, Bayer was already building a permanent plant in a safer location.

In December 1914, Höchst's initial goal for TNT production was 200 tons per month. The Farbwerke based its process on the dinitrotoluene process of its smaller competitor, Griesheim-Elektron, which in August was the only dye company to agree to produce explosives. As the process had peacetime commercial value, Griesheim disclosed it reluctantly, and probably only after the intervention of Haber or Fischer. The first nitration plant went on-line in February, and a second building was added when the War Ministry raised its quota to 400 tons per month. This was achieved in stages by July 1915. By mid-1916, Höchst introduced a new process, which conserved resources by producing TNT directly from the 'mononitro' product, by-passing the 'dinitro' stage that had caused the original controversy. This was a major step towards a dedicated explosives technology. Producing 26,300 tons of TNT and 17,900 tons of other high explosives, the Höchst Farbwerke stood second only to Bayer by the end of the war.[49]

## COMPLETING THE FIRST MOBILIZATION, 1916

By August 1916, the three dye companies (Bayer, Griesheim, and Höchst) had already developed more than twice the capacity to produce high explosives as the entire pre-war German explosives industry. They achieved this by converting idle plant and apparatus, with little new construction (except for government-subsidized nitrate plants). Their scientific and technical capabilities allowed them to make crystallized TNT that was purer (and thus safer) than the explosives industry's product. They also produced the bulk of Germany's picric acid, as well as virtually all of its 'substitute explosives' (mainly trinitroanisol and dinitrobenzene). These compounds were familiar from dye manufacture, so they could be dealt with more effectively, thus minimizing the disastrous explosions and poisonings that plagued explosives firms.[50]

While BASF at first avoided explosives production, they did make large quantities of synthetic ammonia and nitrates, along with other raw materials and intermediates, most of which went to military uses (including chlorine for chemical warfare). By 1916, they had perfected their ammonia oxidation process and were licensing it to other firms in Germany and Austria.[51] The expansion of their Oppau ammonia plant complex took place roughly in accord with pre-war plans, which the war simply accelerated. The same was not true of BASF's decision to build a completely new ammonia plant at Leuna in central Germany, far from Oppau, whose production was suffering repeated disruptions by air raids.[52]

By March 1916, Germany had gained the capacity to produce 60,000 tons of nitrates and nitric acid per month from plants built mostly at BASF, Bayer, Griesheim-Elektron, Agfa, and Zeche Lothringen. Nevertheless, shortages of ammonia left almost half that capacity unused.[53] After months of negotiations, BASF contracted in April 1916 to build a new synthetic ammonia plant with a capacity of 36,000 tons, approximately doubling its total output. This seemed so risky that the company obtained an outright subsidy of 12 million marks from the Reich, plus a long-term loan of 64 million marks, in return for a commitment to share excess profits with the Reich and to share the market with other nitrogen producers.[54] When it began production in April 1917, the Leuna plant would lay the basis for further expansions in explosives production.

As a result of munitions production, the leading dye companies' sales and profits, which dropped in 1914–15, now far exceeded pre-war levels. Even so, the firms would no doubt have gladly reconverted to peacetime production, had the war ended in 1916. Yet as with the Leuna project, military contracts led the dye industry to institutionalize their dual-use technologies and to accept the likelihood of mobilizing again in a future war. The Prussian War Ministry accomplished this through the so-called 'readiness clause', introduced in 1915 in connection with Agfa's new sodium nitrate factories. These were based upon the improved BASF catalytic process for ammonia oxidation, and were thus state-of-the-art, unlike the initial plants built by Bayer and other companies as wartime expedients. The Agfa contracts provided outright grants in return for a pledge that the plants would be held 'in readiness, at the disposal of the military authorities, for a period of ten years after the end of the war'.[55] The company could operate them for commercial purposes in peacetime, but only on condition that it agreed to reconvert to military production within weeks of a resumption of hostilities.

There could be no true return to a peacetime economy for the dye industry. Partly in recognition of the changes brought on by the war, in late 1915-early 1916 the two main competing dye groups plus two independent firms, (all engaged in war production), had established a more centralized organization – the 'expanded IG'. Based on the Bayer-BASF-Agfa pooling agreement this encouraged the much easier exchange of wartime technological know-how among member firms, and which would also prove useful in meeting postwar foreign competition.[56]

To be sure, the dye companies stayed away from propellant production, which was in many ways more an art than a science, requiring highly experienced workers preparing a variety of types for different categories of shell. Although the Germans first stepped up production by investment in private industry, by mid-1916, new state-owned powder works were on the way, including a new Prussian plant at Kirchmöser (Plaue) and a new Saxon plant for field artillery propellants at Gnaschwitz. There was never any question that the new productive capacities would be permanent. In Austria, however, despite increased use of private industry, the government continued to rely upon its state-owned factories, especially Blumau. By June 1916, the Austrians had spent nearly 83,678,000 Kronen (approximately 67 million marks) on these factories, including 44 million crowns at Blumau alone.[57]

Overall, by mid-1916 the Central Powers had reason to be proud of a roughly threefold increase in the production of propellants. Yet as of July, Germany was still producing only 65% of the 10,000 tons per month called for in the current munitions plan, a goal that was not reached until December 1917. Between 1914 and 1917, the limiting factor was the scarcity of nitrogen. With other components, Germany suffered only temporary shortages. Shortfalls in cotton cellulose, in late 1914-early 1915, were solved by substituting rags and then wood cellulose in the form of crepe paper. Significant shortfalls in glycerine did not arise until the autumn of 1916. At this time, following the recommendations of a committee chaired by Emil Fischer, a war corporation sponsored by the Prussian War Ministry rushed into operation a biochemical process for obtaining glycerine as a by-product of sugar fermentation, thus effectively relieving the shortage by early 1917.[58] With sulphates, the situation was more difficult. Due to a monthly shortfall of 6,000 tons of oleum, of 30,000 required, the KCA was obliged to begin confiscating civilian stocks in early 1916. In February 1916, Haber warned the War Ministry that, given the shortage in sulphates, it could not reach its targeted monthly propellant production of 7–8,000 tons before the end of the year (actual production was then a little over 5,000). Even if sulphur production could be increased from domestic sources such as gypsum, propellant quotas could be raised only slowly, and by 'increments'.[59] At least, this was the position of the War Ministry. However, a vast change was soon to follow.

## THE HINDENBURG PROGRAMME AS A 'SECOND MOBILIZATION'

In the summer of 1916, in response to the German failure at Verdun and the massive British offensive on the Somme, Paul von Hindenburg and Erich Ludendorff assumed the *Oberste Heeresleitung* (Supreme Command of the German army, or OHL). In beginning their 'battle of material', they (or better, Ludendorff's able subordinate, Colonel Bauer) worked out an ambitious, far-reaching programme for doubling munitions production, requiring a total reorganization of the war effort.[60] OHL pressured the Austrians to follow suit with a similar increase in production. Amounting in effect to a 'second mobilization', the programme was unrealistic, its consequences, disastrous. Here one can stress only a few of its significant features.

Under the direction of OHL, planning and regulation of munitions production became centralised under a single agency, created in September-October 1916: this was known as the *Waffen und Munitionen Beschaffungsamt* (Weapons and Munitions Procurement Agency, or WUMBA), whose task was to coordinate munitions planning, production, and supply. WUMBA incorporated the KRA, strengthened controls over key sectors, and integrated industrial leaders in the planning process through a *Beirat* (Advisory Council), comprising representatives of key firms, such as Duisberg of Bayer. In November 1916, WUMBA came under the new *Kriegsamt* (War Office), established within the Prussian Ministry of War to centralise control of resources throughout the Reich.[61] In 1917. the Austrians also organized a Munitions Section in their War Ministry, and in September, the two allies undertook their first systematic exchanges of technical information.[62]

Until the autumn of 1916, the dye industry produced large quantities of explosives largely without extensive new construction (except nitrate plants). The new programme would require private industry to construct vast new plant and apparatus, subsidised by the government at huge cost (at least 350 million marks in all branches, as the Prussian War Ministry conservatively estimated in October 1916). The goal was to create permanent, efficient 'readiness plants' such as Bayer's new picric acid plant in Dormagen – a dual-purpose establishment under a fifteen-year contract that permitted commercial use in peacetime, but required quick re-conversion to military production of 2,400 tons (nearly double Germany's entire monthly output of picric acid in August 1916).[63] Many of these – such as two major additions to BASF'S ammonia factory at Leuna – remained incomplete in November 1918. Although propellant output did double (to 12,000 tons) by the autumn of 1918 – more than a year later than planned – this was achieved only by mass-producing an inferior ammonium nitrate/charcoal propellant (about half the field artillery propellant produced in the final months of the war).[64]

Moreover, growing shortages of metal for shell casings, as well as shortages of coal and breakdowns in transportation, became the new limiting factors in shell production. If it is correct (as Feldman argues) that there were only enough shell-casings to utilize 10,000 tons of propellant per month, much of the additional propellant and explosives capacity was wasted, along with the labour tied up in creating it – perhaps one-quarter of the million soldiers recalled for construction work. It may be a final irony of the Hindenburg Program that it kept these men uselessly at home fighting the 'battle of materials', when they could have made a far more significant difference serving in Ludendorff's last offensive on the Western Front in the spring of 1918, once the Russian front disintegrated and before the Americans arrived.[65]

At the end of 1916, in response to German pressure, Austria attempted a similar increase in production, with the goal of doubling output of shells to four million per month. But the long-term results were even more disastrous. Despite an investment of 454.4 million crowns (more than 360 million marks) in armaments between November 1916 and April 1917 – including the expansion of explosives factories at Blumau and three other sites – Austrian munitions production actually declined. The monthly output of shells fell to 0.75 million by 1918, as the Austrian economy collapsed.[66] Overall, it could be argued, the 'second mobilization' may have cost the Central Powers the war.

CONCLUSION

Comparing the German and Allied chemical industries during the war, we must concede the Germans a clear initial advantage. This was so, not because of the superiority of their state-owned facilities, but rather because of their large private chemical firms, whose processes, and to some extent products as well, were readily adaptable to military purposes. The initial dislocations of the war were economically unpleasant for the dye manufacturers, who lost access to much of their world market. By the end of 1914, however, they began to adapt, encountering few major technological obstacles as they did so.

By early 1916, Germany's private explosives companies were producing the bulk of its propellants, and the dye industry, the bulk of its high explosives. Nevertheless, given the dangerous nature of explosives production, which led at times to losses in staff and physical plant, the dye manufacturers were reluctant to make permanent changes to their facilities. The 'second mobilization' initiated by the Hindenburg Programme in the summer of 1916 created a drastically new situation. But the huge requirements of the Program overtaxed the German and Austrian systems of innovation and production, and despite enormous investments in new plant, optimistic targets remained unmet. Ultimately, German and Austrian production lagged behind that of the Allies. In the final months of the war, the Allies were producing three times the German output of high explosives, and nearly twice their monthly average of finished shells. Total British and French propellant production approximated that of Germany, at about 14,000 tons per month; but the Americans were adding another 19,000 tons per month, so doubling the pressure against the Central Powers.[67]

Using Hughes' terminology, the dye companies' ventures into explosives manufacturing – like BASF's production of ammonia and nitrates – had acquired increasing 'technological momentum' as the war became increasingly 'total'. Much of this momentum dissipated with defeat. The provisions of the Armistice and the Versailles Treaty, enforced by the Military Interallied Control Commission, forced the dismantling or modification of most of the readiness plants in private industry, together with all the state-owned explosives and propellant plants. With the break-up of the Austro-Hungarian Empire, the 'technological momentum' of the largely state-owned munitions industry in Austria came to an end. In Germany, however, important elements of the military-industrial system remained in place, and contributed to postwar momentum. Although the Versailles Treaty led to the dismantling of most of the dedicated explosives factories, such as Bayer's Dormagen works – a state-of-the-art 'readiness plant' that had produced 85% of Germany's picric acid and filled nearly four million shells[68] – the treaty did allow many 'dual-use' facilities – including the Haber-Bosch ammonia plants – to remain intact. These were to become vital producers of peacetime fertilizer, if no less vital sources of future military explosives.[69]

After the war, Germany retained its supporting culture of wartime technological momentum – the corporations, professional organizations, and bureaucracies whose collective memories retained the lessons of 1914–18. During the 1920s, the German chemical industry was accorded a strategic importance hardly recognised before 1914.[70] After 1933, the National Socialist regime took care to avoid the mistakes that had brought down the Kaiser's regime, not least of which was a failure to integrate the chemical industry into prewar planning for munitions production. As a military analyst observed in 1937, it will be 'absolutely necessary...not to improvise this arms industry at the outbreak of the war or afterwards, but rather to prepare it fundamentally and carefully in all its aspects.'[71] Accordingly, the chemical industry played a central role in Nazi planning for the next war.[72]

### ACKNOWLEDGEMENTS

This paper has its origins in a larger project on 'The Great War and Modern Chemistry', assisted by the US National Science Foundation. I thank my assistants, Katherine Marcotte-McKay and Andrew Jones of Villanova University, and Susanne Betz of the University of Vienna, for their help with research, and Roy MacLeod for his help in producing a final version.

### NOTES

[1] Thomas Parke Hughes, 'Technological Momentum in History: Hydrogenation in Germany, 1898–1933', *Past & Present*, No. 44 (August 1969), 106–132.
[2] 'Austria' here and later implies 'Austria-Hungary' (up to 1919).
[3] Hew Strachan, *The First World War, vol 1: To Arms* (Oxford: University Press, 2001), 1113.

4 Andrea Theissen and Arnold Wirtgen (eds.), *Miltärstadt Spandau: Zentrum der Preussischen Waffenproduktion 1722 bis 1918* (Berlin: Brandenbürgisches Verlagshaus,, 1998), 153. See also Reichsarchiv, *Der Weltkrieg, 1914 bis 1918 vol. 1: Kriegsrüstung und Kriegswirtschaft* (Berlin: E.S. Mittler & Sohn, 1930), 394.

5 W. J. Reader, *Imperial Chemical Industries: A History* (London: Oxford University Press, 1970–75), vol. I, 131–137, 151, 194–197, 216, 306–309; Wolfram Fischer, *WASAG: Die Geschichte eines Unternehmens 1891–1966* (Berlin: Duncker & Humblot, 1966), 18–26.

6 Alois Jahn, *Das Pulvermonopol in Österreich-Ungarn und die Vorschriften über die Erzeugung und den Verkehr von explosiven Stoffen, Waffen und Munitionsgegenständen* (Vienna: Hof- u. Staatsdr., 1902); Rainer Egger, 'Rüstungsindustrie in Niederösterreich und Heeresverwaltung während des Ersten Weltkrieges', *Bericht über den 16. österr. Historikertag in Krems, 1984* (Vienna: Verband Österreichischer Geschichtsvereine, 1985), 406–421, on 406. Österriechisches Staatsarchiv: Kriegsarchiv (Vienna) (hereafter, KA), Kriegsministerium (hereafter, KM)-Intern, K. 15, Aufteilung der 7.P Abteilung (29 December 1908); KM-Carnegie Manuskript, Fasz. 4, copy 2, Ottokar Pflug, 'Bewaffnung und Munition' (unpubl. ms., 1923), 39–40, 47.

7 R. Norris Shreve, *Dyes Classified by Intermediates* (New York: Chemical Catalog Company, 1922), 261, 434–437, 450–454, 459–463, 494–495, 590–625.

8 Jeffrey Allan Johnson, The Kaiser's Chemists: Science and Modernization in Imperial Germany (Chapel Hill: University of North Carolina Press, 1990), 126, 185–186. For Haber, see Margit Szöllösi-Janze, *Fritz Haber, 1868–1934* (Munich: Beck Verlag, 1998), and Daniel Charles, *Between Genius and Genocide: The Tragedy of Fritz Haber, Father of Chemical Warfare* (London: Cape, 2005 [U.S. ed.: *Master Mind: The Rise and Fall of Fritz Haber* (HarperCollins, 2005)]. On the postwar Allied mythology concerning the provision of ammonia in prewar German military planning, see BASF Unternehmensarchiv (hereafter, BASF UA), Ludwigshafen, G6101, Alwin Mittasch, "Geschichte der Ammoniaksynthese aus den Elementen" (ms., 1919), 394, citing various Allied sources, including Lord Moulton in the *Chemical Trade Journal*, 15 June 1919, 524.

9 Cf. Werner Abelshauser *et al.*, Wolfgang von Hippel, Jeffrey Allan Johnson, and Raymond G. Stokes, *German Industry and Global Enterprise: BASF: The History of a Company* (Cambridge: Cambridge University Press, 2004), 119–127, 142–145,151–157.

10 David T. Zabecki, *Steel Wind: Colonel Georg Bruchmüller and the Birth of Modern Artillery* (Westport, Conn.: Praeger, 1994), 8–12; Bundesarchiv-Militärarchiv, Freiburg/Breisgau (hereafter, BAMA), PH2/87 (Artillerie-Munition...1911–1918), KM to Kriegsminister (25 October 1914).

11 Strachan, *op. cit.* note 3, 1005, 1019.

12 Cf. Ernst Homburg, 'Operating on Several Fronts' (paper in this volume).

13 Cf. Lothar Burchardt, *Friedenswirtschaft und Kriegsvorsorge: Deutschlands wirtschaftliche Rüstungsbestrebungen vor 1914* (Boppard: H. Boldt, 1966), 165–241; Strachan, *op. cit.* note 3, 1016–1019.

14 BAMA PH2/87, Wild v. Hohenborn to the Kgl. Feldzeugmeisterei (hereafter, FZM) (13 August 1914); Strachan seems unaware of this message; cf. *op. cit.* note 3, 993.

15 BAMA, PH2/86, KM. A.D. Nr. 1492.14 g.A4 (9 July 1914).

16 Regina Roth, *Staat und Wirtschaft im Ersten Weltkrieg: Kriegsgesellschaften als kriegswirtschaftliche Steuerungsinstrumente* (Berlin: Duncker & Humblot, 1997), 28–115.

17 Neither of these proved to be as effective as high explosive alone; moreover, the complexity of the fuses hampered production and eventually led to the combined shell's abandonment: BAMA PH2/87, Wandel, 'Denkschrift...betreffend Einführung des Einheitsgeschosses der Feldkanone' (25 September 1911), 3–4; Pflug, *op. cit.* note 6, 59, 62.

18 Margit Szöllösi-Janze, 'Losing the War but Gaining Ground: The German Chemical Industry during World War I', in John E. Lesch (ed.), *The German Chemical Industry in the Twentieth Century* (Dordrecht, NL: Kluwer, 2000), 91–121, on 95–99; Bayer-Archiv (Leverkusen) (hereafter, BAL), 201–20, Carl Duisberg, 'Die Herstellung von synthetischem Salpeter bzw. Salpetersäure...' (typescript, April 1916); BAMA W-10 (Kriegsgeschichtliche Forschungsanstalt des Heeres), 50521 (Chemikalien im Weltkrieg [typescript copy , no date]), KM KRA Zentralstelle für Fragen der Chemie (5 February 1916), 236.

[19] BAL 201-5-4, Pt. 8 (Herstellung und Lieferung von Geschoßfüllungen), Carl Duisberg, 'Die Herstellung von Sprengstoffen und Pulver...', 33.

[20] BAL 201-5-1 (Herstellung und Lieferung von Geschossfüllungen), FZM to Bayer (21 August 1914); Plumpe, *Die I.G. Farbenindustrie AG: Wirtschaft, Technik und Politik 1904–1945* (Berlin: Duncker & Humblot, 1990), 63–66.

[21] BAL, *ibid*, Duisberg to FZM (24 August 1914).

[22] Hoechst Archives (Frankfurt/M.-Höchst) (hereafter, HA) 18/1 - 12 (Weltkrieg I: Salpetersäureanlage), Höchst to FZM (25 August 1914); HA library, Albrecht Schmidt, 'Erinnerungen' (unpublished typescr., early 1920s), vol. IX, 68.

[23] Szöllösi-Janze, *op. cit.* note 18, 93–95, 99–100.

[24] *Johnson, op. cit. note 8, 186–187.*

[25] BAL, Duisberg Papers, Henry T. von Böttinger File, Böttinger to Duisberg (1 October 1914). BAMA W-10/50522 (Hans Frhr. v Wolzogen, 'Die Bewirtschaftung der Chemikalien während des Krieges', typescript, 1937), BASF to KM and FZM (3 October 1914, copy), Bl. 397–399.

[26] Szöllösi-Janze, *op. cit.* note 18, 96–104, 110. HA 18/1 - 12, correspondence (Höchst also made cyanamide fertilizer, but this was unsuitable for explosives production). BAL, Duisberg Papers, Walther Hempel File, Duisberg to Walther Hempel (19 September 1915 and 30 March 1915). Pflug, *op. cit.* note 6, 63–64.

[27] Cited in Abelshauser *et al.*, *op. cit.* note 9, 163.

[28] Bundesarchiv (Berlin-Lichterfelde) (hereafter, BAB), R8128/AW381, *Jahresbericht an den Aufsichtsrat der AGFA, 1915*, 72.

[29] Nordrhein-Westfälisches Hauptstaatsarchiv (Düsseldorf), Regierung Köln, Nr. 2184 (Sprengstoffe und Sprengstofffabriken 1912–1919), Handelsministerium to Regierungspräsident, Köln (5 November 1914); *idem*, Kgl. Gewerbeinspektion Bonn, Tagebuch Nr. 83 (9 January 1914 [= 1915]), 3.

[30] BAMA MSg2/760a (Archivrat Frantz, 'Fabrikatorische Erfahrungen...mit den Ersatzstoffen für Pulver und Sprengstoffe', June 1933) (typescript with excerpts from official documents), 154–158.

[31] *Ibid.*, 158, 174–175. BAL 201-5-4, Duisberg, *op. cit.* note 19, 9–10, 26–27. BAMA, PH2/86, table of shell production to July 1915. British amatol ultimately used only 20 parts of TNT, saving even more toluene; see Wayne D. Cocroft, Chapter 3, in this volume.

[32] Pflug, *op. cit.* note 6, 38, 49, 52, 55.

[33] Pflug, *op. cit.* note 6, 51–52, 56, 59–60, 101–103; KA, KM 1915 7A 48 - 13/2-2 (8 cm Schrapnells Halbfabrikate aus der Schweiz Lieferung). BAL 201-6-3 (Wumba, Produktion), 2 (P-R), Bayer to KM (Vienna) 14 October 1915.

[34] Pflug, *op. cit.* note 6, 64.

[35] Pflug, *op. cit.* note 6, 63–64; James R. Wegs, *Die österreichische Kriegswirtschaft, 1914–1918* (Vienna: Schendl, 1979) (revised and translated Ph.D. dissertation, University of Illinois, 1970), 25–28; KA, KM 1916 7P 31-2/8 and /11.

[36] BAMA PH2/87, *op. cit.* note 10.

[37] BAMA PH3/509, KM to Chef des Feldmunitionswesens, 8 September 1915, 121–122.

[38] See Sebastian Kinder, Chapter 4, *infra*.

[39] Geheimes Staatsarchiv Preussischer Kulturbesitz (Dahlem), I. HA, Rep. 120 (Handelsministerium), BB II a 2, Nr. 1: Pulverfabriken, Bd. 8 (1900–1915), 291–293; BAB, R3101, 1630 (Ausführung des Artikels 172 des Friedensvertrags), Sitzung am 17 August 1921 (minutes dated 20 August 1921), 180–181; UN/League of Nations Archives (Geneva), Commission de Controle Interalliés-Allemagne, 1921–1926, Col. 58, Annexe 6bis. Düneberg produced 75% of Germany's NG propellant during the war, and Reinsdorf (WASAG) most of the rest; only 5% came from Prussia's Plaue factory.

[40] BAMA, MSg2/760a, *op. cit.* note 30, 202–204.

[41] Jeffrey Allan Johnson, 'La mobilisation de la recherche industrielle allemande au service de la guerre chimique', in David Aubin and Patrice Bret (eds.), *Le sabre et l'éprouvette. L'invention d'une science de guerre, 1914–1939 (14–18: Aujourd'hui. Today. Heute,* 6), (Paris: Noesis, 2003), 89–103.

[42] BAL 201-5-1, draft telegram Duisberg to Haber, 5 November 1914; Duisberg to FZM (Garke),13 November 1914.

[43] HA 18/1-17 (Weltkrieg 1, 1915–1921: Ammonnitrat [etc.]).

44 HA. 18/1-18 (Weltkrieg I, 1915–1934), Singer, 'Kriegsproduktion im 1. Weltkrieg', 17 October 1934, 1.
45 Plumpe, *op. cit.* note 20, 83–84; Roth, *op. cit.* note 16. Cf. correspondence in BAB, R1501 (RM d Innern), Wirtschaftliche Mobilmachung, Nr. 18761 (Kriegschemikalien AG, 2 April 1915–22 February 1916); also Bundesarchiv (Dahlwitz-Hoppegarten), R8729 (Kriegschemikalien AG), Nrs. 1–2.
46 Cf. BAL 201-5-1, Duisberg to Operationsabteilung der obersten Heeresleitung (Ost), 30 September 1915; HA 2/001-23, 'Bericht über das Geschäftsjahr der Säurefabrik-Abteilung in 1915', 17–18.
47 BAL 201-6-1 (Waffen- und Munitionsbeschaffungsamt [WUMBA]), Duisberg to Groener, 5 April 1917, Anlage 1, 2; BAL 700-504 (Direktionsabteilung, Kriegsproduktionen 1930–1953), Warnecke to Eichwede, 17 January 1938.
48 BAL 201-019 (Auswirkungen der Leverkusener Explosion am 27.01.1917 auf Kriegslieferungen).
49 HA 18/1-1 (Bomben, 1902–1918), report by Dr. Sohst, January 1915; Dr. F. Hübner, 'Trinitrotuluol-Fabrikation', 5 June 1916; HA 18/1-3 (Gesamtlieferung an Munition), 'Verzeichnis der...hergestellten Sprengstoffe und Kampfstoffe' (no date, but postwar).
50 BAMA MSg2/760a, *op. cit.* note 30, 166–170; BAL 201-5-4, Pt. 7, Table (September 1916). Duisberg, *op. cit.* note 18, 11–15.
51 BASF UA, R1/733, 'Bauliches und betriebliches Geschehen im Werk Oppau von 1911–1939' (typescript, May 1963), 12.
52 Abelshauser et al., *op.cit.* note 9, 167–168.
53 Duisberg, *op. cit.* note 18, 5.
54 Szöllösi-Janze, *op. cit.* note 18, 117–118. The Reich Treasury hoped thereby to secure its large investments in Caro-Frank cyanamide fertilizer production.
55 *Jahresbericht, op. cit.* note 28, 72.
56 Plumpe, *op. cit.* note 20, 96–100; Abelshauser et al., *op. cit.* note 9, 171–173.
57 KA, KM-SR, Zentralstelle für Statistik, 125(113) (Bewaffnung und Munition, 1914–1916: Zusammenstellung der Ausgaben, 30 September 1916), 606–608.
58 BAMA MSg2/760a, *op. cit.* note 30, 10–11, 47–49, 60–63.
59 BAMA W-10/50521, *op. cit.* note 18, 251.
60 Otto Goebel, *Deutsche Rohstoffwirtschaft im Weltkrieg: einschliesslich des Hindenburg-Programms* (Stuttgart, Berlin & Leipzig: Deutsche Verlags-Anstalt, 1930); Gerald D. Feldman, *Army, Industry, and Labor in Germany, 1914–1918* (Princeton: Princeton University Press, 1966), 149–161; Holger H. Herwig, *The First World War: Germany and Austria-Hungary, 1914–1918* (London: Arnold, 1997), 259–266.
61 BAL 201-6-1, *op. cit.* note 47 sect. 1 (Organization), Coupette (FZM) to Duisberg, 17 September 1916; sect. 2 (Sitzungs-und Besprechungsprotokolle), *passim*.
62 Egger, *op. cit.* note 6, 411. KA, Technisches Militärkomitee (TMK), 1159 (Monatsberichte 1917–1918).
63 BAL 201-6-3, 2, 'Vertrag...' ([for Dormagen], 14/15 December 1916, para. 10; 'Auszug aus der Denkschrift der Farbenfabriken vom 3. Oktober 1916'.
64 BAMA MSg2/760a, *op. cit.* note 30, 12, 20–22, 110–120, 158–159; PH3/512, WUMBA to Oberste Heeresleitung, 9 February 1917, 72. Sächsisches Hauptstaatsarchiv (Dresden), Kriegsarchiv (P), Nrs. 74583–74585 (Pulveranfertigung, 1918), monthly production estimates.
65 Cf. Feldman, *op. cit.* note 60, 266–273, 494–495; Herwig, *op. cit.* note 60, 264, 325–326, 334. This is only one of many factors that contributed to the failure of Ludendorff's offensive, such as the million men left in Eastern Europe in 1918 as occupation forces.
66 Cf. Wegs, *op. cit.* note 35, 110–113, 120–123.
67 Benedict Crowell, *America's Munitions, 1917–1918* (Washington, DC: Government Printing Office, 1919), 104; Plumpe, *op. cit.* note 20, 92; BAMA PH3/42 (Major von Ellerts, 'Die Munitionsversorgung des Feldheeres...1914–1918', 1922), Anlage II, 2.
68 National Archives (Kew) MUN 4/7056, *Report of the British Mission Appointed to Visit Enemy Chemical Factories in the Occupied Zone Engaged in the Production of Munitions of War* (Ministry of Munitions of War, Department of Explosives Supply, February, 1919), Vol. 6, 21–22; SUPP 28/216 (Dormagen Filling Plant, C. 4513, 28 March 1922).

[69] Abelshauser *et al.*, *op. cit.* note 9, 184–189; cf. Jeffrey Allan Johnson and Roy MacLeod, Chapter 13 in this volume, which examines limitations on Allied arms controls and disarmament in postwar Germany.
[70] Cf. BAMA W-10: 50521, *op. cit.* note 18; 50522, *op. cit.* note 25; 50530 (=MSg2/760a), *op. cit.* note 30.
[71] BAMA PH2/87, Friedrich Stuhlmann, 'Munitionsbedarf und Rüstungsindustrie' (July 1937), 6.
[72] Cf. Abelshauser *et al.*, *op. cit.* note 9, 273–277.

# MOBILIZATION AND INDUSTRIAL POLICY: CHEMICALS AND PHARMACEUTICALS IN THE FRENCH WAR EFFORT

## INTRODUCTION

As fruitful in its work for peace as in its work for war, through the multiplicity of its operations, through their value and their results, is chemistry not in essence the powerful good fairy of progress and economic prosperity? Nowadays, there are few industries that have no need for its assistance, that have nothing to gain from its efficiency. Not only are the metallurgical industry for the transformation of metals, the glass and ceramic industries for the perfection of their processes, and the perfume industry for the composition of essential oils, unable to do without its support and advice; but also agriculture for its fertilizers, the food industries for product purity, and the therapeutics industry for drugs, are obliged to call on its services. Chemistry, nowadays, is the foundation of all industries. It is the science whose field of work is the broadest.[1]

These words recall how essential may be the chemistry for the whole industry, and explain how important may have been the creation of the 'office des produits chimiques et pharmaceutiques' during the war. This speech, delivered by Paul Kestner to the first session of the Society of Industrial Chemistry (Société de Chimie Industrielle) in March 1918, describes chemistry and its main uses. It also shows how industrial chemistry had become so important.[2] In wartime, the organization of the chemical industry had many strategic consequences. The experience of chemical organization during the First World War proved useful to France.

In France, chemical mobilization was necessary to satisfy both military and civilian needs. Under these conditions, French industry succeeded in overcoming its dependence upon German enterprises and technical devices, and established new methods of production and management. Mobilization for civilian needs, provided a way for the State to promote new organizations, both public and private, once the war was over.

## CHEMISTRY IN WARTIME

In tracing wartime chemical industries, one may distinguish between the organization of production for military needs, involving explosives and poison gases,[3] for which the War Ministry was responsible, and the organization of production for civilian needs. In 1914, the general pattern of the chemical industry in France was distinguished by a low degree of state interventionism, and also by a remarkable lack of concern about the need for state control. Politicians appeared quite divided on the merits of interventionism. On the one hand, political radicals had a programme for promoting solidarity and tariff protection for insufficiently developed businesses (like dyes), or for activities with endemic financial difficulties (like railways). The nature of French parliamentary government also acted to slow down rapid change or development in economic policy. On the other hand, French enterprises and professional unions rivalled for control of the market. Some of them, like the Comité des Forges, were famous.

In the early months of the war, the French government devised a new economic system to make up for lost industrial areas in the north and east occupied by the Germans. On 20 September 1914, Alexandre Millerand explained the terms of this 'economic mobilization'. The rules governing the granting of contracts for military supplies were redefined; and contract holders were regulated, in order to obtain better execution of agreements. In addition, skilled workers sent to the front were recalled to the factories, beginning in October 1914.[4] Some professional unions negotiated military contracts with the government. The Comité des Forges became a main contract holder, and remained so until the end of the war.

The new economic policy of France developed as the war progressed, and successive administrations assisted in the emergence of a war economy. In the summer of 1916, new men come to power: Etienne Clémentel, Albert Thomas, and Louis Loucheur. Clémentel was a Radical who had little influence, but he relied on the support of the administrative services and industrial organizations, and came to wield greater and greater influence. Albert Thomas was a Socialist, and held the post of Under-Secretary of State for Armaments (Sous secrétaire d'État à l'armement).[5] This was a difficult office: Thomas had to manage the production of arms, and had to find solutions to armament factory strikes in 1917. The revolution in Russia also made it difficult for a Socialist to remain in office. Thomas was obliged to find common ground with the ambitious Louis Loucheur, who was in charge of armament manufacturing. Loucheur replaced Thomas in the government headed by Clemenceau, which came to power in November 1917.

With these changes, came new rules and laws to define 'interventionism'. Imports were more and more controlled, and some products were even prohibited. Consortia came to manage the links between State and industry, by collecting information and organizing the purchase and sharing of raw materials and other goods between industries, the first of which was established in 1917. These organizations were initially responsible for managing shortages in raw materials; later they became intermediaries, linking government and industry to organize production, integrate industry, manage prices and share profits. During the last year of the war, government control over the economy reached its height, and the government authorized itself, from February 1918, to make all decisions relating to production, distribution and trade.

## THE FRENCH CHEMICAL INDUSTRY IN 1914

The war deprived the chemicals industry of many products, raw materials and intermediary products. The cessation of trade weakened the French chemical industry generally, but some French firms, led in highly specialized fields — notably in sulphuric acid, nitric acid, gaseous chlorine, soda, phosphorus, mineral salts, superphosphates and electrochemicals. On the other hand, French firms did not manufacture liquid chlorine or bromine, and there was very little production of potassium hydroxide.[6] Organic chemical production was also weakened by a lack of engineers and research, as France had relied upon Germany for the dyes and organic products that were useful in pharmaceuticals. The pharmaceutical industry was wealthy, and France was a leader in exporting branded products.[7]

Military necessity demanded a huge expansion in sulphuric and nitric acid production. Imports increased, while government helped set up new plants. The development of coal distillation processes provided materials for explosives and for dyes. In this context, phenol was very important, as it was useful for both explosives and drugs. The Usines Chimiques du Rhône became the most important producer of phenol.[8] The first use of poison gas in 1915 led the government to begin gas production, and to organize research into manufacturing processes. The French government also created new drugs manufacturing plants to meet military needs. As chemical products may have several uses, it may have been difficult to set a satisfying organisation to manage their production in War time. Cooperation between French government and scientists proved to be a good one as it was successful to provide civilian and military needs. The 'Office des Produits Chimiques et Pharmaceutiques' was devoted to such a mission.

## THE CHEMICALS AND PHARMACEUTICALS OFFICE

The Office des Produits Chimiques et Pharmaceutiques (Chemicals and Pharmaceuticals Office — OPCP), headed by Auguste Behal, was founded by a decree on 17 September 1914.[9] Behal was a pharmacist, a graduate in physics, and of the École de Pharmacie in Paris, and a member of the Académie de Medicine. In 1908, he succeeded Jungfleisch in the chair of organic chemistry at the Pharmacy Faculty in Paris.[10] Behal chose his colleagues from scientists and professors at universities and technical schools like the École de Physique et de Chimie Industrielles de la Ville de Paris. These men were pharmacists or chemists, collecting information and carrying out research in order to offer technical advice to manufacturers. The Office was in charge of supporting chemicals or pharmaceuticals firms that had any production difficulties during the war.

### The Organization and Mission of the Office

The Office was quite a unique organization which was not administered by manufacturers. It may serve to illustrate the pattern of chemical industries in France: the scientists and academics remained in the universities and research institutions, separate from the firms. It may also explain why the chemical industry of France lagged behind that of Germany.

The Office members and Behal seem to have been optimistic about solving the difficulties of the chemical industries. Behal called on a 'sacred union' between the manufacturers, as there was between the politicians.

> Never has so favourable a situation occurred and I must say that it will never happen again: we are faced with the disappearance of the world's main producer, we must convince ourselves that this disappearance will not be temporary, but prolonged. So we are faced with exceptional conditions, extremely favourable conditions; as in politics, we must strengthen the links between all the firms, we have to forget that we are competing, we have to gather together all our production capacity, we have to establish cartels to meet the needs of England, of the Allied nations, and of the neutrals.[11]

Behal invited manufacturers to work and produce together, assuring them that England would offer skilled workers, machines and raw materials. He also insisted on buying raw materials in France, to stop money leaving the country and to create jobs. Any surpluses were sold in England or Russia.

The Office encouraged substitutions for German imports and the modernization of French firms. As Behal wrote:

> The chemical industry almost completely lacks organization. Of course, there are prosperous firms, but their production is not connected, such that they rely on Germany for part of the products they manufacture. We are at a moment when this organization, incomplete and insufficient even in peacetime, must ensure not only the continuation of normal life within the country, but also a new effort, unsuspected until now; when the necessities of hasty mobilization and of resistance to a sudden offensive are depriving the chemical industry of nearly all its means. This is also the moment when the main industrial areas are being occupied and invaded by the enemy.[12]

Several factories belonging to German firms were sequested for production; these included the Aktien Gesellschaft factory in Saint-Fons,[13] which manufactured dyes before the war. It began producing explosives for the French forces; and the Manufacture Lyonnaise des Matières Colorantes, that belonged to Cassella, another German dyes manufacturer. BASF (Neuville sur Saône) manufactured explosives;[14] while Merck (Montereau) manufactured ammoniac sulphate for the War Ministry.[15] Not all of these plants, however, could be exploited; for example, Gebinder Giulini, an aluminium producer, and Auerbach (manufactures theobromine, cocoa alkaloid or coffee and tea alkaloid) ceased production during the war.

Because the same raw materials were used in the manufacture of dyes, explosives, and pharmaceuticals, it was necessary to prioritize. The Office had to know which products were made by each firm, and the capacities of their plants and machinery. The Office then registered all the raw materials available, and organized their redistribution. The Office also managed the redistribution of stocks sequestered from German plants, according to the needs of French firms, thereby helping a number of factories to start up again. At the same time, however, French factories were forced to submit to the requisitions of the War Ministry, which provoked difficulties when such requisitions were not paid for by the government. Many factories were also deprived of skilled workers. The Office asked the War Ministry for payment of the goods requisitioned, for the return of technicians to the factories and for the allocation of combustibles and transport.[16]

The Office also organized the production of dyes and negotiated agreements with the War Ministry to allocate raw materials that were not used by the explosives manufacturers. Another task was to collect information about chemical production. An enquiry at the beginning of 1915, showed that acetone, acetic acid and acetate represented two-thirds of French chemical production. Nitric and sulphuric acids were manufactured by the Service des Poudres, but the quantities were unknown. Acetic-salicylic acid production was deprived of phenol, which was wholly under the control of the War Ministry. Acetic-salicylic acid was used as an analgesic, and was a military, as well as a civilian, necessity. The Ministry also requisitioned raw materials for producing pyramidon (pyramidon was used as an anti-pyretic, a drug that may fight against fevers) and aspirin, normal civilian production of which was insufficient to meet army requirements.[17] The Office was in charge of production, but it also negotiated for the return of skilled workers, and for the supply of raw materials from the International Committee for Supply (Comité International des Approvisionnements), based in London. The Office supported the setting up of new pharmaceuticals plants in order to manufacture, for example, phenol, using coal distillation processes, synthetic phenol, ammoniac sulphate and picric acid.[18]

## CHEMICAL FIRMS DURING THE WAR

The Office collected information about the activities of chemical firms during the war. Thanks to its records, we know that production decreased rapidly during the autumn of 1914 at Établissements Poulenc,[19] except for the production of iodine and arsenobenzol, (drugs for syphilis). Several closed because raw materials or workers disappeared. The main pharmaceutical enterprises had similar difficulties and all of them requested skilled workers.[20] The Hoffmann La Roche plant was unable to purchase codeine and morphine from Switzerland once the Swiss government prohibited exports.[21] Manufacturers also increased the output of products necessary in wartime, such as camphor and hydrogen peroxide. However, the Société de Traitement du Quinquina reduced its quinine salt production: 12,000 kg instead of its usual 30,000 kg, even though its total production capacity was 40,000 kg.[22] Transport difficulties, the lack of skilled workers, and variations in prices were the main problems.

Such difficulties drove firms to ask the Office for help. For example, the Société des Produits Spéciaux de la Société des Brevets Lumière sent a letter to the Office stating that the firm would carry on the manufacture of cryogenine, an antipyretic purchased by the War Ministry,[23] but requesting a deferment for the chemical engineer in charge, and a supply of raw materials. Usually, the Office was able to assist in supplying firms. However, pharmaceutical firms tended to get raw materials only if a drug was considered essential to the war effort. As the war progressed, production increasingly prioritized the bare essentials, such as quinine and antiseptics.

Increasing the production of dyes was another priority. This offered firms an opportunity to improve their processes, with the aim of becoming independent of German products. This would have been possible if the War Ministry had been able to provide raw materials not already destined for use in explosives or gas. However, civilian dyes production remained very small, and the Office set prices to avoid excessive profits: 'The manufacturers will also agree with the Office as to the characteristics of the dyes that will be offered for consumption.'[24] Dyes manufacturers had to register their stocks and sell their dyes in accordance with the authorization of the Office. At the same time, the French Government tried to import dyes from Germany through the agency of the Swiss government. The Institut Pasteur made dyes for military needs and for some private firms. These dyes were more complex than those usually sold, as they were used for bacteriological work.[25] The Office also managed other kinds of production — for example, several plants in the Alps and Pyrenees began manufacturing liquid chlorine at the end of 1915.

The mobilization of firms through the agency of the Office and the support of the War Ministry led to an overall increase in chemical production. The growth in selected drugs, between 1913 and 1917, is set out in Table 1.

To these quantities may be added a large quantity of the pills, phials and tetanus serum made by Institut Pasteur.

As chemical production increased, it became possible to export. Exports brought money, helped trade and increased imports. The Trade and Finances Ministries allowed some exports to begin at the end of 1915. These were not generally subject to restrictions if intended for Allied countries, but controls over pharmaceutical exports were stiffened. Such controls were legitimate when the product was scarce: any exports of essential drugs, for instance, would deprive the army of them. At the same time, certain exports could not be avoided because they were counterpart to the importation of raw materials. For example, France exported mercury salts to Spain,

Table 1. French Pharmaceuticals Production, 1913–1917

|                   | 1913 (kg) | 1917 (kg) |
|-------------------|-----------|-----------|
| Antypirine        | 1900      | 10,000    |
| Chloroform        | 1800      | 25,000    |
| Hydrogen peroxide | 6500      | 500,000   |
| Iodine            | 400       | 4000      |
| Quinine           | 800       | 90,000    |

[Source: Charles Moureu, La Chimie et la Guerre: Science et Avenir (Paris: Masson, 1920), 162].

but later imported the processed mercury back from Spain. The Office promoted this kind of commercial transaction to the Trade and Finances Ministries,[26] with the proviso that military needs were met first.[27]

The scientists and academics in charge of the Office were not completely preoccupied by wartime requirements; they also looked to the future.[28] The success of the Office lay in managing factories, supply, and transport. The results were all the more outstanding, as the organization was not governed by formal rules and regulations: the manufacturers worked with the Office without pressure, they allowed State interventionism for reasons of patriotism. The cooperation of manufacturers was helpful in achieving success; and, once the registration of stocks revealed a surplus, exports were authorized again. The lessons of the war led to the elaboration of a kind of industrial policy in the post-war period. Throughout the war, the Office improved the use of patents, trademarks, and customs regulations, and the production of dyes, pharmaceuticals and organic products, and promoted industrial chemistry studies.

In 1916, the incidence of paludism on the Macedonian front led to new drug needs. The main drug remained sulphate quinine, the production of which relied upon cinchona imports from The Netherlands, and on the factories managed by the army's Medical Service and by private firms (notably, the Société du Traitement du Quinquina). In 1917 and 1918, a decrease in cinchona imports slowed down quinine production, at a time when the demand for the drug was increasing. During the winter of 1917–1918, cinchona imported from Java arrived late in Amsterdam, which meant that French manufacture could not proceed, since production relied not only upon exports allowed by the Dutch government, but also on timely transport between Java and Amsterdam.[29] At the end of 1917, shortages of cinchona from Java and India amounted to about one-quarter of that needed for a year's production. The Dutch and British governments reduced cinchona exports, as they needed it for their own armies. On February 1918, managers of the Société du Traitement du Quinquina finally succeeded in negotiations with Dutch and English traders, but transport and payment difficulties remained. There was no resolution of these issues until the end of the war.[30]

While the supply of cinchona remained insecure, the population of France, and all of Europe, was affected by the great influenza epidemic, which occurred at a time of increasing scarcity in drug stocks. The Office, with the support of the Trade Ministry and the War Ministry, organized a 'new' mobilization. The Office prioritized the most essential drugs, mainly for the army, and at the end of 1918, for soldiers and civilians threatened by influenza. Its success showed that the Office was able to manage the chemical firms. Constant information about the French chemical industry, and the regular inventories of stocks, production and raw materials, helped achieve this goal.

In October 1918, sulphate quinine, like a number of antipyretics, could no longer be manufactured because supplies were disrupted by army movements.[31] The sharing of sulphate quinine was managed by the Office: each pharmacist received a first aid dose, while the manufacturers kept a little stock to be delivered to pharmacists as required. The antipyretics (pyrazolics) manufactured by Usines Chimiques du Rhône were sent to Paris and then distributed to local pharmacists. Production was slow at first, but Usines Chimiques du Rhône worked night and day to increase it.[32]

Other drugs like hydrogen peroxide, soda and ipecac also came under the control of the Office. Because of the influenza epidemic, the 'free' drugs market disappeared. The French army kept an important stock of sulphate quinine, and at the same time, the Office tried to stimulate organic production. It asked several drug manufacturers to distribute the sulphate quinine offered by the army among pharmacists. As the quantities were very small, the Office controlled prices. The influenza epidemic showed how efficient this organization set up by the Office really was. Even if the manufacturers or pharmacists were critical of controls on prices and production, they collaborated with the Office. This collaboration gave them some security in terms of their supply and their business. Many of them also saw it as a moral, and patriotic duty.[33]

In July 1918, Behal left the OPCP. His successor, Ernest Fleurent, was a chemist who promoted links between science and industry. Fleurent recalls that when the Office was founded, 'the war had completely disrupted chemical and pharmaceutical trade.... The Office was founded to sustain the development of production and the creation of businesses that were lacking.'[34] The Trade Ministry recognized the work of the Office, and took the financial decisions that allowed the building of plants. The Trade Ministry called for 'economic independence' and considered that after the war, the Office should be put in charge, and given the financial support of the government.[35]

The Office retained this position through the 1920s, when it became responsible for the regulation of chemical, pharmaceutical and dyes imports. In 1920, the Office was put in charge of the importation, production and distribution of dyes, but its contribution to the management of pharmaceuticals or to the rebuilding of chemical industry was not defined. The peace treaties eventually brought the Office to an end: during the 1920s, the government preferred new types of organization – a national society for dyes, for instance – and public corporations for making potassium hydroxide. But, even though the Office as an organization disappeared, it remained as an administrative service, and re-emerged when the 'Chemicals Office' (Direction des Industries Chimiques) was founded as part of the Ministry of Industry in September 1940. This new organization was a part of a general organization, with technical offices (energy, technical industries, etc.) dedicated to managing relationships between manufacturers and administration and between French and German governments, as was required by the Armistice.

In retrospect, the 'Office des Produits Chimiques et Pharmaceutiques' was the first experience of collaboration between government and industrial chemistry in France. Combining scientists, academics and bureaucrats to offer technical support to manufacturers, it succeeded in managing production of chemical weapons, drugs and dyes). At the beginning of the 20th century, there were only two or three large chemical companies, such as Saint-Gobain and the Usines Chimiques du Rhône, and many smaller manufacturers which had no skills in chemical engineering, or which relied upon German firms for their raw materials. Thanks to the Office, many of these continued manufacturing throughout the war. In the end, the Office contributed to the idea that State interventionism could be successful. It demonstrated that the French government could also be a manufacturer. Indeed, after the war, the state went on to manage the nitrogen, potassium, and oil industries, and many others. The wartime experience of the Office was thus not merely interesting; it set the tone, and the pace, for things to come.

## NOTES

[1] AN F12 7700, Speech delivered by Paul Kestner at the inaugural session of the Society of Industrial Chemistry, 16 March 1918.
[2] P. Cohendet (BETA), *La Chimie en Europe: Innovation, Mutations et Perspectives* (Paris: Economica, 1984).
[3] Charles Moureu, *La Chimie et la Guerre: Science et Avenir* (Paris: Masson, 1920).
[4] John F. Godfrey, *Capitalism at War: Industrial Policy and Bureaucracy in France, 1914–1918* (New York and Hamburg: Berg, 1987).
[5] Thomas was an independant Socialist and did not belong to the Socialist Party (SFIO).
[6] Moureu, op. cit. note 3, 50.
[7] *Tableau général du commerce*, Direction générale des douanes, Paris, Imprimerie Nationale, 1900–1914.
[8] P. Cayez, *Rhône-Poulenc, 1895–1975* (Paris: Armand Colin et Masson, 1988), 57–64.
[9] Office des Produits Chimiques et Pharmaceutiques, files are preserved as AN F12 7698 to 7710.
[10] From *Figures Pharmaceutiques Françaises: Notes Historiques et Portraits, 1803–1953* (Paris: Masson, 1953), 209–214.
[11] AN F12 7700, Behal, 8 October 1914.
[12] AN F12 7698, 'Report on the Chemicals and Pharmaceuticals Office', 30 October 1916, 3–4.
[13] Suburb of Lyon.
[14] Small town 20 km from Lyon, on the Saône.
[15] Small town south-east of Paris.
[16] AN F12 7698, op. cit. note 11.
[17] AN F12 7698, 'Stocks and Productions', 13 March 1915.
[18] AN F12 7698, 'Report', 16 March 1915.
[19] AN F12 7703, 'Pharmaceutical Industry: Stocks and Firms Production', Établissements Poulenc.
[20] AN F12 7703, 'Pharmaceutical Industry: Stocks and Firms Production', Darasse frères, Pharmacie Centrale de France, Comar et Cie, Adrian et Cie.
[21] AN F12 7703, 'Pharmaceutical Industry: Stocks and Firms Production', Hoffmann La Roche.
[22] AN F12 7703, 'Pharmaceutical Industry: Stocks and Firms Production', Société du Traitement du Quinquina.
[23] AN F12 7704, Marius Sestier, 'Pharmaceutical Industry: Stocks and Firms Production: Produits Spéciaux de la Société des Brevets Lumière', 16 November 1914.
[24] AN F12 7706, Pharmaceutical and Chemical Office, memoramdum, 13 April 1916.
[25] Moureu, op. cit. note 3, 167–169.
[26] AN F12 7699, Behal papers, 20 December 1915.
[27] AN F12 7699, Behal papers, 17 December 1915.
[28] AN F12 7698, Members of the OPCP (1916): Arsandaux: professeur de minéralogie à l'école de physique et de chimie; Auger : Maître de conférences à la Sorbonne (matières minérales, colorantes et organiques); Blaise: Maître de conférences à la Sorbonne, professeur à l'école de physique et de chimie industrielles (goudrons de houille et dérivés, produits pour la Guerre); Charabot: professeur à l'Ecole des Hautes Etudes Commerciales; Ders: attaché commercial, régisseur par économie; Detœuf:

préparateur à l'Ecole supérieure de pharmacie (pharmacie galénique, spécialités, objets de pansements, articles de chirurgie); Fauconnier: agrégé de chimie à la faculté de Médecine de Paris; Freundler: Maître de conférences adjoint à la faculté des Sciences (produits pharmaceutiques organiques); Marie: chef des travaux à l'institut de Chimie appliquée (grande industrie chimique); Marquis: chef des travaux à l'institut de Chimie appliquée (matières premières naturelles, alcaloïdes, glucosides, cires, vaselines, essences); Sommelet: professeur agrégé à l'Ecole supérieure de Pharmacie (produits pharmaceutiques minéraux et produits organiques salyciliques).

29 AN F12 7705, Société du Traitement du Quinquina papers, September 1917 and January 1918.
30 AN F12 7710, Chemicals and Pharmaceuticals Office, 7 February 1918.
31 AN F12 7710, Influenza papers, Chemicals and Pharmaceuticals Office, October 1918.
32 *Ibid*
33 As they wrote to the Office when they asked for raw materials and recalled that they worked for the army.
34 AN F12 7699, E. Fleurent, Chemicals and Pharmaceuticals Office, June 1918.
35 AN F12 7710, Chemicals and Pharmaceuticals Office, June 1918.

# FIRST WORLD WAR EXPLOSIVES MANUFACTURE: THE BRITISH EXPERIENCE

## INTRODUCTION

On the eve of war in August 1914, military explosives manufacture in Great Britain was split between the State enterprises at the Royal Arsenal, Woolwich, and the Royal Gunpowder Factory, Waltham Abbey, and around ten private producers. The principal needs of the services were for a propellant and shell fillings, and for other explosives in lesser quantities. Cordite filled the first role; black powder and picric acid or lyddite — a high explosive[1] — the second. It was an industry developed to supply the relatively modest needs of a small regular army, which trained for mobile open warfare and was mainly engaged in policing the empire. The Royal Navy was entrusted with guarding the country against invasion and securing the sea-lanes of the empire, and, although it was rarely in action, its guns consumed large quantities of explosives.

In the autumn of 1914 as the western front campaign in Belgium and northern France opened, the type of war for which the British army was prepared, and which it had envisaged, quickly stagnated into a static war, waged from parallel series of heavily fortified trench systems. To break this impasse, Field Marshall Sir John French, Commander-in-Chief of the British Expeditionary Force, declared in February 1915 that 'the problem set is a comparatively simple one, munitions, more munitions, always more munitions'.[2] In particular, the army required vast quantities of high explosive shells to break through German trench lines.

## PROPELLANTS

From its introduction in the 1890s as the principal British service propellant, an assured supply of cordite was of vital strategic interest to the government. Clive Trebilcock has described the way in which the British government carefully nurtured the development of the cordite industry. By alternating the award of cordite contracts between different firms, it was able to create extra capacity by effectively compelling the companies to erect the necessary plant at their own expense.[3]

Before the war, the yearly demand for cordite stood at around 3600 tons (3657.6 tonnes); the Royal Gunpowder Factory at Waltham Abbey supplied about one third, and seven trade factories the remainder, most of that went to the Royal Navy.[4] Cordite factories were some of the largest and most complex explosives factories. They required large areas of flat land, a reliable water supply, good transport links, a large workforce, and plant to concentrate sulphuric acid and to manufacture nitric acid, as well as facilities to recover spent acids. Units were also needed to produce guncotton and nitroglycerine before the manufacture of cordite could begin (See Figure 1).

In the early stages of the war, cordite production was increased by extensions to existing factories. Plant was also used more effectively with the introduction of shift working. Prior to the declaration of war the largest private explosives producer,

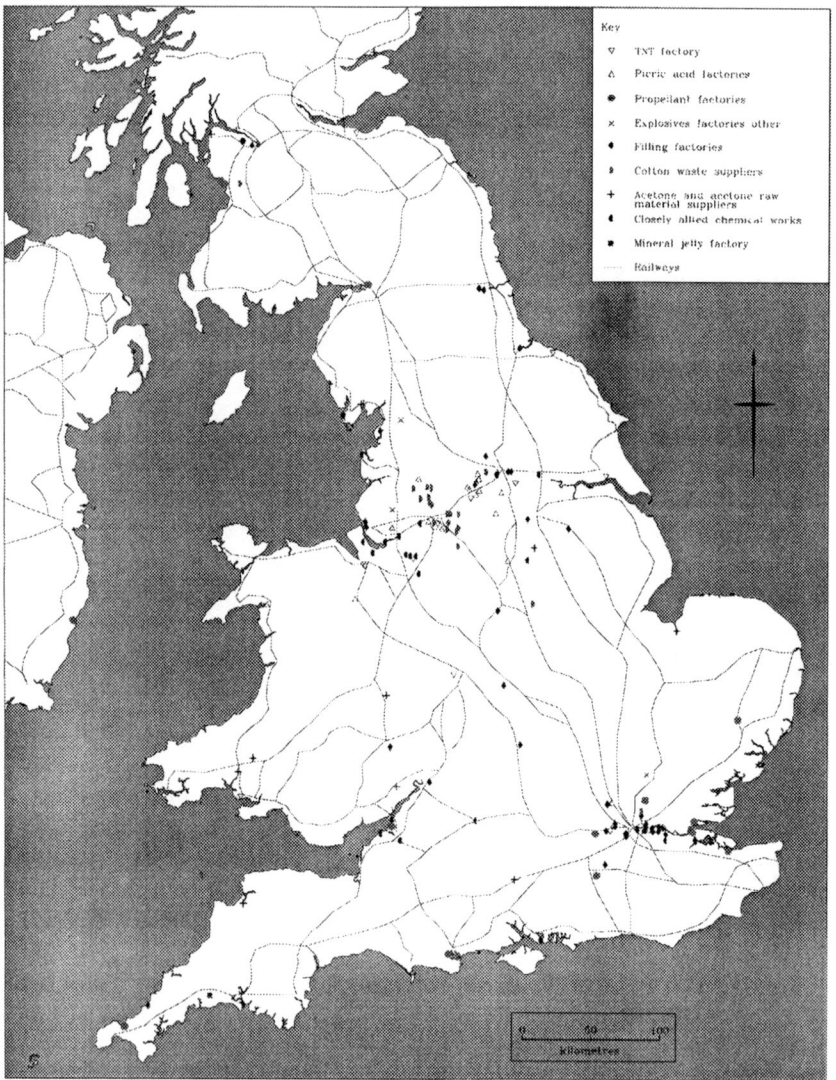

Figure 1. Map showing the distribution of explosives factories during the First World War. © Crown copyright.NMR

Nobel's, had already planned to increase cordite production by building a new factory at Pembrey in South Wales. Its design, entrusted to Sir Frederic Nathan, later informed the design of the new cordite factory adjacent to Nobel's Ardeer factory in North Ayrshire, and later the government factory at Gretna. Nathan's career illustrates the close links that developed between the government service and the private manufacturers. He had joined the Royal Artillery in 1879, and rose to become the Superintendent of the Royal Gunpowder Factory at Waltham Abbey, before joining Nobel's in 1909. He returned to government service in 1915, and took charge of the construction of the new Royal Naval Cordite Factory at Holton Heath, Dorset, before his great organizational skills were harnessed as Controller of Propellant Supplies at the Ministry of Munitions.[5]

A key part of the cordite manufacturing process was the incorporation of the cordite paste — a mixture of highly nitrated (insoluble) guncotton and nitroglycerine — using acetone as the solvent. Britain's reliance on cordite made a secure supply of acetone critical. Before the war, acetone was derived mainly from the destructive distillation of wood, and much was imported from Austria and the United States of America. One of the most innovative processes devised during the war (and one with the most far-reaching consequences) was the synthetic manufacture of solvents using a bacterium to ferment acetone and butanol from starch sources, including potatoes, maize and chestnuts. The process has been closely associated with Chaim Weizmann, although it was the product of the pre-war work of around fifteen scientists working in research laboratories set up around the turn of the century.[6] Remarkably, part of the acetone plant at the former Royal Naval Cordite Factory at Holton Heath survives (See Figure 2) and is now protected as a scheduled monument. Study of its fabric has revealed the careful attention paid by the construction engineers to the pipework's seals, an essential feature to maintain the sterility of the operation.[7]

As a further step to conserve supplies of acetone-based cordite — due to its stability and uniformity of effect, the preferred choice for naval gunnery — the Research Department at the Royal Arsenal at Woolwich devised a new type of cordite for land service. This was known as cordite RDB. In place of highly nitrated insoluble guncotton, soluble nitrocellulose was substituted, and ether-alcohol was used as the solvent. This resulted in scarcity and higher prices for ether-alcohol, produced by the alcohol distilleries. Despite this, and although it was more expensive to manufacture, cordite RDB was accepted for use in May 1915, and put into production as a war emergency measure.[8]

However, this innovation had consequences. Notably, it increased demand for glycerine. To maintain ballistic performance, it was necessary to raise the nitroglycerine content to 42 per cent compared to the 30 per cent content of pre-war cordite MD.[9] Sections of the Royal Gunpowder Factory were converted to manufacture cordite RDB, and plans for the new cordite factory at Gretna were redrawn to include additional facilities for the production of the new type of cordite. Yet another adjustment — in this case, to conserve supplies of cordite for use in larger guns — was the substitution of single-based propellants for use in rifle cartridges,

Figure 2. Royal Naval Cordite Factory, Holton Heath, Dorset, remains of the fermentation vessels used for the production of acetone (BB94/16943) © Crown copyright.NMR

a large quantity of which were imported from America.[10] New factories were also built at Irvine in Scotland and at Henbury, near Bristol, to manufacture nitrocellulose powders.

## HIGH EXPLOSIVES

In contrast to the case of cordite, whose manufacture was well developed and understood, the capacity to produce large quantities of high explosive had to be built up almost from scratch. Before August 1914, the types of operations in which the British Army was predominantly engaged, called for anti-personnel shrapnel shells, exploded by a blackpowder charge. The smaller numbers of high explosive shells used by the army were filled from the mid-1890s with lyddite. Soon after the outbreak of war, trinitrotoluene (TNT) — which had been in use in Germany since the turn of the century — was recommended as the preferred type of shell filling. TNT, although less powerful than lyddite, used smaller amounts of raw materials in its manufacture, which made it cheaper to produce. It could also be used in a less than pure form, and further savings in raw materials could be made by diluting TNT with ammonium nitrate to form 'amatol', which produced a relatively cheap shell filling.[11]

In the production of high explosives, Britain was in a strong position. Her coal by-products industry and town gasworks were able to supply a large proportion of the raw materials for the country's lyddite and TNT requirements. Nevertheless, this took prodigious quantities of coal. To produce a typical weekly output of TNT in wartime, for example, required 720 tons of toluene, which in turn required 600,000 tons of coal. Geographically, the production of high explosives and their precursors was concentrated in the north of England, between Manchester and Leeds, in an area closely associated with the manufacture of synthetic textile dyes and other coal tar derivatives. Another significant group of factories was located along the river Thames, close to London (see Figure 1).

Before the war, the manufacture of TNT was restricted to two private companies, the Clayton Aniline Company and Nobel's, the latter having a capacity of just ten tons per week.[12] In comparison with other areas of munitions supply (especially shells, whose manufacture was often organized by locally self-appointed committees) the supply of high explosives was in November 1914 put into the hands of a newly-appointed Committee on High Explosives, under the chairmanship of the lawyer, Lord (Fletcher) Moulton, FRS.

Moulton quickly instigated a survey of private chemical works capable of producing, or being adapted to manufacture, high explosives. In addition to these private concerns, the government itself also entered into bulk high explosives manufacture for the first time. On Moulton's initiative, an exemplary national TNT plant was constructed adjacent to Messrs. Chance's factory at Oldbury, near Birmingham, supervised by chemists from the Research Department at Woolwich.[13] Following established practice at the Royal Gunpowder Factory, the Oldbury TNT factory was open for private producers to inspect what was considered to be best practice. By May 1915, it was producing 100 tons (101.6 tonnes) of TNT per day.[14]

At the beginning of the war, the manufacture of TNT was poorly understood. The main risk was thought to come from its flammability, and, as a result of the urgency to increase production, some factories were built in inappropriate urban locations. Almost inevitably, this had tragic consequences. For example, a factory in a former textile mill at Ashton-under-Lyme, Greater Manchester, and a factory constructed in a former caustic soda plant at Silvertown, in London, both blew up with heavy loss of life. Less visible as catastrophes were the fatalities and sickness caused by TNT poisoning.

One of the most important technical challenges put to the British explosives industry was the need to devise methods to increase the ratio of ammonium nitrate to TNT in 'amatol'. British manufacturers soon managed to raise the proportion to 80 parts ammonium nitrate to 20 parts TNT, resulting in significant savings in raw materials. German industry, desperately short of toluene supplies, never succeeded in increasing the proportion beyond 40 parts ammonium nitrate.[15] A new industry developed to supply ammonium nitrate, led by the great alkali concern of Brunner Mond and Company, located in the north-west of England.[16] Their chief chemist, Dr Francis A. Freeth, developed a process for the industrial manufacture of ammonium nitrate, based on the work of Dutch chemists, which was patented in 1910.[17]

Through the technical expertise of Brunner Mond's research workers and plant managers, the production of ammonium nitrate was raised from around 1000 tons (1016 tonnes) per year, to an output of over 200,000 tons (203,200 tonnes) between and 1915 and 1918.

## THE FILLING FACTORIES

During the first months of the war, with the exception of explosives production, the coordination of munitions supply suffered from the absence of strong leadership and direction. This culminated in May 1915 in the 'shells scandal', when it was claimed in *The Times* that the shortage of high explosives and shells had contributed to the lack of progress and heavy loss of life on the Western Front.[18] Partly in response to this criticism, a newly formed coalition government under Herbert Asquith created the Ministry of Munitions in June 1915, which was to be responsible for all areas of munitions supply and manufacture. Under the new ministry, directed by David Lloyd George, a system of national factories was planned. The selection of sites, design of factories, and manufacture of process plant and machinery, together with the recruitment and training of a labour force, was a complex task. Nevertheless, by spring 1916, the first of the new national factories was in production, in readiness to equip Kitchener's New Armies and in preparation for the Somme offensive of that summer. Most of the new factories were working by 1917, and by the end of the war, the Ministry of Munitions was operating around 200 factories, manufacturing everything from aircraft, explosives and shells to boxes and concrete slabs.[19]

To ensure a steady supply of shells for the front, one of the most pressing needs was for 'filling factories' — where the various components and explosives were brought together for assembly. Under strong central direction, the emergence of standardized factory designs and plant might be expected. In practice, considerable latitude was given to individual preference. At Chilwell, in Nottinghamshire, Viscount Chetwynd, a former steelworks manager with no previous experience of handling explosives, took a novel approach to the design of a factory to fill shells with amatol. After visiting a number of French explosives factories, and unimpeded by convention, he adapted coal-crushing, stone-pulverising, sugar-drying, paint-making and sugar-sifting machinery, and used porcelain rollers usually found in flour mills to grind TNT. Contrary to normal practice in the explosives industry, he built multi-storey mills similar to those used for milling flour. Despite being a government factory, Chilwell so closely identified with Chetwynd that a monogram of crossed 'C's was applied to lamp posts and the ironwork of buildings' balconies (Figure 3). The factory filled prodigious quantities of munitions, including over nineteen million shells, 25,000 sea mines, and 2500 aerial bombs.[20] It was also the scene of the worst accidental explosion of the war, in 1918, when 134 people were killed.

At Greenford, West London, a similar arrangement was entered into, whereby A.G.M. Chalmers was asked to undertake the design, construction and management of a factory for filling gas shells. There, H.M. Office of Works designed the

Figure 3. Chilwell, Nottinghamshire, National Filling Factory No.6, balcony of the officers' mess and main offices showing Chetwynd's crossed 'C's monogram, but beneath a King's crown (AA96/3561).
© Crown copyright.NMR

buildings, while Chalmers was responsible for the machinery and its arrangement, and later for the management of the factory.[21] By 1918, the Ministry of Munitions operated twenty-two national filling factories, together with other specialized filling factories — for example, designed for the needs of the Trench Warfare Department. The policy of permitting strong local or individual control led to a multitude of factory designs, often resulting in plants that were closely designed around a specific production process, with little flexibility for adaptation.

## CONTINENTAL AND EMPIRE CONNECTIONS

During the late 19th century, the development of the British explosives industry was heavily reliant on foreign expertise. This know-how was brought by individuals, such as the Hungarian Oscar Guttmann, but was also represented in factories established by continental companies — most notably by Nobel's, but also smaller concerns producing patent explosives. This dependence on foreign specialists continued throughout the war. Some links — such as the continuing employment of German workers by the Chilworth Gunpowder Company, a firm partly owned by the Vereinigte Rheinisch-Westfälische Pulverfabriken — finally ceased in 1915, when the links within the Nobel Dynamite Trust were also cut.

Despite the efforts of Britain's chemists, the country was short of 'chemical engineers' with the skills to build the new munitions factories, and of factory chemists, and their assistants, to manage the factory sections and to undertake quality control. Fortunately, Britain was able to make up part of this shortfall by chemists drawn from the empire.[22] Amongst their number was the chemical engineer Kenneth B. Quinan, who made the single greatest contribution to the British war effort. Quinan was an American, who before the war had been employed by the Cape Explosives Works in South Africa. He designed some of the largest government plants erected during the war, including the factory at Queensferry in North Wales, which was capable of producing staggering amounts of explosives, including 500 tons (508 tonnes) of TNT and 250 tons (254 tonnes) of nitrocellulose per week.[23]

Even larger was the cordite factory he designed on the English and Scottish border close to the village of Gretna. As the largest explosives factory in the empire, this covered 9000 acres (3642.3 hectares) and stretched for seven miles (12 km) and employed nearly 20,000 people.[24] Gretna had a maximum production capacity of 1000 tons (1016 tonnes) of cordite per week and, given economies of scale, was able to produce cordite at a price around 25 per cent lower than before the war.[25] Such factories were not only remarkable for their scale, but also for the use of innovative chemical technology. At Gretna and Queensferry, this included the manufacture of sulphuric acid by the contact process developed by continental acid producers.[26] At Gretna, too, Quinan introduced his patent guncotton stoves, a number of which survive. They combined the twin advantages of a faster drying time with improved safety, as there was less guncotton in a stove at a given time, as compared with older designs.

Continental expertise was also drawn upon to build the smaller raw materials plants, which were also critical to production. In early 1915, a distillation plant owned by Shell, which was capable of producing toluene from Borneo petroleum, was brought from Rotterdam and re-erected at Portishead, near Bristol. Soon afterwards, an almost identical plant was constructed at Barrow-in-Furness in Cumbria. During the war, these two factories produced almost the same amount of toluene as the entire British coal-gas industry. Under Shell's chief engineer, W.R. Aveline, company chemists, who included Dutch citizens, also assisted in the construction of the nitration and TNT plants at Oldbury and Queensferry.[27]

Another example of resourcefulness in drawing on expertise was the National Ammonium Perchlorate Factory, established in 1916 at Langwith in Derbyshire, which similarly relied upon technology supplied by Carlson's of Stockholm.[28] Another, although unfinished, factory was intended to produce nitric acid, which was essential for the production of most explosives, and which was throughout the war produced from sodium nitrate from Chile. This source was vulnerable to interception by German U-boats, and also occupied valuable shipping space in a three-month round journey.[29] To lessen this dependence, plans were drawn up to build a synthetic ammonia plant at Billingham, Teesside, using the Haber-Bosch process devised by German chemists before the war. However, the development was incomplete at the time the war ended.[30]

## ALTERNATIVE PERSPECTIVES

The production of munitions in Britain during the First World War is comparatively well documented. Sources include the official twelve volume *History of the Ministry Munitions* published in the early 1920s, individual factory histories, and unpublished records held by the Public Record Office, along with secondary histories derived from these. The Ministry of Munitions, and later the Department of Scientific and Industrial Research, commissioned William Macnab to document some of the most innovative processes devised during the war.[31] Nevertheless, sources and resources have not been exhausted; and questions can change and enlarge. Many local archives remain unexplored, and often contain material not found in national archives. Other sources, beyond what may be regarded as traditional areas of documentation, may also be used to increase our understanding of production processes, factory architecture, working conditions, and the social context of manufacturing activity.

Few wartime participants remain alive, but the sound archive of the Imperial War Museum, London, holds many interviews with former munitions workers. Some of the most valuable untapped sources of information are contemporary photographs. These were taken for a variety of reasons; some to document construction work, others consciously commissioned as a historical record.

One of the most important collections was produced in anticipation of the foundation of a National War Museum (later renamed the Imperial War Museum). Its subcommittee was established to record the contribution of women to the war effort, and photographers were appointed to record their work.[32] Official statistics generally reveal few facts about factory life, beyond perhaps the percentage of female staff employed. Photographs can be used to uncover valuable information on working conditions, the roles and activities undertaken by men and women, their ages, clothing, and machinery and plant, which rarely survive as historical artefacts (Figure 4).[33]

The images of large numbers of women in British factories contrast with the French experience, where numerous wounded and convalescent servicemen were employed in the munitions industry.[34] Contemporary photographs are also often the only evidence of the appearance of munitions factories, especially where they were hastily erected in timber and quickly dismantled after the end of the war.

The construction of new munitions factories represented just one element of the militarization of Europe's landscape. There was a grim symmetry, hinging on the Western Front, of munitions factories, depots and supply lines delivering armaments to the killing fields.[35] It is only now, after the end of the Cold War, that much of this land is being returned to civilian use, thereby giving access to many previously restricted installations, and creating opportunities to study their physical fabric.[36] This approach is able to open up new sources of information, avenues of research and perceptions about this great enterprise over and above what is available from traditional documentary sources.

Even in the countries with best documentation, factory plans and drawings of plant may be lacking. This may be partly remedied by archaeological survey, such

Figure 4. In this view of the top floor of the Cunard's Shellworks, Liverpool, the women are working on 4.5-inch shells. In the foreground the operator is cutting a groove at the base of the shell ready to accept the copper driving band. On the adjacent lathe the operator is using a gauge to test the depth of the cut and on the centre bench the dimensions of the shell are being checked with another gauge (BL24001-21). © Crown copyright.NMR

as the recent identification and record of wartime extensions to the Chilworth gunpowder works in Surrey. While some structures or groups of buildings leave distinctive footprints, ascribing functions without supporting documentation is often problematic. Among the features that munitions manufacture brought to 20th-century factory design was the application of scientific management to factory layout and the improved organization of the workforce. Whereas, before the war, a skilled worker might have carried out a number of tasks, the policy of dilution broke down the manufacturing process into a series of repetitive tasks that could be carried out by unskilled, and often female, labour. Because of these developments, spatial analysis has become one of the most rewarding ways of studying large factory complexes. Plans may reveal the incremental growth of a factory. Even where wartime factories represent a single phase of activity, simple block plans may be used to study organizational structure and production processes (see Figure 5). Moreover, as surviving original drawings of structures and plant within factories are comparatively rare, careful recording work in the field can often allow the manufacturing process to be reconstructed. Some of the most revealing evidence

# FIRST WORLD WAR EXPLOSIVES MANUFACTURE

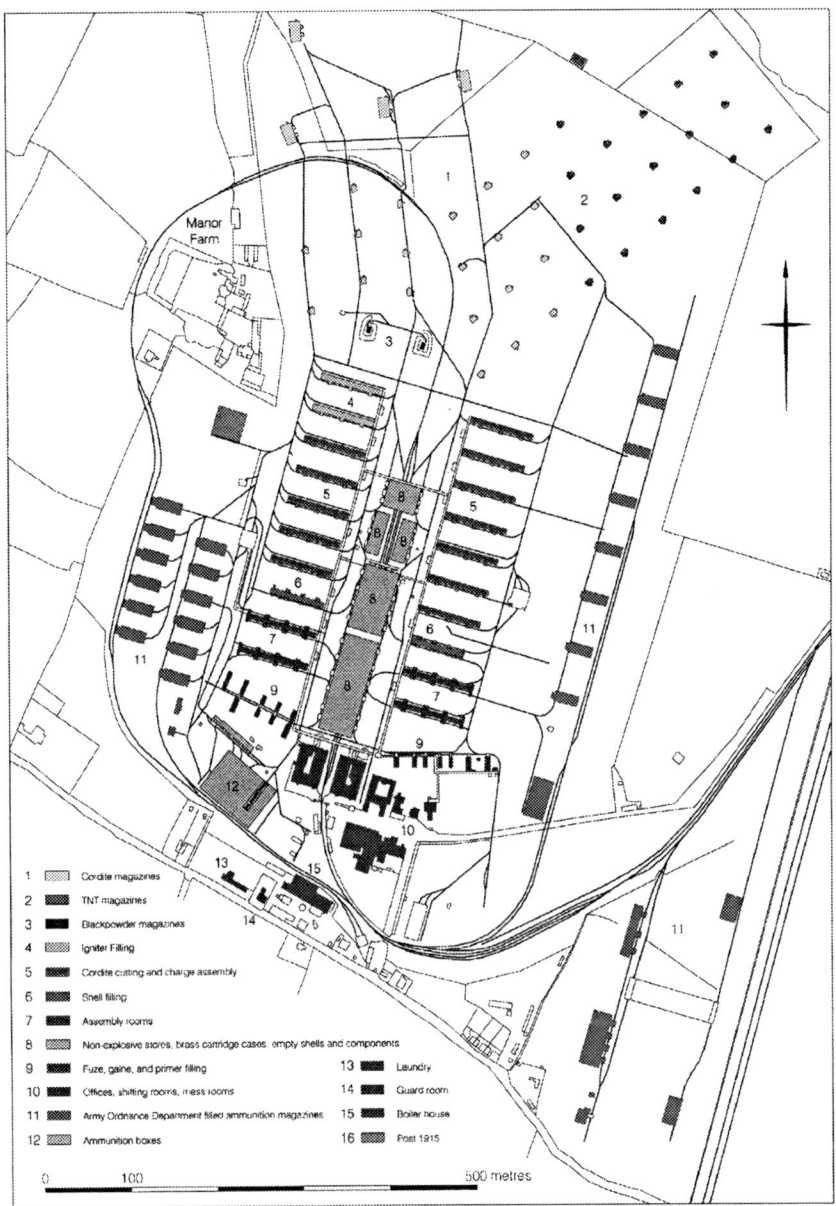

Figure 5. Quedgeley, Gloucestershire, National Filling Factory No.5, block plan illustrating the main activities of the building within the factory. Site plan redrawn from Ordnance Survey, Gloucestershire, 1923, 25-inch, Sheets XXXIII.10 and XXXIII.14. © Copyright.NMR

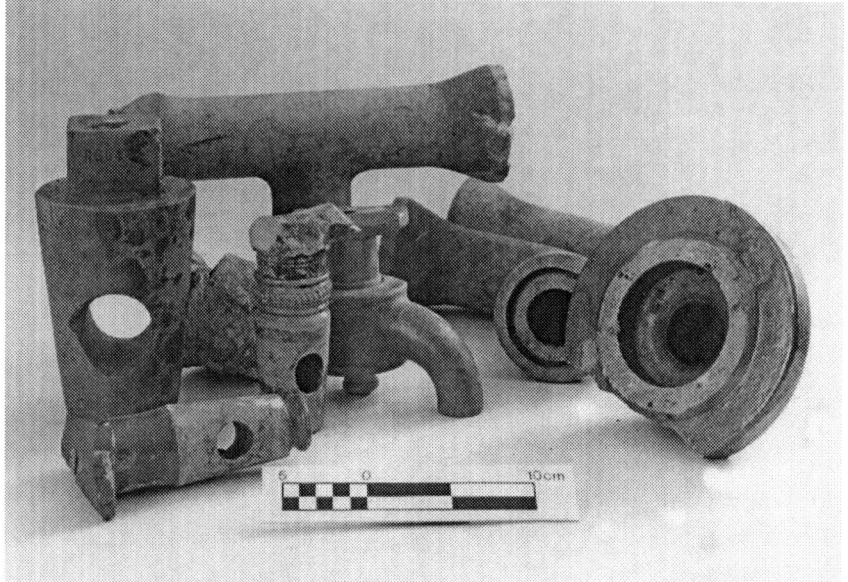

Figure 6. Royal Gunpowder Factory Waltham Abbey, Essex, examples of excavated earthenware taps and pipes used in the acid factory. Before the development of acid resistant metals, the chemical industry was reliant on suppliers of specialist earthenwares (BB94/7995). © Crown copyright.NMR

of former production activities is to be found in dumps of industrial earthenware, an adequate supply of which was in itself a limiting factor in the expansion of chemical production (see Figure 6).

In addition to chemists, wartime government departments employed teams of architects, some of them eminent in their profession. At the Ministry of Munitions, the head of the design team of the Explosives Supply Department, which was responsible for design of HM Factory, Gretna, was Raymond Unwin. At HM Office of Works, Frank Baines was the principal architect, and amongst his work was the lyddite factory at Rotherwas in Herefordshire. The designs they produced — especially for the public or accessible face of these factories — sought to convey a deliberate image. At the beginning of the 20th century, many prestigious factories — such as Marconi's works erected in 1913 at Chelmsford, Essex — were fronted by grand and ornate administrative ranges, expressing the international standing of the company. During the war, neo-Georgian buildings became what could be regarded as the official style. It was more restrained than pre-war styles and, in contrast to contemporary buildings in an Arts and Crafts style, many of its elements such as sash windows could be prefabricated as standardized units. Nevertheless, the construction of imposing entrances and offices represented a considerable investment in skilled labour, construction time, and materials.

Their design raises the question of what these buildings, built at time when the country was locked in a deadly struggle, actually represented. Some of the new

Figure 7. His Majesty's Factory, Avonmouth, near Bristol, office building of a factory initially built to produce picric acid, it was later converted manufacture produce mustard gas. The windows are modern replacements (BB97/117). © Crown copyright.NMR

factories, such as the Royal Naval Cordite Factory at Holton Heath, were intended to be permanent establishments; grand style was a reflection of the pride and patriotism inherent in a 'national' factory. In most instances, however, new factories were erected to fulfil short term needs, with little thought to future demands. Even in such cases, such as at the picric acid factory at Avonmouth, Bristol (see Figure 7), there was an impressive neo-Georgian frontage — a style closely associated with the Liverpool School of Architecture, which produced facades behind which factory owners hid the unpleasant realities of the early 20th century.[37]

The style had a particular resonance in the area of munitions, for it was the same architectural style that was associated with the expansion of the Board of Ordnance sites a century earlier, when the country was locked in war with revolutionary France. Outwardly, architectural standards were maintained to the war's end. The National Machine Gun Factory at Burton-on-Trent, Staffordshire, whose construction was approved in 1917, was given an imposing three-storey range resembling a country house; but steel framing, concrete and slab filling were used to speed up construction. Some private manufacturers who benefited from wartime expansion also enhanced their factories with impressive gatehouses, medical centres and administrative blocks, such as those built at Kynoch's Witton cartridge factory, Birmingham, in 1915. Elsewhere, factories were erected quickly, using timber or corrugated iron. Surprisingly, given its comparative cheapness and ease of use, concrete was rarely employed.

## LEGACY

Discussions about the future of the explosives industry began well before the end of the war, with the growing realization that there would be massive over-capacity.[38] With large remaining stocks of explosives, and the belief that a major war in Europe was unlikely in the next ten years ('the 10-year rule'), there could be no justification in maintaining the large network of explosives factories, and most of the new factories were destined for closure.

The government relinquished its control of armaments production, and the Cessation Act wound up the Ministry of Munitions in 1921. Government explosives production and research retreated to Waltham Abbey, Holton Heath, and the Royal Arsenal, with its enlarged Research Department. Nearly all the national factories were closed; some sites were soon cleared, and returned to agricultural use. There was a similar picture of closures amongst the trade factories, and by 1920 most had amalgamated to form Nobel Industries Limited, which was absorbed into ICI in 1926. Under a drastic restructuring scheme, blackpowder manufacture was consolidated at a handful of sites, and commercial chemical explosives manufacture was concentrated at Ardeer in Scotland. Specialist explosives plants offered few opportunities for conversion to peaceful purposes, but other types of munitions factories with large, flexible, covered, open spaces (such as filling and projectile factories), as well as other engineering concerns, were put to new uses.

In particular the emerging automotive industry benefited from the legacy of newly-built munitions factories; elsewhere the orderly layouts of these factories were easily adapted to trading estates with premises for small businesses.[39] Of more enduring use have been the residential estates, many of which were originally built to house munitions workers.

The victors also had opportunities to inspect enemy plant. Of particular interest was the German synthetic ammonia factory at Oppau. Despite the obstructions placed in their way, the visit convinced the directors of Brunner Mond that they should acquire the part-finished artificial nitrate plant at Billingham, thereby forming the basis of the modern chemical industry on Teeside.[40] It was also as a result of contact with German wartime practice that the Royal Naval Cordite Factory began experiments in the 1920s with wood (in the form of paper) as a cellulose source in place of imported cotton. The most important legacy of the vast munitions programme, however, was the experience that it gave government experts who became responsible for armaments production during the 1930s.[41]

In the context of modern conservation agendas in England, this work on the physical remains of the industry has played a central role in informing English Heritage's assessment and designation programmes for the gunpowder — and, more recently, the chemical industry — including chemical explosives.[42] Parts of the legacy are accessible and interpreted to the public, notably within the displays at Waltham Abbey's Royal Gunpowder Mills. But these can give only a limited impression of the scale of the national endeavour, the ingenuity and organization upon which it drew, and the experience of the many thousands whose contribution to the war to end all wars was the labour, and sometimes the lives, they gave to munitions production. For our generation, the physical remains of locations, sites, buildings and artefacts continue to provide distinctive insights not available from other sources.

## ACKNOWLEDGEMENTS

This paper is based on fieldwork and research carried during the early 1990s by the Royal Commission on the Historical Monuments of England, which merged with English Heritage on 1 April 1999. For more detail, illustration and supporting references, see Wayne Cocroft, *Dangerous Energy: The Archaeology of Gunpowder and Military Explosives Manufacture* (Swindon: English Heritage, 2000), especially chapter 6 'The Great War 1914–18'. I am grateful to my colleagues Paul Everson for his helpful comments on the text, and Philip Sinton, who prepared the figures.

## NOTES

[1] When used as a shell filling in British service, Picric acid was known as lyddite, after Lydd in Kent, where it was first tested. In other countries, it was variously called melinite, shimose, pertite and picrinite.

[2] V.B. Lewes, 'Modern Munitions of War', *Journal of the Royal Society of Arts*, LXIII (1915), 821.

[3] R.C. Trebilcock, 'A "Special Relationship"—Government and Rearmament, and the Cordite Firms', *Economic History Review*, Series 2 (19), (1966), 364–379.

[4] Public Record Office (Kew) (afterwards, PRO), MUN7/555, 'Reduction in Output at Royal Gunpowder Factory Waltham Abbey and Question of Future of Factory', 1918–1919.

[5] William Rintoul, 'Frederic Lewis Nathan, 1861–1933', *Journal of the Chemical Society*, 1 (1934), 564–565.

[6] Robert Bud, *The Uses of Life: A History of Biotechnology* (Cambridge: Cambridge University Press, 1994), 37–45; and Roy M. MacLeod, 'The Chemists go to War: The Mobilization of Civilian Chemists and the British War Effort, 1914–1918', *Annals of Science*, 50 (1993), 455–481.

[7] Wayne Cocroft, *Dangerous Energy: The Archaeology of Gunpowder and Military Explosives Manufacture* (Swindon: English Heritage, 2000), 158–159.

[8] PRO SUP5/436, Establishment: Gretna Committee (W.3504), 1919–1920.

[9] John Reader, *Explosives* (Harmondsworth: Penguin, 1942), 112.

[10] HMSO, *History of the Ministry of Munitions* (London: HMSO, 1922), Vol. 10, 3.

[11] Reader, *op. cit.* note 9, 125; MacLeod, *op. cit.* note 6.

[12] PRO MUN7/36, Minister's Queries on Report of the Committee on the Silvertown Explosion, 23 February 1917–9 October 1919; Alan Cartwright, *The Dynamite Company: The Story of the African Explosives and Chemical Industries Ltd* (London: Macdonald, 1964), 136; Luit Haber, *The Chemical Industry, 1900–1930* (Oxford: Clarendon Press, 1971), 189.

[13] Hampshire County Record Office (Winchester), 109M91/RM9, 'Memo on HE by Lord Moulton', 16 June 1915.

[14] Anon., 'Birmingham and the Production of Munitions No. 1', *The Engineer*, 29 March 1918, 270–271.

[15] S. Levy, *Modern Explosives* (London: Pitman & Sons, 1920), 12.

[16] John Watts, *The First Fifty Years of Brunner Mond and Company* (Winnington: Brunner Mond and Company, 1923), 52–54.

[17] Francis Freeth, 'Ammonium Nitrate and TNT in World War I', *New Scientist*, 270 (1962), 157.

[18] *The Times*, 14 May 1915; See MacLeod, *op. cit.* note 6.

[19] PRO MUN5/146/1122/8, 'List of Factories operated by the Ministry of Munitions in March 1918'.

[20] M. Haslam, *The Chilwell Story: VC Factory and Ordnance Depot* (Nottingham: Boots Company, 1982), 1–70.

[21] PRO MUN5/154/1123.3/39, 'History of NFF Greenford, Middlesex (Chemical Shell)', November 1918.

[22] Roy MacLeod, 'The Industrial Invasion of Britain: Mobilising Australian Munitions Workers, 1916–1919', *Journal of the Australian War Memorial*, 27 (1993), 37–46.

[23] Levy, *op. cit.* note 15, 92.

[24] HMSO, *op. cit.* note 10, 3.

[25] PRO MUN7/555, *op. cit.* note 4.

[26] D. Hardie, and J. Davidson Pratt, *A History of the Modern British Chemical Industry* (Oxford: Pergamon Press, 1966), 99.

[27] P.G.A. Smith, *The Shell that Hit Germany the Hardest* (London: Shell Marketing, 1919), 5–11; see also Ernst Homburg in this volume.

[28] PRO MUN 5/294, HM Explosives Factory, Langwith, Nottinghamshire, January–November, 1918; Personal communication, Whaley Thorns Heritage Centre to author.

[29] Robert Robertson, 'Some War Developments of Explosives', *Nature*, 107 (2695), (1921), 524.

[30] Tony Travis, 'The Haber-Bosch Process: Exemplar of 20th-Century Chemical Industry', *Chemistry and Industry*, 15 (1993), 581–585.

[31] Anon., 'The Manufacture of Explosives', *Nature*, 109 (2739), (1922), 541–544.

[32] Gareth Griffiths, *Women's Factory Work in World War I* (Stroud: Alan Sutton, 1991); Gaynor Kavanagh, *Museums and the First World War* (Leicester: Leicester University Press, 1994), 138–143.

[33] Wayne Cocroft and Ian Leith, 'Cunard's Shellworks, Liverpool', *Archive*, 11 (1996), 53–64.

[34] Guy Hartcup, *The War of Invention: Scientific Developments, 1914–18* (London: Brassey's, 1988), 55.

[35] Nicholas Saunders, 'Excavating Memories: Archaeology and the Great War, 1914–2001', *Antiquity*, 76 (2002), 101–108.

[36] See Marie-Luise Buchinger, *Stadt Brandenburg an der Havel Teil 2: Äussere Stadtteile und eingemeindete Orte* (Worms am Rhein: Wernersche Verlagsgesellschaft, 1995), 212–219; Malcolm Bowditch and Lesley Hayward, *A Pictorial Record of the Royal Naval Cordite Factory Holton Heath* (Wareham: Finial Publishing, 1996); Cocroft, *op. cit.* note 7, ch. 6; Sebastian Kinder, 'Brandenburg an der Havel Der Industriestandort Kirchmöser von der Pulverfabrik bis zum Ausbesserungswerk der Reichsbahn', *Brandenburgische Denkmalpflege*, 1, (2000), 4–16; and Stuart Foreman, 'Nitroglycerine Washing House, South Site, Waltham Abbey Royal Gunpowder Factory, Essex', *Industrial Archaeology Review*, 23 (2), (2001), 125–142.

[37] Simon Pepper and Mark Swenarton, 'Neo-Georgian Maison Type', *Architectural Review*, 168 (1002), (1980), 92.

[38] PRO MUN7/555, *op. cit.* note 4.

[39] Paul Collins and Michael Stratton, *British Car Factories from 1896: A Complete Historical, Geographical, Architectural and Technological Survey* (Godmanstone: Veloce Publishing, 1993), 46, 68, 162, 173, 196, 200, 219, 237.

[40] Francis Freeth, 'Explosives for the First World War', *New Scientist*, 402, (1964), 275; Travis, *op. cit.* note 30, 585.

[41] PRO CAB102/627, D. Mack Smith, 'Construction of the Filling Factories'. unpublished narrative, ca. 1945/6.

[42] Gould Shane, *Monuments Protection Programme: The Gunpowder Industry* (Combined Steps 1–3 Report, typescript report for English Heritage, 1993); David Cranstone, *Monuments Protection Programme: Chemical Industry* (Steps 1 Report, typescript report for English Heritage, 2002).

SEBASTIAN KINDER

# TRANSFORMING A VILLAGE INTO AN INDUSTRIAL TOWN: THE ROYAL PRUSSIAN POWDER PLANT IN KIRCHMÖSER (BRANDENBURG)

## INTRODUCTION

The appearance of German cities did not change substantially, if at all, in the wake of the First World War. With the exception of some East Prussian towns that were destroyed, their physical structure was not visibly influenced by the course of the war. However, the severe disruption of cities during the Second World War has overshadowed the massive changes some towns experienced in the First World War, due to the demands of the war economy. Those that became the location of new factories were especially affected. The study of such places can provide interesting insights into the practical implementation of military programmes. They can also show how these programmes were realized, and what constraints they faced. This is particularly important since detailed information on German mobilization, and on the Hindenburg Programme of 1916, is still very fragmentary.

This paper describes and discusses the only state-owned powder plant built during the First World War in Germany. It considers this particular factory site in the context of the general pattern of German powder plants. It aims to show the major constraints that arose during the construction of the plant and within its production processes, and to draw parallels with contemporary programmes elsewhere in Germany.

## GERMANY'S MUNITIONS INDUSTRY ON THE EVE OF THE FIRST WORLD WAR

In the late 19th and early 20th centuries private firms dominated the German munitions industry. State-owned powder plants, on the contrary, played only a minor role within the industry. Indeed, if measured by their output, such plants produced between a third or a quarter of the guncotton powder produced by private powder plants during the war, and only about five per cent of the nitroglycerine produced towards the end of the war.[1]

When the Second Reich was founded in 1871, the munitions industry was extremely prosperous in Germany for two reasons. First, the wars against Austria, Denmark and France, by increasing the demand for munitions, had led to an enormous expansion of the industry. Second, the foundation of the Reich was followed by a rapid process of industrialization that required more explosives for coalmines, quarries and tunnel construction. German munitions and explosives corporations therefore shifted their production from the needs of the military to the demands of other industries.

47

*R. MacLeod and J.A. Johnson (eds.), Frontline and Factory, 47–60.*
© 2006 *Springer.*

Moreover, poor transportation networks and customs frontiers had restricted powder plants to regional markets before 1871, but started to expand into other German regions.[2] This, however, led to greater competition among the leading firms. In the 1870s, the producers of munitions and explosives faced an emerging over-production of gunpowder, despite the continuously high demand for explosives. To overcome the problems of high competition and over-production, most of the private firms merged, step by step, and finally created cartels and syndicates.

In 1873, nineteen Westphalian powder plants joined together in the *Vereinigte Rheinisch-Westfälische Pulverfabriken*,[3] and in 1878, founded a powder cartel. This effectively determined prices in the German munitions industry, and it later developed into a syndicate that controlled not only prices, but also the production output of every plant and supplier contract. Almost all German explosives plants joined this syndicate. With its headquarters in Cologne, the *Vereinigte Rheinisch-Westfälische Pulverfabriken* had most of its production plants in western and southern Germany. The most important northern German powder plant, Wolff & Co in Hamburg, and the famous Duttenhofer plant in Rottweil (Württemberg), remained outside the syndicate until 1890, when they merged under the name '*Vereinigte Köln-Rottweiler Pulverfabriken*'. Before 1918, they represented the most influential powder plants in Germany.[4]

In addition to these privately owned powder firms, there were a few state-owned plants in the kingdom of Prussia, which served the needs of the Prussian army. These Royal Prussian powder plants were located in Spandau, a suburb of Berlin; and Hanau, close to Frankfurt am Main. The two plants could guarantee only to meet the basic demands of the peacetime Prussian army. In wartime, higher demand was met by buying additional gunpowder and other munitions from the privately owned plants. Other state-owned plants were located in the kingdoms of Saxony (the Royal Saxonian powder plant in Gnaschwitz) and Bavaria (the Royal Bavarian powder plants in Dachau and Ingolstadt).[5]

During the First World War, Germany's powder plants were barely able to produce the amount of munitions required. A major problem arose from the lack of raw materials and components. Germany had no natural deposits of some of the materials, and was cut off from suppliers. Therefore, it had to find substitutes and synthetic processes, of which the Haber-Bosch process is perhaps the most famous. For example, Germany used coal for the production of toluol and phenol, and potatoes for the production of alcohol. Saltpetre — an essential raw material for the production of explosives — was not naturally available in Germany, and had to be imported.[6] During the war the lack of saltpetre became a major constraint on German munitions production.

### THE NEED FOR NEW POWDER PLANTS

A few weeks after the outbreak of the war, it became obvious that Germany's strategy, based on the Schlieffen Plan, had failed. This strategy was intended to bypass a multi-front war by beating France in the west within two months, and by then sending troops to the Eastern front. This failure resulted in intensive trench warfare. Thereafter, movement along the front seemed possible only by the use of vast amounts of munitions. But the rising demand for explosives was hampered by major shortages of raw materials. Mobilization plans proposed for a monthly production output of 200 tons, which did not nearly meet the need. By the autumn of 1914, German powder plants were producing 1000 tons per month.[7] The plants were able to raise their output almost continually. Nevertheless, during the four years of the war, their output lagged behind demand.[8]

Only a few weeks after the Battle of the Marne, the General Staff and the Prussian War Ministry realized that the immense and unforeseen demand for gunpowder and explosives could not be met by the existing state-owned and private apparatus. In response, in September 1914, they decided to build an additional plant in Plaue/Kirchmöser, and to expand existing plants in Ingolstadt, Gnaschwitz, and Dachau (Figure 1).[9] How the War Ministry influenced the expansion of the private plants is still unknown, but information suggests that they had to raise production by a given percentage or up to a given amount. How individual firms reacted to these often poorly planned demands is a field that needs further investigation.[10] Most private corporations hastily started temporary facilities, which were intended to be dismantled after the end of the war. The state-owned plants present a rather different

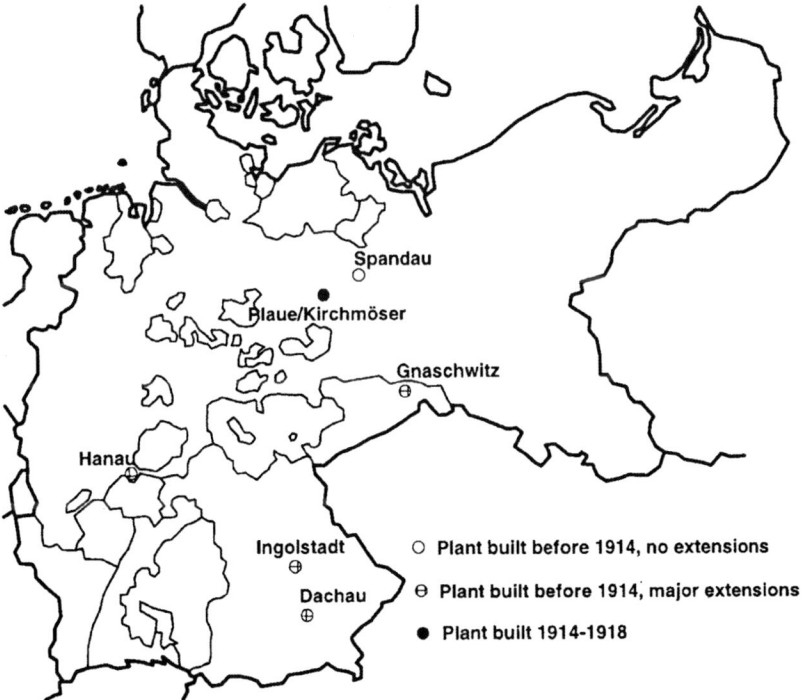

Figure 1. German State-owned Powder Plants, 1914–1918

picture. The extensions of these plants, and the construction of the Kirchmöser plant, were designed not to be temporary, but to be permanent facilities.

After summer 1914, the German explosives industry suffered not so much from a lack of production facilities, as from a lack of components. Guncotton, the basic material in gunpowder, was usually made of lint. After the outbreak of the war, German producers started to use rags and cellulose instead. Eventually, other raw materials were substituted — turpentine, obtained from resin, for example, was used instead of camphor. In 1916, 30,000 workers collected about 3500 tons of resin from German forests. The production of gunpowder also required ether, for which alcohol was needed. Alcohol was obtained in large quantities from potatoes, which then became scarce on the food markets.[11] The production of glycerine, for TNT, required sugar that also came at the expense of food.[12]

In the summer of 1916, following the battle of the Somme, Erich Ludendorff and Paul von Hindenburg replaced General von Falkenhayn. The goal of their so-called 'Hindenburg Programme' — the second phase of the intensification of the war — aimed to double munitions output in six months.[13] But this entailed a doubling of the production of nitrogen, which initially was not available in sufficient quantities. The invention of the Haber-Bosch process, and its implementation by the BASF works, could have solved the problem.[14] However, due to severe wartime shortages and the huge scale of the undertaking, the Hindenburg Programme was never fully completed. The programme intended to extend existing state-owned plants, but its primary focus was on the private firms. Plans were developed for permanent 'readiness plants', but these were only partly constructed by the end of the war. These plants were designed to be easily converted to peacetime uses. In case of a new war, they could be converted to military production. To ensure their readiness, they were tied by long term contracts to the Ministry of War.[15]

## FINDING A SUITABLE LOCATION

While three powder plants experienced major extensions after September 1914, only one completely new state-owned powder plant was built during the war — the Royal Prussian powder plant in Kirchmöser. The search for a suitable location, and the building and operation of the plant, were undertaken and controlled by the *Feldzeugmeisterei* in Berlin.[16] In a secret letter, dated 28 September 1914, to the *Königlich Preußischer Regierungspräsident* in Magdeburg the *Feldzeugmeisterei* underlined the need for an additional powder plant.[17] The *Regierungspräsident* was asked to suggest a suitable location in his province before 10 October 1914. This deadline — a period of less than two weeks — highlights the priority of the project. The letter contained a list of ten criteria for the site, requiring that it be:

1. On or near a waterway in central Germany navigable for ships of least 350 tons loading capacity for easy access.
2. On or near a hilly area of about 350 hectares with one-quarter wooded. The hills should provide protection in case of explosions and the woods should prevent espionage.

3. Next to a main railway line of at least four kilometres for easy access to major transportation routes.

It should also have:

4. The availability of at least 8000 to 10,000 cubic metres of drinking water per day for production and for the workers.
5. No built-up areas within at least 250 metres of three sides of the site, in order to establish a safety zone to protect people from emissions and explosions.
6. Suitable building ground, allowing for the fast erection of buildings.
7. The capacity to accommodate 600 to 800 workers with their families in the surrounding area, and 50 to 60 workers on the site.
8. A major city with building companies and well-established industries.
9. Coalmines.
10. State-owned land in order to keep land costs low.[18]

Similar letters were sent to the *Regierungspräsident* in Merseburg and to the *Herzöglich-Anhaltinische Staatsregierung* (State Government of the Duchy of Anhalt). This shows that the Ministry of War intended to build the new factory in a core area of Germany, far away from the German borders and the fighting fronts. Only two days later, the *Regierungspräsident* in Magdeburg asked every *Landrat*[19] in his province to suggest a location in his county.[20] In a secret letter dated 10 November 1914, the Ministry of War informed the *Regierungspräsident* in Magdeburg that the commission responsible[21] had decided to suggest a site near the village of Möser.[22] The plant had to start production within six months. For the first time, this letter gave information regarding the intended products: guncotton powder, nitroglycerine powder, and other, various sorts of powder.

The site near Möser fulfilled almost all of the requirements.[23] It was situated west of the city of Brandenburg on a large peninsula within the lakes of the river Havel. This river provided access to the industrial centres of the west (e.g. the Ruhr District) and east (e.g. Upper Silesia) as well as to Berlin and the main coastal ports (Hamburg and Stettin). The peninsula was covered with woods, it was hilly, and it offered even more space than the required 350 hectares. The village lay directly on the main Berlin–Magdeburg – Hanover railway line. Owing to its location on a peninsula, there was plenty of water. The village itself was far enough from the site so that inhabitants would not have to fear explosions and emissions. Furthermore, the number of inhabitants in the surrounding area (about 300) was rather low. The building ground on the proposed site was suitable, except for some swampy parts which could easily be by-passed. Workers could find accommodation not only in Möser village itself, but also in other surrounding villages, and especially in the small town of Plaue north of the site and on the other side of the river. Only two locational demands could not be met: coal mines and state-owned property. However, it was easy to send coal by ship or train to the site, and the land was not in agricultural use and was, therefore, cheap.

In a letter to the Deputy General Command of the IV Army Corps, dated 26 November 1914, the Ministry of War explained that because of the high number of private landowners on site, it would be necessary to expropriate the land and to hand it over to the Ministry of War.[24] The Kaiser gave his approval on 29 November 1914.[25] One day later, the land was transferred to public property. Altogether, 563.3 hectares were expropriated for the new plant, for which 1,154,480 Marks were paid to the owners as compensation — the equivalent of 0.20 Mark per square metre. The area consisted predominantly of woodland, but also included public paths and a lake.

## PLANNING AND BUILDING OF THE KIRCHMÖSER POWDER PLANT

Once the location was decided — even before the land became public property — construction planning began. On 12 November 1914, the *Neubaubüro* (Construction Office) and the *Neubauamt* (Construction Department) were set up at the Artillery Construction Offices at the Spandau powder plant. The *Neubaubüro* was responsible for the general layout of the plant and the provision of machines, while the *Neubauamt* dealt with the design of buildings. Both institutions were led by military architects.[26]

As it would normally take far too long to build a plant of such dimensions various steps were taken to start production within six months. The land survey, the design of the buildings, the delivery of building materials, and the construction were all carried out at the same time. Altogether seven companies were chosen, most of them based in Berlin.[27] Each in turn selected local sub-contractors for the delivery of materials and the provision of labour.

The layout plan of the plant, based on estimates about the required number, size and ground plans of the buildings, was accepted by the end of November 1914. The plan was handed over to the *Militärbauamt*, which started to design the various buildings. Because it was still not very clear how many buildings would be required, and what purposes they would serve, the *Militärbauamt* designed about twelve basic types of production buildings, while the building companies began the factories and warehouses. Once the actual uses of the buildings were fixed — which happened mainly in the spring and summer of 1915 — the interiors were adjusted to actual needs.

The general layout of the plant was typical for an industrial plant of its kind. It ignored most of its environmental conditions, and was, by a rectangular street network, divided into plots for the various factories and additional infrastructures. The site was confined by the lakes, and included the area north of the village and of the Berlin–Magdeburg railway. In the southern part, a lake was included for sewage purposes. The whole area was enclosed by a fence that had been souvenired from the fortress in Liège (Belgium) at the beginning of the war. A main road connected the northern and the southern gates of the plant, and together with several east–west roads, the site was divided into various parts dedicated to single factories. In an extension of the north–south main road, a bridge was built to connect the plant with the main Berlin–Magdeburg highway (later the Reich Road No. 1).

Coming from this road and crossing the bridge, one entered a square with some of the main buildings of the plant: the main administration, the officers' mess and the northern gate. Altogether the plant was divided into twelve parts (see Figure 2): the main administration, a workshop group, a power station and waterworks, the guncotton factory, the guncotton powder factory, the laboratories, the TNT factory, the nitroglycerine powder factory, the mixing group, storage buildings, the nitroglycerine factory, and, finally, a fireworks laboratory for the production of fusers, which was administrationally independent from the actual powder plant.

All production parts of the plant were equipped with additional buildings for the storage of raw materials and immediate and final products. The chess pattern layout of the plant allowed each part of the plant to be assigned one or more building plots. In retrospect the plant could, therefore, have been easily extended. In addition, the layout plan included tracks, a small channel for a port, a sewage plant, and a shooting range. Furthermore, north-west of the plant, a site had been reserved for a future housing estate.

Construction involved a number of problems. In the beginning, the building companies had difficulties finding enough trained workers. From January to March 1915, weather conditions were very poor, and a long frost period caused major

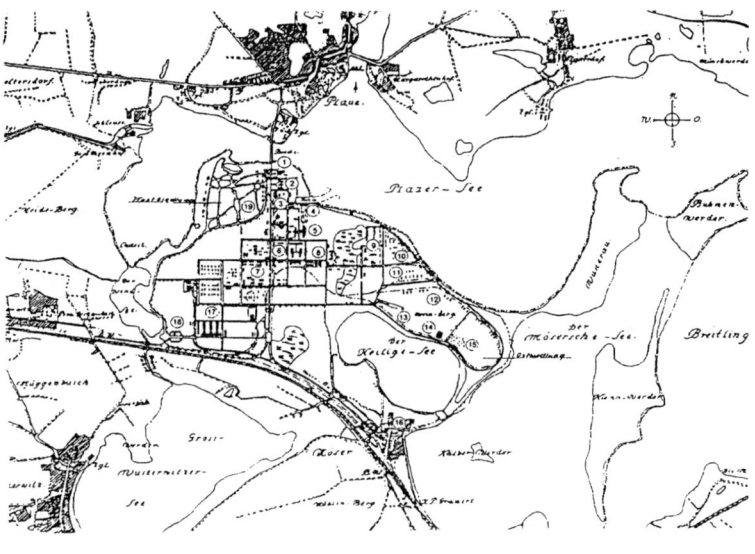

| | | | |
|---|---|---|---|
| 1 | Officers' mess | 11 | Mixing group |
| 2 | Main administration | 12 | Powder storage |
| 3 | Workshop group | 13 | Shooting range |
| 4 | Port | 14 | Nitroglycerine facotry |
| 5 | Power station | 15 | Eastern housing estate |
| 6 | Guncotton factory | 16 | Kirchmöser (village) |
| 7 | Guncotton powder factory | 17 | Fireworkslaboratory (production) |
| 8 | Laboratories | 18 | Fireworkslaboratory (administration) |
| 9 | TNT factory | 19 | Western housing estate |
| 10 | Nitroglycerine powder factory | | |

Figure 2. Layout of the Royal Prussian Powder Plant at Kirchmöser

delays. The difficult terrain, partly swampy and very sandy, complicated the transportation of building materials. Further, the Berlin–Magdeburg railway line was often closed to non-military transports and the waterways were frozen. In addition, the building site lacked accommodation facilities for the workers, many of whom had difficulty getting to the site during the winter months.

Despite these problems, the site was connected with the main railway at the end of January 1915. The building of the power station demonstrates the difficulties. In mid-February 1915, the foundation of the huge building was laid, and on 7 May 1915, the installation of the cauldrons began, although the building site was not yet connected by road or rail. At that time, the building had neither a roof nor windows and doors. Construction workers and fitters had to work at the same time. Nevertheless, the building was finished by 1 July 1915, and most other buildings were finished the same month, so that the installation of machines could start immediately. Within a year, more than 300 buildings had been built as well as fifteen kilometres of roads, thirty-two kilometres of rail tracks, seventy-seven kilometres of pipelines and ten kilometres of fences.

On 12 May 1915, almost exactly within the specified six months after the location had been decided, a testing factory began to produce one ton of guncotton powder per day. This factory was used to train workers, since other powder plants, affected by call up orders, were not able to spare men. By January and February 1915, it became obvious that, while building materials were sufficient, more construction workers were needed. To engage workers from the surrounding region turned out to be time consuming, and numbers rose only slowly, up to 3000 by the end of 1915. Another problem was that most of the workers worked in the agricultural sector and were now needed on farms. The local mayor, therefore, asked the county administrator to use prisoners of war.[28] This case caused an intense discussion, which eventually included the Ministry of War.

In March 1915, the Ministry decided not to use prisoners of war for construction, since the Hague Convention did not allow this, although it would be possible to use them later in powder production. Furthermore, the Ministry concluded that the local farm workers would not go back to the land, where they received lower wages. Instead, the Ministry suggested the use of migrant workers from occupied Russian Poland. From May 1915 onwards, therefore, migrant workers were used to overcome the shortage of construction labour.[29]

From photographs taken in 1917 or 1918, and from postcards, we know that the workers in the production process were partly prisoners of war, and partly workers from various parts of Germany (especially women). Labour market shortages seem to have been so intense, that it was necessary to recruit female workers from other parts of Germany. Although the administration was able to bypass the shortage of construction workers by using migrant workers it proved impossible to finish the entire plant in the envisaged time. Therefore, the construction of the so-called east factory (nitroglycerine) was suspended, while the construction of the west factory (guncotton and guncotton powder) was prioritised. With the implementation of the Hindenburg Programme in 1916, however, the factory was finally completed and brought to full capacity by mid-1918. The TNT factory was the last section to take up operation in July 1918. This is mirrored in the production output of the plant.

Figure 3 illustrates how the output of the various product groups developed, according to the construction of the plant. The plant started with the production of guncotton and guncotton powder. In the wake of the Hindenburg Programme and the extension of the plant, the production of nitroglycerine and nitroglycerine powder started in 1916, and TNT was produced from the summer of 1918. The sequence of production follows an obvious pattern. The plant began with the production of the technologically easiest products, which placed lower demands on the skills of the workers. The production of nitroglycerine powder and TNT demands much greater skills. This organizational strategy became necessary when we recall that most workers at the Kirchmöser plant were not trained for this kind of work.

Another major problem arose from the absence of accommodation. Although we have only fragmentary information on the absolute number of workers we can estimate there to have been 3000 construction workers and between 6000 and 7000 production workers. The surrounding villages were so small that they could accommodate only some of these. This problem was solved with the erection of twenty provisional huts on the plant site. Workers usually lived in the huts until they found a flat or a room on the outside. The accommodation situation improved during the war when a housing estate was developed north-west of the plant, offering villas and semi-detached houses for the managers, and terraced houses for the workers.[30] In addition, a small garden city was built for workers on the other side of the river, north of the town of Plaue.[31]

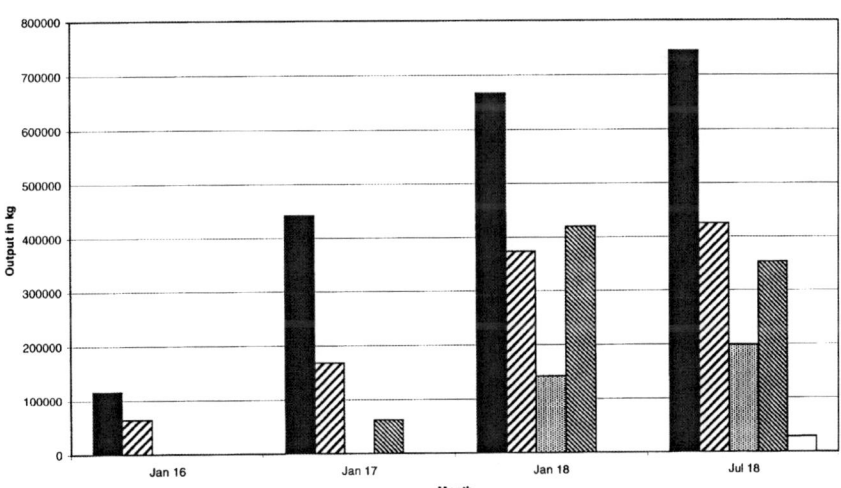

Figure 3. Output development at the Royal Prussian Powder Plant in Kirchmöser, 1916–1918

South of the powder plant, in 1916, a fireworks laboratory started to produce fuses for shells and torpedoes. The fireworks laboratory and the powder plant were separate and independent plants, because they came under different departments of the Ministry of War.

The buildings for production and storage were designed in a rather simple manner. All were made of red brick, the traditional building material of the Prussian heartland, with facades of alternating brick walls and long vertical windows that stretched up to the eaves. The roofs were wooden, and had a small inclination of 15° to 20° in order to let shockwaves escape upwards in case of an explosion.

As compared with the factory buildings, greater attention was given to the design of the administration buildings, which were all given Neo-Renaissance features, in conformity with style used in most official buildings of the Wilhelmian era. These features included porticoes, columns, and flower ornaments, as well as high roofs with windows. Such buildings often had three wings, and formed a quadrangle. The main administration building, situated at the northern entrance to the plant, was a similar building with a central portico, displaying the Prussian eagle and the Prussian motto '*Suum cuique*'. Other typical buildings were the laboratories, with three wings and a central portico; and the administration complex of the fireworks laboratories. The latter consists of two, two-storey buildings facing each other with identical façades. The northern building housed the administration of the fireworks laboratory, while the southern building was reserved for dining halls and social infrastructure. Both buildings were connected by gatehouses, so that the whole complex created a closed square where one could feel lost under the great dimensions of the buildings. This impression was underlined by the design of the façades, with their huge columns and rich ornament. Other buildings having detailed façades were the swimming baths for workers and the officers' mess for the military employees. The interiors of all these buildings were often of a rather simple design, lagging behind the design of the façades.

## THE POSTWAR STORY: KIRCHMÖSER AS A CENTRE OF GERMAN RAIL WORKS

After the end of the war, munitions machinery was appropriated by the Allies, plants were dismantled, and a large part of the German munitions industry was handed over to the newly-founded Polish state. At Kirchmöser, all machines were sent to France, Belgium, and Serbia, and all buildings were to be destroyed, except the power station.[32] However, in April 1919, all the state-owned powder plants were transferred to the responsibility of the Treasury, and in November, the National Assembly of the Weimar republic decided that they should be adapted to civil production.[33] Accordingly, in 1920, the Kirchmöser plant was given to the Ministry of Transport, and to the newly-founded *Deutsche Reichsbahn* (German Rail).

The *Reichsbahn* faced major problems. A large number of its railway carriages, and goods wagons and locomotives were lost to the victorious powers. The overhaul of remaining, damaged locomotives took an average of 105 days. The Reichsbahn decided to reorganize its repair facilities, and Kirchmöser became the first place where this happened. The central part of the *Eisenbahnwerk Brandenburg-West* (Railway Works Brandenburg-West) — as the Kirchmöser plant was called in the coming years — were the Locomotives Works, on the site of the former fireworks laboratory. For the first time in Germany, Fordist-Taylorist mass production techniques were used to overhaul locomotives. Delegations from all over the world came to visit the modernized works. In addition, the Central Works of the *Reichsbahn* were established at the site of the former powder plant. They were the major supplier of various materials for the railway.

Moreover, Kirchmöser became the location of three out of four research departments of the *Reichsbahn* and of the Central School, where the upper grades of the *Reichsbahn* civil servants were trained. In the following years, the various *Reichsbahn* plants and institutions attracted many people. The number of inhabitants rose quickly, and reached more than 5000 in 1939 (compared with 1000 in 1918). The *Reichsbahn* completed the housing estate of the former powder plant and developed a second housing estate east of Kirchmöser, close to the old village.

In 1942, the Ministry of Armament decided to move the Kirchmöser plant to Kramatorsk, a town in the Ukraine, and to use the site for munitions again. When informed of this decision, the president of the *Reichsbahn* said that without the Kirchmöser works, the *Reichsbahn* would have a broken neck.[34] This underlines the extraordinary importance that Kirchmöser had acquired during the 1920s and 1930s. The site was finally taken over by the Flick Corporation. The machines and equipment of the railway works never reached the Ukraine. Instead, in late April 1945, the Soviet Army occupied Kirchmöser, and established a tank production line on the former site of the fireworks laboratory. This production line occupied about one-third of the whole area, and was in use until 1993, when the Russian Army finally withdrew.[35]

After 1945, parts of the former railway works were used by the *Reichsbahn*, but the plant never regained its former importance. Since the early 1990s, Deutsche Bahn has privatized the remaining production lines. Today, 80 per cent of the site is derelict, but the city of Brandenburg, as well as the Federal Land, have begun to develop the area for business uses. Since 1993, the site has been listed as an architectural monument. It still typifies the former Royal Prussian powder plant, with its proud buildings. Despite the *Reichsbahn* and the Soviet Army, the site retains much of its overall character.

CONCLUSION

The small village of Kirchmöser is a rare example of how a village or town can be changed dramatically by the demands of an industrial war. For hundreds of years, Kirchmöser was a village remote from the centre of world history. But within a few months, it became an industrial town, and in the following decades, it became nationally important, directly experiencing the ups and downs of German political history.

Historians have said that, in the wake of the failure of the Schlieffen Plan, German mobilization was poorly planned and hastily completed.[36] This mobilization is said to have produced inefficient and temporary facilities.[37] The example of Kirchmöser suggests a somewhat different picture. The plant was planned directly in response to the failure of the Schlieffen Plan, and is therefore part of the wave of mobilization. From the beginning, however, it was not designed to be temporary. This is obvious from the size and quality of the buildings. Furthermore, it cannot be assumed that the plant was inefficient, given the impressive expertise of leading powder plant engineers and architects from Spandau and Hanau, and the efforts made to equip the new plant with state-of-the-art technology.

On the other hand, the construction history of the powder plant reveals all the typical problems of early mobilization. These problems were at first associated with shortages of labour. However, these shortages were quickly overcome, and even though the plant was planned and constructed hastily, it was functioning in a surprisingly short time. Technical expertise and thousands of workers produced a major facility within only six months from the date of the decision to locate the works in Kirchmöser. As we have seen, the original plans were too ambitious, and the plant was finally finished only in the wake of the Hindenburg Programme. By this time, its completion faced few major problems. Obviously, the War Ministry had learned much from the experiences of 1915.

## NOTES

[1] Wolfgang Seel, 'Preußisch-Deutsche Pulvergeschichte', Folge 2: Rauchloses Pulver, Teil 8, *Deutsches Waffen Journal*, 9 (19), (1984), 1135. During the pre-war years, state-owned powder plants produced no nitroglycerine powder at all, while the output of guncotton powder was almost equal to that of the private corporations. It is important to note that reliable data on the output of the various kinds of powder and munitions is still very fragmentary, since the German Military Archive for the First World War was destroyed in 1945. The above-mentioned figures are therefore estimates only.

[2] *Ibid.*, 976.

[3] Hans Mohn, 'Ein Beitrag zur Geschichte der deutschen Pulverfabriken' (Unpublished PhD dissertation, Würzburg, 1924–1925), 27–36.

[4] Olaf Mußmann, 'Komplexe Geschichte: Systemtheorie, Selbstorganisation und Regionalgeschichte. Von der Papiermühle zur Pulverfabrik — Ein historischer Längsschnitt der Gemeinde Bomlitz', (Unpublished PhD Dissertation, Hannover, 1994), 168–172.

[5] Information on the location of other State-owned plants remains incomplete. What survives of the German Military Archive and some secondary literature indicate only the five locations that are mentioned. It would seem that every German kingdom would have its own powder plant, but we have no hints as to their location in the German kingdom of Württemberg. All sites have been cleared during the 20th century, leaving scarce reminders of the former plants. The Spandau site was cleared in the 1970s to build a BMW plant; and Hanau's powder plant site has been used for the construction of a nuclear power plant. In Ingolstadt, Audi bought the site for car production; while the site of the Dachau plant was used by the Nazis as a concentration camp. Gnaschwitz in Saxony is the one and only site where explosives are still produced, although the remains of the original plant are few, due to demolition after 1919.

[6] Gotthold Schaefer, 'Die Bewirtschaftung von Pulver und Sprengstoff im Kriege', in Friedrich-Wilhelms-Universität zu Berlin (ed.), *Jahrbuch der Dissertationen der Philisophischen Fakultät der Friedrich-Wilhelms-Universität zu Berlin*, Dekanatsjahr 1921–1922 (Berlin: Selbstverlag der Friedrich-Wilhelms-Universität zu Berlin, 1922), 174.

[7] Seel, op. cit. note 1, 1134

[8] Ibid., 1134.

[9] Ibid., 1134. The sites of the existing Royal Prussian powder plants in Spandau and Hanau were not suitable for major extension, which may have been a reason for building an additional plant in Prussia.

[10] The study of individual explosives firms is made difficult by the fact that most of them were dissolved after the war, or destroyed in the Second World War. Their archives, also, have been lost.

[11] Schaefer, op. cit. note 6, 158. The production of, for example, 5000 tons of guncotton required 75,000 tons of potatoes. Especially in the winter of 1916–1917, a high proportion of the potato harvest was used for alcohol production, resulting in the so-called *Kohlrübenwinter*, 'swede winter', when starving people had to substitute swedes for potatoes. This winter became an enduring memory in many German families.

[12] In summer 1916, an average German adult received 1983 kcal, in winter 1917, only 1100 kcal. This malnutrition caused deficiency diseases and even death, in some cases. In 1917, 300,000 people died from malnutrition in Germany. This provoked enormous social and political tensions, known as the guns-versus-butter debate. See Anne Roerkohl, 'Die Lebensmittelversorgung während des Ersten Weltkrieges im Spannungsfeld kommunaler und staatlicher Maßnahmen', in Hans Jürgen Teuteberg (ed.), *Durchbruch zum modernen Massenkonsum, Lebensmittelmärkte und Lebensqualität im Städtewachstum des Industriezeitalters* (Münster: Coppenrath, 1987), 47–60.

[13] Gunther Mai, *Das Ende des Kaiserreichs: Politik und Kriegführung im Ersten Weltkrieg* (München: Deutsche Taschenbuch Verlag, 1987).

[14] While the needs of the munitions industry could almost be met by substituting raw materials and implementing new processes, the provision of fertilizers for agriculture became a major problem, leading to famine and discontent. This is seen as one of the major factors that led to the German November Revolution in 1918.

[15] See the essay by Jeffrey Johnson in this volume.

[16] The Feldzeugmeisterei was the Supply Department of the Prussian War Ministry, and was responsible for the production of weapons and munitions.

[17] Staatsarchiv Magdeburg/Landeshauptarchiv (Magdeburg), Acta des Königlichen Landrats-Amtes II. Jerichowschen Kreises zu Genthin betreffend die Errichtung einer Sprengstoff-Fabrik bei Möser, Rep. C30 Jerichow II, A No. 2246, Feldzeugmeisterei Berlin to Königlicher Regierungspräsident, piece Bh.Nr.3386.9.14 A IV, 29 September 1914. The Königlich Preußischer Regierungspräsident in Magdeburg was the highest representative of the Prussian province of Saxony (not to be confused with the Kingdom of Saxony). This province is today part of the German Federal Länder of Sachsen-Anhalt and Brandenburg.

[18] Ibid.

[19] A *Landrat* was the head of a county (*Landkreis*). The Prussian provinces were divided into various counties.

[20] Gunter Michael Wittke, 'Vorgeschichte und Geschichte des Werkes für Gleisbaumechanik Brandenburg-Kirchmöser in der Zeit von 1914 bis 1932: Ein Beitrag zur Betriebsgeschichte für das Werk für Gleisbaumechanik Brandenburg-Kirchmöser' (Unpublished Master's thesis, Humboldt-University Berlin, 1989), 42.

[21] The commission included representatives of the Königliches Militär-Kriegsbauamt 5 (Royal Military Building Authority), the Ministry of War, and the Artillery Command, as well as a military architect from the powder plant in Hanau.

[22] Staatsarchiv Magdeburg/Landeshauptarchiv (Magdeburg), Acta des Königlichen Landrats-Amtes II. Jerichowschen Kreises zu Genthin betreffend die Errichtung einer Sprengstoff-Fabrik bei Möser, Rep. C30 Jerichow II, A No.2246, Kriegsministerium Berlin to Königlicher Regierungspräsident, piece Nr. 6110.O.P., 13 November 1914.

[23] In 1916, the village Möser was renamed Kirchmöser since there was a nearby town with the same name. In 1952, Kirchmöser was incorporated into the city of Brandenburg/Havel.

[24] Staatsarchiv Magdeburg/Landeshauptarchiv (Magdeburg), Acta des Königlichen Landrats-Amtes II. Jerichowschen Kreises zu Genthin betreffend die Errichtung einer Sprengstoff-Fabrik bei

Möser, Rep. C30 Jerichow II, A No.2246, Kriegsministerium Berlin to Deputy General Command, piece Nr. 903/11.14.B.5, 26 November 1914.

[25] Staatsarchiv Magdeburg/Landeshauptarchiv (Magdeburg), Acta des Königlichen Landrats-Amtes II. Jerichowschen Kreises zu Genthin betreffend die Errichtung einer Sprengstoff-Fabrik bei Möser, Rep. C30 Jerichow II, A No. 2246, Staatsministerium Berlin to Kriegsministerium Berlin, 29 November 1914.

[26] Private archive of Axel Schulze (Kirchmöser), Kriegserfahrungen der Technischen Institute, Pulverfabrik Plaue, by Mil. Baumeister Hilpert, piece Nr. Pf. Pl. 2261.15.I, 20 December 1915.

[27] The only exception was the building company Phillip Holzmann & Co. with its headquarters in Frankfurt/Main. Holzmann played an important role in many public and politically initiated building projects of the Second Reich, e.g. the construction of the Bagdadbahn (Berlin to Baghdad Railroad).

[28] Staatsarchiv Magdeburg/Landeshauptarchiv (Magdeburg), Acta des Königlichen Landrats-Amtes II. Jerichowschen Kreises zu Genthin betreffend die Errichtung einer Sprengstoff-Fabrik bei Möser, Rep. C30 Jerichow II, A No. 2246, Bürgermeister to Landrat, 15 February 1915.

[29] Staatsarchiv Magdeburg/Landeshauptarchiv (Magdeburg), Acta des Königlichen Landrats-Amtes II. Jerichowschen Kreises zu Genthin betreffend die Errichtung einer Sprengstoff-Fabrik bei Möser, Rep. C30 Jerichow II, A No. 2246, Kriegsministerium to Feldzeugmeisterei, 31 March 1915.

[30] The construction of the housing estate was not finished by the end of World War I but it was completed by the Deutsche Reichsbahn (German Rail) in the 1920s.

[31] The garden city in Plaue was designed by Paul Schmitthenner and developed by a community of workers, one of the leading figures of the so-called 'Stuttgart School' of architecture, a conservative school that developed as a major opponent of the Bauhaus in the 1920s and 1930s. It became a leading school in Nazi Germany.

[32] Wittke, *op. cit.* note 20, 35–36.

[33] Seel, *op. cit.* note 1, 1139.

[34] Deutsche Reichsbahn, '50 Jahre Reichsbahnbetriebe Kirchmöser: 23 Februar 1920–23 Februar 1970' (Kirchmöser: Deutsche Reichsbahn, 1970), 17.

[35] Personal observations of the author.

[36] See essay by Jeff Johnson in this volume.

[37] *Ibid.*

# WARTIME CHEMISTRY IN ITALY: INDUSTRY, THE MILITARY, AND THE PROFESSORS

## INTRODUCTION

On 31 May 1914, less than a month before the assassinations in Sarajevo, Italian chemists met in Turin for a conference, convened by the Associazione Chimica Industriale, to celebrate the life of Ascanio Sobrero (1812–1888).[1] Sobrero was an early chemistry student of T.J. Pelouze in Paris and Justus von Liebig in Giessen, who had spent his life as a professor of chemistry at the University of Turin. In 1846–1847, he made his name with the development of nitroglycerine, and later acted as a consultant for the dynamite plant that Alfred Nobel established in Avigliana, a few kilometres from Turin.[2]

The timing of the 1914 celebrations was regarded as significant, although the participants could have anticipated neither the events of Sarajevo, nor their aftermath. To the opening ceremony, the Associazione attracted Italy's Prime Minister, as well as leading Italian chemists, and representatives from several private plants that produced explosives, since the government's gunpowder monopoly had been lifted in 1869.[3] For at least one major Italian manufacturer, the celebrations were equally timely. Italy's major private manufacturer of explosives — the *Società Italiana Prodotti Esplodenti* (SIPE), based in Milan — thought fit to print a lavish volume, devoted to the life and works of Sobrero, the history of explosives, and, above all, to its own successes.[4] These were graphically represented in a diagram showing that, while during the period from 1892 to 1904, SIPE's explosives production oscillated between 900 and 1000 tons, production reached 2600 tons in 1912.[5] The 2600 tons produced by SIPE in 1912 amounted to about 60 per cent of the total production of explosives by Italian private firms in 1911.[6] As to the causes of this dramatic expansion, the SIPE offered an obvious explanation. Its report reproduced a letter in which, in December 1912, the Italian Minister for the Navy congratulated the firm on the fine job its TNT had done during the Italian-Turkish war. That year, Italy had conquered Libya.

If colonial wars were a major impulse towards increased explosives production, another factor was the expansion of the Italian railway and road systems, especially in the Alps, where several new tunnels were being constructed. Accordingly, the monument to Sobrero emphasized peaceful undertakings, rather than the arts of war: it portrayed a worker who, thanks to modern explosives, opened up new routes through the mountains by breaking up hard rocks.

In 1912, the combined effects of colonial wars and tunnel openings, plus the normal demands of the mining and hunting industries, sustained about 130 explosives factories in the peninsula (fewer than a handful of which were owned by the State, and run by the military). These employed an estimated workforce of 3000, one-third being women.[7] According to SIPE, Italy's colonial campaigns pushed firms to invest in the explosives business: their 1913 volume included a section devoted to a detailed, self-celebratory description of the recently completed, 'state of the art' explosives plant at Cengio, a little village on the Bormida river, between Piedmont and Liguria. The plant — occupying over 600,000 square metres, surrounded by 25,000 poplars — was designed to produce nitroglycerine, dynamite, picric acid, guncotton, collodion cotton and TNT. It could employ 400 workers and, during the war, it would expand to accommodate 6000 people.[8] Big orders from the Navy in 1906, and from the Ministry of War in 1910, accounted for the main increases in the pre-war production of SIPE and its Cengio facility.[9]

That the years preceding Sarajevo were momentous for private producers of explosives is confirmed by the fact that another firm, Bombrini Parodi-Delfino, inaugurated in 1913 a new big plant at Segni, near Rome, for the production of ballistite and dynamite, with in-house facilities for nitric acid and glycerine. This plant, too, would undergo gigantic expansion during the War.[10]

If one combines this information with some recent, general assessments of the situation of chemistry and chemical industries before the war,[11] one gains the impression that, around 1914, the long sought-after 'take-off' of the Italian chemical industry — for which, for fifty years, academic chemists like Stanislao Cannizzaro, as well as industrialists and investors, had been lobbying — was at last in sight. However, several weaknesses remained. Early twentieth century observers noted that chemical laboratories in Italian universities attracted less than one-tenth of the resources devoted to similar laboratories in Germany, and complained that Italian industry employed far too few university-trained chemists for Italy to aspire to an independent role in Europe. This, according to the same observers, conspired with the nation's lack of raw materials, and the government's reluctance to tax imported chemicals (especially from Germany), to keep Italy's chemical industry in a backward state.[12]

Coming at a moment of uncertain 'take-off', and structural weakness, the Great War acted as a powerful catalyst, and a tremendous test. It is a test we can use to assess the roles played by industry, the military, and the professors in the history of Italy's wartime chemical industry.

## MUNITIONS AND THE MOBILIZATION OF ITALIAN INDUSTRY, 1915–1918

The effects of the war on munitions and explosives production after 24 May 1915, when Italy declared war on Austria, were impressive in many respects.[13] A report prepared in 1918 under the supervision of Giacomo Ciamician, professor of chemistry at the University of Bologna, declared that, during the war, the average monthly production of explosives had multiplied 17.7 times. Wartime improvements in plant and processing ensured that, by 1918, the production of explosives could, if needed, be brought up to twenty-six times that of 1914. Over the same period, the production of explosives intended for munitions, and ordered by the Ministries of War and the Navy, increased 13.7 times over 1914. Yet, despite this huge effort, Italy had to rely heavily on explosives imported from France, Britain, and the United States: in 1918, about half of Italy's demands in explosives were imported, with a 200 per cent increase over 1916.[14] The key problems were the lack of raw materials, and the weak roots of chemistry in the country's industrial system.[15]

In fact, Italy's mechanical industries during the war considerably out-performed its chemical industry. On entering the war in 1915, the Italian army had 2.47 million rifles, and the two plants run by the army managed to make another 2.42 million in three years. In the meantime, 36,567 machine guns (including 5537 for aircraft) were made by Italian private industry, which were added to the 700 foreign-built machine guns already held by the army in 1915. During the next three years, an additional 7000 machine guns were purchased abroad.[16]

Rifle ammunition was produced in two publicly owned plants run by the army, and reached a daily production of 3.4 million cartridges in 1918, with a total wartime production of 3,616 million pieces. In the meantime, domestic private firms supplied the army with 700 million artillery shells, 7.3 million shells for trench mortars, and 880,830 bombs for aerial use. Because the Italian chemical industry lagged behind the metallurgical and mechanical industries, in 1918 there were more cartridges and shells made than filled.[17]

## HOW WERE THESE RESULTS OBTAINED?

The driving forces behind Italy's war effort were two: an apparently endless flow of government and Allied money; and the administrative provisions introduced by special wartime legislation for 'industrial mobilization'. These were first approved in June 1915, and were followed in July by the creation of a Sottosegretariato (Under-Secretariat) (after June 1917, a Ministero (Ministry) per le Armi e Munizioni (for Arms and Munitions).[18]

These measures, conceived and carried out until 15 March 1918 by General Alfredo Dallolio, were meant to confer special powers aimed at intensifying cooperation between the military and private industry; a cooperation that, as we have seen, was already developing at an accelerating pace well before the war. The measures established a central committee for arms and munitions, cooperating with seven (later, eleven) regional committees, all answering to the Ministry of War.[19] The committees, in which civilians representing industry and labour sat next to the military, became powerful instruments for attaining the goals of wartime production. They proved less effective in securing sound administration of the huge economic resources that Italy and the Allies invested in the war.

The central committee had powers to declare certain private plants 'auxiliary' for the purposes of wartime production. 'Compensation', offered to these firms in exchange for their services, was negotiated directly by the central committee, with the agreement of the Ministero della Guerra (Minister of War), the Ministero della

Marina (Minister of the Navy), and the Ministero del Tesoro (Treasury). In most cases the compensation was regarded — at the end of the war — as overwhelmingly remunerative for the firms. A special parliamentary commission was appointed to investigate the matter in 1920.[20]

The Ministero della Guerra central committee provided private firms with raw materials free of charge — a very delicate matter, involving the need for coordination with other Allied purchasing. So, for example, in July 1915 a permanent military mission was established in the United States with branches in Baltimore, Philadelphia and New York, as well as in Washington, DC. In financial matters, the military mission was assisted by the Italian and American Treasuries.[21]

At home, the regional committees established in 1915 focused on technical, labour, disciplinary, and security issues. These were also sensitive in a country that had held its first 'universal suffrage' elections as recently as 1913, when Socialist groups (some of which were against the war) gained seats at the expense of the Liberals.[22]

The Ministero della Guerra also had power to give contracts for supplies to private firms that were not declared 'auxiliary'. This combination of legislative tools proved effective for domestic 'industrial mobilization', if we are to judge by the estimates of munitions and arms production. A separate assessment must be made of the munitions plants belonging to the army, on which detailed studies are wanting. However, judging from the information available on the munitions plant run by the army in Bologna, mobilization in the public sector generated only slightly less impressive results than in the private sector.

The army's Bologna plant, called the *Laboratorio Pirotecnico*, was established in 1862 and was the largest in Italy. In 1897, it employed 1055 people (about 60 per cent of whom were women), which rose to 1500 by 1914. Before the war, the *Pirotecnico* was by far the largest factory in the city and the only one involving extensive chemical know-how and operations. During the war, the number of workers multiplied several times, reaching 10,160 in 1918 — a large number for a city of 150,000 people.[23] Using special automatic machines for filling cartridges invented by Captain Luigi Stampacchia, the Bologna plant was credited with a daily wartime production of between one and two million cartridges. The plant also included laboratories devoted to research on new munitions, such as luminous bullets for shooting at night.

Besides the big army plant, Bologna hosted twenty-three private firms declared 'auxiliary' for the production of military supplies, and another dozen that received contracts under wartime legislation.[24] Some of these were already pivots of the local industrial system before the war, a role that many would afterwards retain or expand.

The core of Italy's war effort was thus a system of coordination between public plants, private firms and the military, ensured by central and regional committees. These committees favoured intertwining military and private firms, which either already supplied the military, or which could be converted quickly to wartime production. In total, about 1500 firms or institutions (including several gas companies, and a few university laboratories) were declared 'auxiliary', and another 350 firms obtained contracts without being so declared.[25] The large number of firms benefiting from the system seems to indicate that, under the pressures of war, industrial development was fostered across a comparatively wide spectrum of firms, firm sizes, and locations. The more so because — although only the largest, most powerful firms had representatives sitting in the committees — they had to involve a large number of sub-contractors if they wanted to meet the challenging production targets and short-term deadlines imposed by the war.

According to an estimate by the economist Luigi Einaudi, over the period 1914–1930, war expenses in Italy amounted to 65,204 million of gold, pre-war liras; almost 30% of which [19,209] — equal, roughly, to the country's peacetime GDP for 1914 — was financed by Britain and the United States. After the war, these loans were either extinguished by Germany's war reparations, or else never repaid. Again, according to Einaudi, given the low efficiency of the Italian tax system, and frequent political turmoil, the government managed its war debt mainly through loans imposed upon the Italian population. Postwar inflation enabled the government to repay only a fraction (in real terms) of the sums it had borrowed.[26]

Considering Italy's weak industrial development at the beginning of the twentieth century, historians agree that the country's mobilization — including the mobilization of the chemical industry — was on the whole a 'technical success'.[27] Historians also agree, however, that this success was achieved through measures that, by emphasizing the role of the State and weakening the responsibilities of entrepreneurs and workers, weakened the pivots around which a healthy economy is supposed to turn.[28] According to others, these same measures prefigured a model of economic and social relations that Fascism would soon revive and exploit to its own advantage.[29]

## WARTIME CHEMISTS

That chemistry and chemists had a central role to play in the war was a frequent claim voiced emphatically by Italian chemists through the press, and at the many public and private meetings that accompanied the country's mobilization. An outspoken propagandist of this idea was Ettore Molinari, who combined a chemical education at the Polytechnic of Zurich, the University of Basel, and the Polytechnic of Milan, with personal leanings toward the anarchist movement. Molinari was the author of a successful chemistry textbook that went through several editions, a couple of which were in English translation.[30] During the war, Molinari was the chief chemist at SIPE's Cengio plant. At a meeting of the Italian Society for the Advancement of Science (SIPS) held in Milan from 2–7 April 1917, he neatly phrased his view of the role of chemistry in the war:

> The problem of the war, from whichever perspective you consider it, is overwhelmingly a chemical problem. Armed defense and offence are directly connected with it, and so is the performance of the remaining social organism in its economic, industrial, commercial and civil functions.

> The problems that war has brought to the surface all bear, directly or indirectly, on applied chemistry, which is in turn the daughter of...theoretical chemistry.[31]

Molinari was personally against the war — which he regarded as a dramatic, but temporary interruption in the process by which industrial societies would attain their proper goals of modernity and social reform. However, his description of the problems faced — first by the Germans and the Austrians, and then by the Allies — in the production of explosives reveals his commitment as a chemist deeply involved in the war effort. In Molinari's view, the war was demonstrating what clever observers had known perfectly well beforehand: chemistry was a key asset in all advanced industrial societies, and equally important in peace and in war.

Throughout their careers, chemists like Molinari moved between industry and academic appointments. In both capacities, they took advantage of contacts with international networks. These networks were reshuffled (but not dismantled) by the war — as shown in the case of the German directors who were expelled from Italian chemical plants in 1915,[32] and in the experience of the Alsatian chemists who were employed at Cengio throughout the war.[33] Personnel problems, indeed, figured prominently in Molinari's description of the 'chemical problem' that Italy and other countries faced. He blamed the lack of well-trained chemists on the long term neglect of higher education in the natural sciences, for which, in his view, the Italian government was directly responsible.

In the absence of a systematic survey of Italian chemists $c.$1915–1918,[34] it is hard to distinguish in Molinari's complaint the real dimensions of the personnel problem that Italian chemical industry faced during the war. Molinari's view was that poor industry-university relations were even more of a problem than Italy's small output of chemistry graduates.[35] Many of his colleagues concurred that the teaching of applied or technological chemistry in Italian universities was weak. A report published in 1916 by Raffaello Nasini — a well-known chemist and public figure at the University of Pisa — confirmed that, in the twenty or so universities scattered throughout the peninsula, there were only four chairs of applied chemistry — or six, including electrochemistry. Furthermore, according to Nasini, applied chemistry chairs had small and badly equipped laboratories, usually with only one assistant. Worse, given the traditional prominence of theoretical chemistry in universities south of the Alps, applied chemistry chairs were often held by professors who had no real interest in the subject, nor deep ties with industry, and who fled back to theoretical chemistry chairs as soon as a chance was offered. Again, according to Nasini, the recent introduction of chemical engineering as a course of study in engineering schools was no remedy. Because of the deep-rooted traditions in Italian engineering schools, the few chemical engineers so far trained tended to excel in mathematics, rather than in applied or industrial chemistry. This contributed to scepticism on the part of Italian industrialists about the advisability of appointing academic chemists in industrial contexts, even in the special circumstances created by wartime production.

Contemporary reports liked to emphasize data that showed how, during the last year of the war, SIPE employed in its several plants fifty chemists with university degrees, twenty-two engineers, two architects, and another thirty-seven chemists without university training, most of them working in Cengio.[36] Observers regarded these numbers as showing the unprecedented degree of interconnection between pure and applied chemistry brought about by the war. Similar judgments were evoked in 1920, by a visit to the recently established, well-equipped research laboratory set up by the Società Bomprini Parodi-Delfino at its Segni plant: a laboratory that most Italian universities had good reason to envy. In Segni, besides the usual positions in chemical operations, and in process and product testing, eight chemists worked daily on research under the supervision of a Professor Sernagiotto, and benefited from the regular advice of academic chemists, including luminaries such as Nasini, and Emanuele Paternò, of the University of Palermo.[37]

Given the limitations of scholarship, it is hard to say how well Italian chemists performed during the war, apart from their unquestionably important contribution to the achievement of remarkable targets in explosives production. Only future studies — arguably focusing on single plants and chemists, as well as on national and comparative assessments — will provide an answer.

## WARTIME ACADEMIA

On the margins of industrial mobilization, grew an expanding belt of other wartime initiatives. Among these were proposals by university professors, in conjunction with politicians, the military, and industrialists, aimed at using wartime conditions to remedy the country's shortcomings, traditionally blamed on delayed industrialization and poor higher education. The professors' efforts were meant to assert new roles and find new resources for the endemically indigent system of higher education and research, while at the same time offering support to the military and industry. Two such initiatives shed a particular light on scientists' attitudes towards the war, and reveal additional features of Italy's 'national system' at work.

On 10 June 1916, a year after Italy's declaration of war on Austria, and seven months after a preliminary conference in November 1915, the Ministero per l'Agricoltura, l'Industria e il Commercio (Ministry for Agriculture, Industry and Commerce, later renamed the Ministero per l'Industria, Commercio e Lavoro, or Ministry for Industry, Commerce, and Work) set up a Permanent Committee for Chemical Industries. The composition of this Committee illustrates its main goals, and intended strategy. The Committee initially comprised eight people. Three were chemistry professors at the universities of Bologna, Palermo, and Pisa. Four were officers belonging to technical institutions of the public administration in Rome, and had themselves held temporary appointments in academia. Another was a chemist from a leading chemical and pharmaceutical manufacturer, based in Milan. One of the professors chaired the Committee; another (who happened to be the chairman's brother-in-law) was put in charge of the office created inside the ministry to support the Committee and put its initiatives into effect. When it was decided to appoint six more members to the Committee, they included five officers from the technical corps of the public administration, and one representative of SIPE.

The chairman of the committee was Giacomo Ciamician, professor of chemistry at the University of Bologna. Ciamician had a Ph.D from Giessen, and was well known internationally. He was repeatedly nominated for the Nobel Prize in Chemistry between 1905 and 1921, and was a member of the Italian Senate from 1910. During the war, he sat with the Constitutionalists in Bologna's local assembly.[38]

The committee chaired by Ciamician has received little attention from historians. Its most visible output was an annual report and a directory of Italian chemical industries, the first of which was published in 1918. This report was something new, and much needed in the Italian context. Yet, it was a somewhat modest initiative, when compared with the high expectations of new links between higher education, industry and the State that the war was supposed to foster.

The second initiative born within academia became somewhat better known, and has received more attention. This initiative targeted the core of the 'system' sketched above: it was aimed at the Ministero della Guerra and, especially, the Sottosegretariato (Under-Secretariat), later the Ministero per le Armi e Munizioni (Ministry for Arms and Munitions). The promoter was Vito Volterra, a brilliant, well-known and well-travelled mathematician, a founding member and a leading personality of the Italian Society for the Advancement of Science (SIPS), a frequent nominator for the Nobel Prize in physics and, from 1905, a member of the Senate in Rome.[39]

Volterra's initiative was inspired by a similar step taken in France by his colleague and friend, the mathematician Paul Painlevé, who served as the French War Minister and then Prime Minister from March through November 1917. In a letter to the Italian War Minister on 27 January 1917, Volterra submitted a plan for the mobilization of Italian universities. Volterra estimated that, of the 400 or so university laboratories scattered around the peninsula, more than 100 could be easily converted to carry out research of interest to the military, with a view to converting them to industrial research at the end of the war. Since most of the student population was serving at the front, Volterra argued, the intended conversion would not disrupt the little that was left of the educational mission of university laboratories during the war.

The conversion of university laboratories to military purposes was to be coordinated by a central office, attached to the Ministry for Arms and Munitions, called the Ufficio Invenzioni, and placed under the direction of Volterra himself. The Ufficio was established in March 1917. Its quick realization was a result of the spirit of 'mobilization' created by the war, as well as a reflection of Volterra's own qualities — including his credentials as an early proponent of Italy joining the war on the side of the Allies with his national and international renown; and his connections with the military, reinforced by his decision in 1915 (aged 55) to join the army as a volunteer. Later restyled the Ufficio Invenzioni e Ricerche, Volterra's wartime creation is regarded as the first manifestation of the Italian Consiglio Nazionale delle Ricerche (National Research Council), eventually established under Mussolini in November 1923.[40]

In his letter to the Minister of War, Volterra mentioned the high value he accorded chemistry, as second only to 'mechanics, precision instruments, etc', and which he put before 'hygiene, physiology, biology, etc'.[41] Once established, chemistry still came second (preceded by physics, and followed by mineralogy, materials science, radiotelegraphy, mechanics, electrotechnics, and physiology) in the list of the fields of competence the Ufficio soon acquired.[42]

A report for the period from October 1917 to March 1918, preserved among Volterra's papers, offers a glimpse of the kind of chemistry that the Ufficio carried out.[43] During the semester, some sixty chemical 'inventions' or new processes were submitted to the Ufficio for examination, in addition to an unspecified number of proposals and questions submitted by the other branches of the same office. Judging from the report, the chemical research carried out on the Ufficio's own initiative was developed mainly at the University of Pisa. Here, chemists working for the Ufficio focused on quite advanced (and theoretical) topics, such as spectrograph research — carried out with quartz and prism spectrographs, on minerals such as bauxite, scheelite, and manganese — and on radioactive substances, including an (inconclusive) study to establish the presence of uranium in granites from Sardinia.

The same report listed several hundred 'inventions' in any field that the Ufficio, using its scattered facilities, had managed to test, review and occasionally recommend to the military. Among these, many dealt with aeronautics and airships — Volterra's favourite fields of investigation during the war.

Despite Volterra's abilities, larger personnel and financial problems shaped the life of the Ufficio. Consider that the man in charge of reviewing chemical inventions for the Ufficio, Raffaello Nasini (who, as we know, became a consultant for Bomprini-Parodi-Delfino after the war), was also in charge of the office established by Ciamician at the Ministero per l'Industria. Consider also that — according to a note written by Volterra — as late as June 1918, the 'inventions' office had no independent budget, and its head was still asking the Ministry for a dozen proper employees, since most of the (few) staff attached to the office were shared with other institutions.

Cooperation between university, industry and the military, set up so quickly in the early fervour of wartime was, however ambitious in its intentions, often stunted in its outcome. This is certainly the impression one gains from a report describing poison gas research at the laboratory of pharmaceutical chemistry in the University of Bologna from January 1918 to August 1919. The Analytical Chemistry Laboratory, as it was called, was set up in forty-five days with money from the Ministero per le Munizioni (Ministry for Munitions) and the support of local artillery authorities. Under the direction of Professor Pietro Spica, a pharmaceutical chemist, the laboratory at its apogée employed six chemistry graduates in uniform, with a supporting staff of five, all working in three rooms made available by the University. In nineteenth months, the laboratory produced 160 reports describing

199 analyses carried out on poison gases, gas masks, and explosives seized from the enemy and submitted by the military for examination. The laboratory's activities culminated in September 1918, with the production, by Dr. Giuseppe Velardi, of samples of 'yperite' (mustard gas), regarded 'as pure as the one used by Germans', and prepared by an 'easy and economical procedure'.[44]

## CONCLUSION

Few have studied Italian industrial performance during the war, and fewer still the role of chemistry. One must therefore be cautious when trying to assess what the Italian case has to say along the interpretative lines suggested by the editors of this volume. It is tempting, however, to make at least one point, to encourage comparisons, and to stimulate further investigation.

This point has to do with the metaphor of the 'core' and the 'belt' used in describing the performance of different sectors at war. Judging from our story, one might say that, just as contemporaries saw the mechanical industries performing better than the chemical industries, the 'core' of the Italian system — made up of the military and the comparatively large number of private firms engaged in wartime production — performed better than the 'belt' of the system, where the champions of technological innovation and higher education — the professors, in our two examples — took the initiative.

One might as well argue, however, that in industrial societies (whether at war or in peacetime) the strength of a 'national system of innovation' resided precisely *in the belt, as well as in the core* of the system. If so, in our case, the comparatively weak performance of the belt and of the champions of innovation drawn from academia should be further investigated, as their experience may shed light on some oft-blamed, but little-understood, long term weaknesses displayed by Italy *vis-à-vis* scientific and technological innovation.[45]

It may turn out, however, that a prominent feature of the system, and a symptom of its major weakness, was precisely what some of the protagonists (and some historians) regard as a strength: that is, the comparatively easy access to state resources enjoyed by some university professors *qua* members of the Italian Senate. Because that comparatively easy access to public money was seldom accompanied by a wider appreciation of the need for closer interaction between higher education and industrial performance, easy access to public money could favour the proliferation of badly-staffed, poorly-financed, State-controlled bodies such as that which emerged from the otherwise well-intended, pioneering efforts of Ciamician and Volterra.

The initiatives of these professors, and the agencies that the war helped them to establish, could hardly compensate for the shortcomings of a national culture afflicted by an illiteracy rate that, in 1911, still totaled as much as 38 per cent of the population above six years of age.[46] Nor could those new, fragile agencies completely compensate for the far too limited provision, in Italian society at large, of widespread, effective, and easy channels of communication between the educational system, industry, and the military — channels already limited by the low circulation of scientific and technological literature.[47]

Viewed from this perspective, the 'technical successes' of Italy's mobilization, like Italy's wartime exercises in 'institutional engineering', look like performances of elites dancing on the quicksand of under-development. In this sense, the Italian case invites us to consider that broad social and institutional factors such as literacy rates, and the circulation of scientific and technological information in the culture at large, must be included in any final assessment of the performance of the Italian 'national system', whether during the war or afterwards.


ACKNOWLEDGEMENTS

I wish to thank especially Roy MacLeod, Jeffrey A. Johnson, Ernst Homburg, Erik Langlinay, Seymour H. Mauskopf, Mario Morselli, and John K. Smith for their comments on earlier versions of this paper.


NOTES

[1] Associazione Chimica Industriale di Torino, *Onoranze Centenarie ad Ascanio Sobrero* (Turin: Unione Tipografico-Editrice Torinese, 1914). The Associazione Chimica Industriale di Torino was created in 1899. For a survey of the chemical industry in Italy during the period, including information on educational and professional institutions: Paolo Amat di San Filippo, 'The Italian Chemical Industry from 1861 to 1918', in Ernst Homburg, Anthony S. Travis and Harm G. Schröter (eds.), *The Chemical Industry in Europe, 1850–1914* (Dordrecht, Boston, London: Kluwer, 1998), 45–57. See also the important survey by Vera Zamagni, 'L'Industria Chimica in Italia dalle Origini agli Anni '50', in Franco Amatori, Bruno Bezza (eds.), *Montecatini 1888–1966: Capitoli di Storia di una Grande Impresa* (Bologna: Il Mulino, 1990), 69–148. For the history of chemistry in Italy, see Antonio Di Meo (ed.), *Storia della Chimica in Italia* (Rome: Theoria, 1989), Introduction. On the arms industry, see Luciano Segreto, *Marte e Mercurio: Industria Bellica e Sviluppo Economico in Italia, 1861–1940* (Milan: Angeli, 1997); and Fabrizio Onida and Gianfranco Viesti (eds.), *L'Industria della Difesa: l'Italia nel Quadro Internazionale* (Milan: Angeli, 1996). For a recent survey of work on the 'chemists' war', see Jeffrey Allan Johnson, 'Chemical Warfare in the Great War', *Minerva*, 40 (1), (2002), 93–106.

[2] On Sobrero, see *Atti del Convegno in Celebrazione del Centenario della Morte di Ascanio Sobrero* (Turin: Accademia delle Scienze, 1989), and especially the articles by Gaetano Di Modica, 'Vita e Opere di Ascanio Sobrero', 5–13, and Luigi Cerruti *et al.*, 'Chimica e Cultura Chimica ai Tempi di A. Sobrero: Un Caso Esemplare: Sintesi e Struttura dell'Indaco', 39–71.

[3] See Amat di San Filippo, *op. cit.* note 1, 52.

[4] Ettore Molinari and Ferdinando Quartieri, *Notizie sugli Esplodenti in Italia* (Milan: Ulrico Hoepli Editore, 1913).

[5] See the diagram showing SIPE's overall production of explosives in the period 1890–1912, in Molinari and Quartieri, *op. cit.* note 4, 239.

[6] *Ibid.*, 278.

[7] Molinari and Quartieri, *op. cit.* note 4, 278.

[8] *Ibid.*, 251–265; on SIPE's wartime production and workforce, see Angelo Cappadoro, 'Visite agli stabilimenti industriali', *Giornale di Chimica Industriale ed Applicata*, 7 (2), (1920), 366–375, 371.

[9] *Ibid.*

[10] *Ibid.*, 366–368.

[11] Luigi Cerruti and Eugenio Torraca, 'Development of Chemistry in Italy, 1840–1910', in David Knight and Helge Kragh (eds.), *The Making of the Chemist: The Social History of Chemistry in Europe, 1789–1914* (Cambridge: Cambridge University Press, 1998), 133–162.

[12] These arguments were reiterated in academic and professional literature. A classical example is Giacomo Ciamician, 'I Problemi Chimici del Nuovo Secolo', in R. Università degli Studi di Bologna, *Annuario*, (1903–1904), 17–58, esp. 46, 48. On Ciamician, see the literature in note 38, below.

[13] See John Gooch, 'Italy during the First World War', in Allan R. Millett and Williamson Murray (eds.), *Military Effectiveness*, Vol. 1 *The First World War* (Boston: Allen and Unwin, 1988), 157–189. On the Italian army before the war, see Piero Del Negro, *Esercito, Stato, Società: Saggi di Storia Militare* (Bologna: Cappelli, 1979), and John Gooch, *Army, State, and Society in Italy, 1870–1915* (Basingstoke: Macmillan, 1989). For a recent collection on the war in Italy, see Giampietro Berti and Piero Del Negro (eds.), *Al di Qua e al di là del Piave: l'Ultimo Anno della Grande Guerra. Atti del Convegno Internazionale, Bassano del Grappa, 25–28 Maggio 2000* (Milan: Angeli, 2001).

[14] Ministero per l'Industria, Commercio e Lavoro, *Annuario per le Industrie Chimiche e Farmaceutiche*, Anno I° (Rome: Ministero per l'Industria, 1918), 19.

[15] Carlo Rostagno, 'Lo Sforzo Industriale dell'Italia Nella Recente Guerra', *Rivista d'Artiglieria e Genio*, 65, (1926), 2074–2097; 66, (1927), 35–54, 246–288, 368–407. On wartime chemistry, see 253–265.

[16] The figures in this and the following paragraph are taken from Carlo Montù, *Storia della Artiglieria Italiana* (Rome: 1948), Part 4, Vol. 10, 555–562.

[17] *Ibid.*, 558.

[18] On Italian industrial mobilization, see Luigi Tomassini, *Lavoro e Guerra: La Mobilitazione Industriale Italiana, 1915–1918* (Naples: Edizioni Scientifiche Italiane, 1997). See also Massimo Mazzetti, *L'Industria Italiana nella Grande Guerra* (Rome: Stato Maggiore dell'Esercito, Ufficio Storico, 1979), and Gooch, 'Italy during the First World War', *op. cit.* note 13, 157–189, esp. 163.

[19] The information in this paragraph is taken from Archivio Centrale dello Stato, Aldo G. Ricci and Francesca Romana Scardaccione (eds.), *Ministero per le Armi e Munizioni: Decreti di Ausiliarità* (Rome: Ministero per i Beni Culturali e Ambientali, 1991), Introduzione, 7–15.

[20] Camera dei Deputati, *Relazioni della Commissione Parlamentare d'Inchiesta per le Spese di Guerra*, 2 vols., (Rome: Camera dei Deputati, 1923). The papers of the commission are preserved in the archives of the Camera dei Deputati, Rome; an Inventory in Camera dei Deputati, Biblioteca, *Quaderni*, 7 (2001). A multi-volume publication containing essays devoted to an analysis of the works of the commission is: Carlo Crocella and Filippo Mazzoni (eds.), *L'inchiesta parlamentare sulle spese di Guerra (1920–1923)*, 3 vols., Rome: Camera dei Deputati, 2002.

[21] Archivio Centrale dello Stato, *op. cit.* note 19, 24.

[22] On the politics of the First World War in Italy, see Piero Melograni, *Storia Politica della Grande Guerra, 1915–1918* (Milan: Mondadori, 1998). See also Vincenzo Calì, Gustavo Corni and Giuseppe Ferrandi (eds.), *Gli Intellettuali e la Grande Guerra* (Bologna: Il Mulino, 2000).

[23] Massimo Mazzetti, *L'Industria Italiana nella Grande Guerra* (Rome: Stato Maggiore dell'Esercito, 1979), 65, 82.

[24] As evinced from the two volumes published by the Archivio Centrale dello Stato. See Ricci and Scardaccione, *op. cit.* note 19, and Francesca Romana Scardaccione (ed.), *Ministero per le Armi e Munizioni: Contratti* (Rome: Ministero per i Beni Culturali e Ambientali, 1995).

[25] These data are extrapolated from the lists published in Ricci and Scardaccione, *ibid.*, 277–435, and Scardaccione, *ibid.*, 411–422.

[26] Luigi Einaudi, *La Condotta Economica e gli Effetti Sociali della Guerra Italiana* (Bari and New Haven: Laterza and Yale University Press, 1933), 37–39, 41, 56–57.

[27] *Ibid.*, 387.

[28] *Ibid.*

[29] For a survey of the interpretations of Fascism, and its roots in Italian history, see the papers collected in Angelo Del Boca, Massimo Legnani, Mario G. Rossi (eds.), *Il Regime Fascista: Storia e Storiografia* (Rome and Bari: Laterza, 1995).

[30] Ettore Molinari, *Treatise on General and Industrial Inorganic Chemistry*, second edition, translated from the fourth revised and amplified Italian edition by T.H. Pope (London: Churchill, 1920). A brief sketch of Molinari's life, with information on his involvement in the anarchical movement, is in P. C. Masini, 'Il Giovane Molinari', *Volontà: Rivista Anarchica Bimestrale*, 29 (1976), 469–476. Part of Molinari's personal and scientific papers are now preserved in the Biblioteca Civica 'Angelo Mai', Bergamo.

[31] Ettore Molinari, 'Lo Sviluppo di Alcune Grandi Industrie Chimiche in Rapporto alla Guerra', *Annali di Chimica Applicata*, 4 (1917), 13–41, 13.

[32] Raffaello Nasini, 'La Chimica Italiana nel Momento Attuale', *Annali di Chimica Applicata*, 5, (1916), 125–142, 131.

[33] Molinari, *op. cit.* note 31, 38.

[34] A much needed prosopography of Italian chemists *c.* 1915–1918 is made difficult by the circumstance that, at that time, a national register of chemists did not exist. Steps leading to its creation were taken by a Socialist politician, Francesco Zanardi, who had a degree in pharmacy, and who was mayor of Bologna during the war. For information on Zanardi's efforts, see Nasini, *op. cit.* note 32, 126.

[35] Molinari, *op. cit.* note 31, 37–39.

[36] Cappadoro, *op. cit.* note 8, 373.

[37] *Ibid.*, 368.

[38] On Ciamician: G.B. Bonino, 'Ciamician, Giacomo', in *Dizionario Biografico degli Italiani*, Vol. 25 (Rome: Istituto della Enciclopedia Italiana, 1960), 118–122, and Albert B. Costa, 'Ciamician, Giacomo Luigi', in Charles C. Gillispie (ed.), *Dictionary of Scientific Biography*, (New York: Scribner, 1970–1990), Vol. 3, 279–280. On Ciamician as a nominee for the Nobel Prize in chemistry, see Elisabeth Crawford, John L. Heilbron and Rebecca Ullrich, *The Nobel Population, 1901–1937* (Berkeley: Office for History of Science, University of California, 1987), Index. On the circumstances in which, in the end, Ciamician failed to get the Prize, see Robert Marc Friedman, *The Politics of Excellence: Behind the Nobel Prize in Science* (New York: Times Books, 2001), 34.

[39] On Volterra, see Edoardo Volterra, 'Volterra, Vito', *Dictionary of Scientific Biography*, Vol. 14, 85–88. On Volterra and the First World War, see Judith R. Goodstein, 'The Rise and Fall of Vito Volterra's World', *Journal of the History of Ideas*, 45 (1984), 607–617; Giovanni Paoloni (ed.), *Vito Volterra e il suo Tempo (1860–1940)* (Rome: Accademia dei Lincei, 1990), 91–106; and Giuliano Pancaldi, 'Vito Volterra: Cosmopolitan Ideals and Nationality in the Italian Scientific Community between the *Belle Epoque* and the First World War', *Minerva*, 31 (1), (1993), 21–37. See also the anonymously published 'Matériaux pour une Biographie du Mathématicien Vito Volterra', *Archeion*, 23 (1941), 325–359. On Volterra as a nominator for the Nobel Prize in physics, see Crawford, Heilbron and Ullrich, *op. cit.* note 38.

[40] On the origins of the Italian national research council (the CNR), see Luigi Tomassini, 'Le Origini', in Raffaella Simili and Giovanni Paoloni (eds.), *Per una Storia del Consiglio Nazionale delle Ricerche* (Rome and Bari: Laterza, 2001), Vol. 1, 5–71.

[41] Accademia dei Lincei, Rome, Volterra's papers, Volterra to the Minister of War, 27 January 1917.

[42] According to a letter in which the Minister for Arms and Munitions invited its several branches to avail themselves of the new facility. See Ministero della Guerra, Sottosegretariato per le Armi e Munizioni, 'Costituzione dell'Ufficio Invenzioni', circular letter from Rome, 16 March 1917, in Volterra papers, Archives of the Accademia Nazionale dei Lincei, Rome.

[43] Ministero per le Armi e Munizioni, Ufficio Invenzioni e Ricerche, Volterra Papers, 'Relazione a S. E. il Sottosegretario di Stato. Riassunto dell'attività dell'Ufficio nel Semestre Ottobre 1917 Marzo 1918', typescript in the Volterra papers, Archives of the Accademia Nazionale dei Lincei, Rome.

[44] Pietro Spica, 'Il Laboratorio di Chimica Analitica dell'UMCG (Ufficio Materiale Chimico di Guerra)', *Atti del Reale Istituto Veneto di Scienze, Lettere ed Arti*, 80 (1920–1921), 917–1005, 919. On wartime research in chemistry at the University of Bologna': Andrea De Nittis, 'Chimica di Guerra a Bologna,' 1915–1918 (Università di Bologna: Tesi di Laurea, 2001).

[45] For comparisons of the situation around 1900, see Robert Fox and Anna Guagnini, 'Life in the Slow Lane: Research and Electrical Engineering in Britain, France, and Italy, *c.* 1900', in Peter Kroes

and Martijn Bakker (eds.), *Technological Development and Science in the Industrial Age* (Dordrecht: Kluwer, 1992), 133–153.

[46] See Carlo G. Lacaita, *Istruzione e Sviluppo Industriale in Italia, 1859–1914* (Firenze: Giunti, 1973), 32. See also Carlo M. Cipolla, *Literacy and Development in the West*, (Harmondsworth: Penguin Books, 1969), ch. 5.

[47] For an assessment of popular scientific culture in Italy *c*. 1900, see Paola Govoni, *Un Pubblico per la Scienza: La Divulgazione Scientifica nell'Italia in Formazione* (Rome: Carocci, 2002), ch. 3.

# MUNITIONS, THE MILITARY, AND CHEMISTRY IN RUSSIA

INTRODUCTION

The Great War imposed immense strains on all countries in the conflict. In many respects, however, Russia was the least prepared of all of the Great Powers to cope with these pressures. The failure of the state to wage war led directly to the overthrow of Tsar Nicholas II in early 1917. Until recently, however, little work has been done by western historians on the conduct of Russian science and technology during the war, and the Russian experience has not been adequately integrated into more general histories.[1] The most complete study in English of Russia's war is Norman Stone's study of the eastern front.[2] Stone covers a vast range of topics, but does not discuss the activities and organisations of Russian science. Nor do other historians of science. For example, L. F. Haber's study of gas warfare contains almost no information about the Russian experience.[3]

Similarly, Russian historians have focused more on the February and October Revolutions of 1917 than on the war itself. As one scholar has noted, most have seen the war as a 'disintegration leading to a disgraceful end.'[4] In a similar vein, studies of Russian science have largely ignored the war.[5] Only recently have scholars begun seeing the war as a critical moment in the establishment of the new Soviet State.[6]

Perhaps a principal reason for this general neglect lies in the fact that, until recently, archival information about wartime events in Russia remained classified by government security. This has been a particular problem for historians of science and technology. One finds in the literature, for example, only scattered references to the production of munitions and other chemical products.[7] This may be explained partly by the fact that General V. N. Ipat'ev, the central figure in the production of wartime explosives and other chemicals, continued to play a leading role in Soviet chemistry until forced to flee to escape a purge of economic officials in 1930. Only with the collapse of the Soviet Union in 1991 have historians had an opportunity to consult archives, and better understand Ipat'ev's contribution.[8]

Like the other allies, Russia entered the war with shortages of munitions, explosives, and other chemical products. Large orders were placed abroad, but these took time to arrive. Meanwhile, Russia's domestic output was called upon to supply a growing share of the military's needs. The domestic effort in chemical products was led by Ipat'ev, who headed the Commission for the Preparation of Explosives (CPE) [in Russian, *Komissiia po zagotovke vzryvchatykh veshchestv*], created in early 1915. This was one of a myriad of committees and councils that were intended to bring order to the chaotic system of military procurement. Other agencies included the Commission for the Preparation of Asphyxiating Gases (CPAG) [*Komissiia po izgotovleniiu udushaiushchikh sredstv*], the War-Chemical Committee (WCC) [*Voenno-khimicheskii komitet*] of the Russian Physical-Chemical Society (RFCS) [*Russkoe fiziko-khimicheskoe obshchestvo*], the Chemical Section of the Supreme Chief of the Army Sanitary and Evacuation Service (SCASES) [*Verkhovnii nachal'nik po Sanitarnoi i Evakuatsionnoi chasti*], the Chemical Section of the Central War-Industry Committee (WIC) [*Voenno-Promyshlennyi Komitet*], the Chemical Sections of the regional War-Industry Committees, the Chemical Section of Zemgor,[9] and many others. This paper will examine a number of Russian chemists and their organizations, in their attempt to deal with wartime crises, and will show how their partial success helped pave the way for the eventual establishment of the Soviet chemical industry. This paper does not aim to provide a definitive analysis of Russian chemistry and the Russian chemical industry during World War I, but rather is intended as a preliminary sketch to probe the complex contours and interactions of the military and science during the war years.

PRE-WAR PLANNING FOR MUNITIONS AND MILITARY SUPPLIES

Before 1914, Russian military and naval armaments were produced under contract from the Ministry of War by domestic factories using prototypes and designs licensed or purchased from foreign firms. The military conducted little research and development, being content to follow the lead of other countries, principally France and Germany. In the 1890s, for example, the Russian Army converted to smokeless powder, using a French process, while the Russian Navy developed a similar type of powder for their large-caliber guns.[10] In 1914, most Russian factories producing goods for military consumption were state-owned. The problem was that, when pushed, state factories could not produce enough, nor fast enough. In 1914, there were only three State-run arsenals producing cartridges and shells – at Tula, Lugansk, and St. Petersburg – and five other factories that produced explosives and gunpowder. All suffered from a lack of investment, and failed to keep up with technological advances. For example, about 70% of all industrial machinery purchased in Russia in 1913, for example, was of foreign origin, and most came from Germany.[11]

The argument for restricting the production of military goods to state-owned plants was strategic. The Russian military was concerned that privately-owned factories in Russia, largely owned by foreigners (principally German) would be unreliable in times of war, and would eventually report to foreign governments. In addition, the Russian government feared that private companies would demand excessive payment for urgently needed supplies.

Despite these concerns, given the limited capacity of the state factories, the Ministry of War did turn to foreign suppliers in times of great demand. This happened often, despite a law mandating domestic production. During the 1890s, the introduction of a new type of rifle impelled the military to place massive orders with Belgian and French firms. Once sufficient quantities were obtained to equip most of the standing army, the foreign orders ceased and yearly replacements were met by the state-owned factories.[12] However, Russian technological backwardness and delay sometimes forced the Ministry to give contracts to privately-owned factories even when they evaded domestic production regulations. For example, the Ministry contracted with a private factory for the manufacture of TNT using domestic toluene. However, the factory obtained its toluene from a German-owned plant in the Russian city of Revel, which fractionated the toluene from crude benzene imported from Germany. The Ministry apparently knew this was the case, but decided to turn a blind eye in order to get the product, instead of devoting time and effort to ensuring that the domestic production regulations were followed.[13]

During the Russo-Japanese War (1904–6), domestic producers were unable to meet the demand for military goods, and the Russian government placed extensive orders with foreign suppliers. For example, the War Ministry placed abroad orders worth 2.6 million rubles in 1903, 16.9 million in 1904, and 73.1 million in 1905. Foreign orders for the Naval Ministry increased from 10 million rubles in 1903 to 68 million in 1905.[14] The fact that, during the war with Japan, neither foreign nor domestic orders were delivered in time, does not seem to have deterred the Russian military, as it revised its planning for a future war in Europe. Apparently, the military continued to believe that supplies could be obtained from abroad when needed. This policy had the advantage of keeping down costs, since the state could avoid spending capital on new plant. The financial position of the government was always precarious, and any way to economize was greatly favoured. Although the Russian chemical industry grew rapidly in the two decades before the war, almost all pharmaceuticals were imported, as were large quantities of other chemical essentials, from raw materials and intermediates to finished products.[15]

The agency charged with responsibility for obtaining munitions for the Russian army was the Main Artillery Administration [Glavnoe Artilleriiskoe Upravlenie, or GAU]. The GAU was formally under the War Ministry, but enjoyed considerable autonomy. In 1914, it was led by General D. D. Kuz'min-Karavaev, but Grand Duke Sergei Mikhailovich, Chief Inspector of the Artillery, actually wielded control. The Grand Duke's royal pedigree made him a rival even to the War Minister (V. A. Sukhomlinov). Bureaucratic in-fighting complicated efforts to equip and organize the artillery. At the beginning of the war, the GAU was a dysfunctional organization, with many overlapping organizations. However, it appears that leading military, as well as civilian authorities did not realize the extent of the problem.[16] In the summer of 1914, The Russian military felt fatally overconfident. Indeed, Sukhomlinov stated: 'In a future war, the Russian artillery will never have to complain of a lack of shells. The artillery is supplied with a large amount of equipment and is assured of a properly organized delivery of shells'.[17] Even earlier he also had declared: "If you want peace, be ready for war. Russia...wants peace, but she is ready."[18]

## MUNITIONS AND CHEMICALS DURING THE EARLY MONTHS OF THE WAR

Like the other Great Powers, Russia had no inkling of what lay ahead at the outbreak of the war on 1 August 1914.[19] Initial mobilization was astonishingly rapid and successful. However, ominous trends emerged in the first few weeks. When, as early as August, the Chief of Staff, N. N. Ianushkevich, warned the Minister of War about the unexpectedly large expenditures of shell, and requested more. Sukhomlinov responded, "Couldn't measures be taken to economize...?".[20] Russian field commanders were indeed forced to economize, but continued to use shells at a rate that threatened to exhaust stockpiles in less than a few months.

Part of the problem was a classic underestimation of the equipment that mobilisation would require. In a plan adopted in 1910, the General Staff estimated that a full-scale European war would require Russia to field an army of 4.5 million men. This would require 4.5 million rifles, plus 700,000 replacements each year. When war began, the General Staff decided to mobilise far more soldiers than were called for in the 1910 plan. Thus, it turned out that Russia needed not only five million rifles simply to equip the first mobilisation, but also 5.5 million for later mobilizations, and 7.2 million to replace losses. This totaled 17.7 million rifles, or more than eleven million above original estimates.[21] The number of cartridges needed to supply these rifles was correspondingly huge.

In addition, the 1910 plan assumed that a European war would last less than six months. During the first few months, the military would use stockpiles, while domestic production was mobilised and foreign orders were placed.[22] During the Russo-Japanese war, the Russian army used 1,276 artillery pieces and fired 918,000 shells, or just over an average of 700 shells per gun. In planning a future war, the Ministry decided to stockpile a reserve of 1,000 shells per gun. However, this number would give continuous fire at the maximum rate for no more than about 100 minutes. While the Ministry duly met their stockpile targets, their preparations turned out to be ludicrously inadequate.

On 9 September 1914, the Minister of War called a meeting of sixteen large enterprises, both private and state, involved in shell manufacture, and placed orders for 6.65 million shells, to be delivered at a rate of 1.5 million shells per month. However, these enterprises were unable to produce more than 500,000 shells per month with existing capacity, so the Ministry advanced them 10 million rubles to increase production to one million shells per month by October 1915, with continued expansion of production after this date until the full amount was achieved. The shortfall would be met by foreign orders (totaling 9 million shells), rather than by placing orders with other domestic producers (especially small producers) or by further expanding domestic production.[23] Large orders were placed at this time with British, French, American, and Japanese firms.[24]

These foreign orders did not, however, solve the shell shortage. Foreign firms were swamped with orders, and could not expand output fast enough to meet demand. By September 1915, British firms, hard pressed to supply British needs, had shipped only 5,000 of one million shells ordered by the Russian Ministry.

As late as June 1916, only 875,000 of the 9.1 million 3-inch shells ordered from the USA and Canada had arrived. In addition, the closing of Russia's western border reduced import access to the far northern port of Arkhangel'sk [Archangel] and the port of Vladivostok. Even when shells and other military goods were actually delivered, heavy demands on the transportation system often meant that supplies sat on the docks, exposed to the weather, for long periods of time.[25]

As the need for more shells and high explosives became apparent during the first few months of the war, the Russian military authorities began to consider how much of these could be provided by domestic production. In September, 1914, a commission of the GAU, including A. V. Sapozhnikov, a professor of physical chemistry at the Artillery Academy, was sent to the southern Russian coal-mining region of the Donets Basin to investigate the possibility of obtaining toluene from coke by-products.[26] At this time, most of these by-products were burned as fuel in coke furnaces. The commission decided that expanded domestic production was not possible, and Sapozhnikov and others were sent to the USA to oversee the placing of foreign orders.[27] In November, however, the GAU sent a second delegation to make another assessment.[28] V. N. Ipat'ev – a general of artillery and professor of organic chemistry at the Artillery Academy – L. F. Fokin and other engineers came to the opposite conclusion. Ipat'ev reported that, on the contrary, the plants in the region could easily and rapidly be converted to produce toluene. If the private companies did not want to participate, then he urged GAU to build a factory itself, on the grounds that the 'greatly increased output of valuable products would more than pay for the small cost of construction'.[29]

Ipat'ev was a firm believer in a domestic chemical industry, and saw this as an ideal opportunity to make Russia's industry independent of foreign suppliers.[30] However, he had to overcome the skepticism of the War Council. One general asked, 'What kind of a guarantee can you give, General Ipatieff, for the completion of this project in such a short time?' Ipat'ev replied, 'I am not a capitalist, Your Excellency. I cannot bind my guarantee with cash; I can offer only my head as security.'[31] The War Council did not immediately approve his plan, which Ipat'ev saw as evidence that 'the mood in general was one of pessimism with a lack of confidence in our own forces and a feeling of inferiority in the face of German technology.'[32] In Stone's words, 'The Artillery Department could not imagine that Russian industry would be capable of manufacturing war-goods, and therefore had recourse to foreign suppliers, who were supposed to be cheaper and more efficient, and who had also been supplying Russia for many years past."[33]

In January 1915, the GAU set up another Special Commission to assist in the production and delivery of munitions, under the direction of Grand Duke Sergei Mikhailovich. Six months later, this was replaced by a permanent committee (the Special Council for the Coordination of Measures to Guarantee the Supply of Munitions and Other Material to the Army).[34] Meanwhile, on 6 February 1915, the GAU set up a Commission for the Preparation of Explosives (CPE), with Ipat'ev as chairman, charged to coordinate the production of raw materials.[35] The staff of the Commission included engineers L. F. Fokin, V. Iu. Shuman, O. G. Filippov, and

A. A. Solonin; S. P. Vukulov (Director of the Scientific-Technical Laboratory of the Naval Ministry); I. I. Andreev (Director of the Scientific-Technical Laboratory of the War Ministry); and several others.[36]

GAU ordered the CPE to increase production of explosives from the 3,000–5,000 puds (one pud is equal to 36 pounds) [[48–80 tons] per month then being produced in the state plants, to 60,000 puds [964 tons] per month, by calling upon private industry. At this time, in early 1915, there was no survey of available chemical resources and factories, so the CPE surveyed materials and equipment needed to produce explosives. The CPE had limited authority to impose its decisions on industry, particularly those factories owned by foreigners, but Ipat'ev felt comfortable about being able to produce this amount, despite initial resistance. At first, only one private company cooperated with the Commission. Some did not want to expand production because they feared the consequences of overcapacity after the war. Others were reportedly reluctant 'to violate the rights of German concessionaires who might subsequently hold them legally liable'[37] Incredibly, even the Ministry was concerned about the legal ramifications of Ipat'ev's plans. 'Providing materials actually seemed to them of only secondary importance. It took a great deal of persuasion to show them that since we were now at war with Germany, with the future unpredictable, a 200,000-pud [3,214 tons] yearly output of crude benzol [benzene] was much more important than mere legal complications in the future'.[38]

Despite these difficulties, explosives production rapidly increased, fulfilling Ipat'ev's expectations.[39] By September 1915, his initial goal of 60,000 puds per month [964 tons] of explosives was reached.[40] While only 6,000 puds [96 tons] of explosives were produced in February 1915, explosives production jumped to 51,000 puds [820 tons] by July and 85,000 puds [1366 tons] in October. Most of this increase came from the private factories, which increased their output by over fifty times by the end of 1915, while the state factories only doubled their output.[41]

While the CPE had great success with explosives, however, it was unable to alleviate the shortage of shells. State plants increased production from 229,400 shells in January 1915 to 989,000 in June, nearly meeting targets. But the targets were unrealistically low, and by the spring of 1915, actual demand tripled to three million shells per month. The military managed to hide the seriousness of the situation until Russian troops suffered staggering defeats all along the front in Galicia and Poland in May 1915, and were forced to retreat deep inside Russian territory. These losses were blamed on the lack of shells. One observer noted, 'The Russian artillery was practically silent'.[42] After months of enthusiasm for the war effort, the Russian public's view of the war sharply changed to anxiety.[43] The Russian public had continued to support the government up to this time, despite the embarrassing defeats during the initial month of the war at Tannenberg and Masurian Lakes, where the Russians had lost more than 250,000 men and many guns due to the mistakes and cowardice of their generals.

In the wake of the setbacks in the spring of 1915, the CPE was ordered to increase explosives production from 60,000 to 165,000 puds per month [964 tons to 2,651 tons]. To do this, huge amounts of sulfuric acid (as well as other materials) were needed. In 1914, factories in Russia produced about 200,000 tons of sulfuric acid per year [1.25 million puds], as contrasted with 2.5 million tons in the USA and 1.8 million tons in Germany.[44] The main raw material was pyrite, iron disulphide, which is burnt to sulfur dioxide in the manufacture of sulfuric acid. Before the war, Russia imported substantial quantities of pyrite from Scandinavia and Turkey (over five million puds [about 94,000 tons] in 1914), since these could be shipped cheaply by sea.[45] Consequently, many of Russia's sulfuric acid factories were built along the Baltic coast. The main source of Russian pyrite (about nine million puds [about 170,000 tons] in 1914) was the Ural Mountains, and transportation costs to the centers of population were high.[46] With the war, the supply of imported pyrites came to an abrupt halt, even from Sweden. The situation was particularly critical in Petrograd and the Baltic region, which in any case relied upon imports for a wide range of materials. These regions were forced to rely exclusively on domestic sources of pyrites, greatly raising the cost of producing sulfuric acid, as well as adding to the strains on the transportation system.[47]

In February 1915, the CPE began a survey of sulfuric acid stockpiles, and turned to increasing the production of sulfuric acid by requisitioning pyrites at their source. By April 1915, supplies of sulfuric acid had almost vanished in Petrograd. Ipat'ev estimated that 35,000 puds [562 tons] of concentrated sulfuric acid a month were needed to maintain explosives production in the capital and requisitioned 80,000 puds [1,286 tons] of acid from the Urals in a desperate attempt to keep the factories operating.[48] On 14 May 1915, the CPE issued a proposal 'to expand the production of sulfuric acid at state factories and also to encourage the construction of new factories by all means'.[49] However, the retreat of the Russian army during the summer resulted in the loss of factories in Poland, and factories in Riga were evacuated. These two areas produced more than 500,000 puds [8,036 tons] of sulfuric acid per month. The loss of these factories threatened to cripple explosives production. By July 1915, the production of sulfuric acid in Russia dropped to 700,000 puds [11,250 tons] per month.

The CPE had to expand production at significantly fewer factories. Following the losses in Poland and Riga, Ipat'ev convened leading pyrite mine owners and asked them to draw up estimates for increasing production and their lowest prices for two-year contracts. At the same time, he proposed to the Special Committee for Defense that the CPE be given the task of organizing the production of mineral acids using Russian pyrite instead of foreign sources. The Commission requested powers to construct twenty new factories and to renovate old ones, with a goal of doubling sulfuric acid production within a year. Finally, Ipat'ev proposed that the CPE have a central office in Petrograd and six regional offices, with powers to oversee distribution between defense and civilian factories.[50]

The Special Committee for Defense approved, and in June 1915, a congress of owners of pyrite mines and factories gave their support, setting a price of 30 kopeks per pud for pyrites. There followed a rapid increase in pyrites mined and delivered. By January 1916, production of sulfuric acid had risen to 1 million puds [16,072 tons] per month, and by March, stood at 1.3 million puds [20,893 tons] per month.[51] By this time, according to Ipat'ev and Fokin, sulfuric acid was no longer the limiting factor in explosives production.[52]

By early August 1915, the CPE controlled the production and distribution of all sulfuric acid in Russia. Military needs were given priority. However, the diversion of large quantities of acid to explosives precipitated a black market.[53] In September 1915, Ipat'ev informed the CPE that the War Ministry had established, under his direction, a commission to regulate prices. This included representatives of the CPE and the Central War-Industries Committee. At the first meeting of the new Commission, Ipat'ev reported that while the majority of factory owners were satisfied with the agreement, the CPE gave each regional office authority to set prices for supplies that exceeded military needs, which resulted in increased profits for the factories.[54]

By the autumn of 1915, having organized distribution and regulated prices, the Commission began to work with factory owners to increase production. The Commission proposed two methods. First, it advocated increasing the quantities of pyrites mined and the number of factories constructed. It secured support from the War Ministry to improve transportation of pyrites to the factories. Next, the Commission proposed to introduce new methods for preparing sulfuric acid. One such method, widely practiced in the United States and Germany, used sulfur gas collected as a by-product at copper smelting factories.[55]

There were some who objected to the CPE'S measures. For example, A. E. Makovetskii, an instructor of chemistry at the Petrograd Technological Institute, petitioned the Special Council for Defense to increase production targets much more than those of the CPE. Makovetskii believed that 20–22 million puds [32,144–35,358 tons] per year was insufficient to meet military and civilian needs, and proposed 80–90 million puds [128,576–144,648 tons] per year instead. Ipat'ev pleaded with his military superiors to continue plans for a balanced increase. In this case, as in many others, Ipat'ev succeeded in deflecting opposition.[56] He later wrote:

> One does not need to be a prophet to predict the terrible conditions for the Russian chemical industry if the largest proportion of its available materials were directed to the production of such a large amount of sulfuric acid. There is no doubt that Russia needs a large amount of sulfuric acid, not only during war time, but also in peace time. But it is necessary also to fulfill this grandiose program gradually, not hurting the other branches of the chemical industry.[57]

Also, in October 1915, the Chemical Section of the Moscow WIC complained to the Special Council for Defense about the measures being taken by the CPE in regards to the production of sulfuric acid. Ipat'ev defended himself in a long report to the Special Council, apparently to their satisfaction.[58]

The CPE also directed research towards overcoming problems with other raw materials. For example, when imports of Chilean nitrates – used in the production of nitric acid – were cut off, Ipat'ev directed the search for domestic sources. A leading part in the search for natural resources was taken by the Imperial Academy of Sciences, which in 1915 established a Commission for the Study of Natural Resources [Komissia po izucheniiu Estestvennykh Proizvoditel'nykh Sil Rossii (KEPS)], under the direction of Academician V. I. Vernadskii.[59] When only small reserves of nitrate were located, Ipat'ev and the CPE turned to experiments to determine if ammonia could be produced as a by-product of the coking of coal. During the summer of 1915, small scale experiments in Petrograd were so successful that the War Ministry gave Ipat'ev funding for a large factory. This factory went ahead, but became operational only in early 1917.[60]

The CPE was not the only agency set up to organize the manufacture of chemical products for military and civilian needs. For example, in the wake of the shell crisis in 1915, a group of Moscow industrialists, having been excluded from direct involvement in provisioning the army, decided to set up their own organization, the War-Industries Committee (WIC). Eventually, a number of provincial War-Industries Committees were also established to act as facilitators, distributing orders for military supplies to a wide range of small and medium-sized factories.[61]

Both the Central War-Industries Committee and the provincial War-Industries Committees had numerous sections. The Chemical Sections often worked harmoniously with the CPE, but sometimes conflicts arose. One occurred soon after the shell crisis in the summer of 1915. High explosives required vast quantities of toluene. The CPE had earlier acted by converting coke ovens in the Donets Basin region of South Russia to recover by-products that had been burned off as waste. But this hardly produced enough to reach the targets set. The Chemical Section of the Central WIC decided that the demand could be satisfied only by building a large number of new coke ovens with the recovery process, as well as totally rebuilding old ones. Ipat'ev, chairman of the CPE, opposed the building plan. In a report to the Chemical Section, Ipat'ev argued that the CPE had already done everything that could be done to increase toluene production in the Donets Basin, and that it would be counterproductive to build new ovens due to the huge costs involved. Ipat'ev proposed that the WIC send an engineer to the Basin to report on what could be done within a year. The engineer duly reported that production could not be increased without a huge investment in factories and ovens – the construction of which would require massive investment in manpower and supplies.[62]

However, the Chemical Section ignored the engineer's recommendation, and proposed the construction of 2,000 new ovens and the rebuilding of old ones during the next ten months, at a cost to the state of up to 45 million rubles. The Chemical Section sent a budget to the Special Committee for Defense, where it was opposed by Ipat'ev, on the grounds that it would disrupt construction already underway and drain money from other urgent needs. Ipat'ev later noted that 'It was necessary to have much energy in order to prove the danger of such experiments and to convince the military officials and the people who proposed them that these grandiose proposals could disrupt that which was created on a firm, business-like basis'.[63]

Ipat'ev recognised the need for toluene, but he also believed that delays in building new ovens and factories would only delay actual production. In addition, such a massive plan would disrupt the chemical industry. In a way that became typical of the CPE (and the later Chemical Committee of the GAU), Ipat'ev and his colleagues approached the problem in a different way. They decided to produce benzene and toluene from petroleum by means of pyrogenetic aromatization – a stop-gap measure until the ovens and factories of the CPE could be finished. By early 1915, even before the full impact of the shell crisis had exploded in public view, N. D. Zelinskii (director of the Central Laboratory of the Ministry of Finance and a former professor of chemistry at Moscow University) and chemists at several laboratories were conducting experiments for the CPE on the pyrogenesis of oil using various catalysts.[64] When the experiments showed success, the chemists moved to scale up the process. Even though yields were low, the cost was in line with the cost of imports, and so could substitute. The first factory-scale experiments were conducted by the CPE at the City Gas Plant in Kazan. Although these were successful, the amount of toluene produced was tiny (about 1,000 puds [16 tons] each month). However, the experience served as a model for other factories that could produce much larger quantities.[65]

To increase production, the Central WIC had several chemists (including A. E. Porai-Koshits, a professor of chemistry at the Petrograd Technological Institute) prepare a proposal for a factory that could produce 4,000 puds [64 tons] a month of benzene and toluene from petroleum. In October 1915, the Baku WIC agreed to construct a factory and signed an agreement with the GAU to begin delivery within seven months. However, once the plant was finished, it was unable to begin production. Ipat'ev visited the plant in early 1916, and sent one of the engineers who had supervised the Kazan plant to offer technical assistance. This engineer brought the plant on line, and by the autumn of 1916, it was in full operation. Ipat'ev believed that further improvements could be made and, in early 1917, sent another engineer to Baku.[66]

The efforts of the CPE and other chemistry organizations to help produce the materials needed for munitions and explosives were very successful. After a slow start during the first months of the war, Russia was able to mobilize its industrial resources during the course of 1915, mainly by increasing production at private firms. By the end of 1915, Russian industry was producing sufficient numbers of shells to supply the needs of its army, with the exception of very large shells (from 203–305 mm). The production of these very large shells remained a problem for Russian industry throughout the war and even as late as 1916 only small numbers of these shells were being manufactured.[67] The increase in production of other artillery shells was very dramatic, however. The numbers of shells provided to the army from domestic sources increased from 150,000 each month in 1914, to 950,000 each month in 1915, to 1,850,000 each month in 1916.[68] According to figures of A. A. Manikovskii, who was the head of the GAU during the war, the Russian army used 72.3 million shells of all kinds. Of these, 56.6 million were produced in Russia, with 15.6 million (about 21%) imported from various countries.[69]

## POISON GASES AND GAS MASKS

While the CPE devoted great attention during 1915 to chemical munitions, the spectre of gas weapons soon overshadowed even the horrors of high explosives. At the end of May 1915 and continuing intermittently for several months, the Germans attacked Russian troops near Warsaw with various types of gases.[70] The initial attacks cost more than 1,100 gas fatalities of a total of 9,000 gas casualties.[71] The result was to force the Russians to mobilize their efforts in chemical warfare, both in defensive (gas masks) as well as in offensive (the production and delivery of gases).

Even before the German gas attacks on the Western Front, the Russians had speculated about the possible use of gas for offensive use. At a meeting on 26 January 1915, General A. A. Zabudskii, head of the newly-founded Central Scientific-Technical Laboratory of the War Ministry, considered the use of 'suffocating and intoxicating gases in shells,' although others attending the meeting rejected this, since 'such methods can be regarded as inhumane and have not been previously used by the Russian army'. However, they left open the possibility of using these gases 'in case of the enemy's gross abuse of such methods.' General Zabudskii decided to begin investigations so that gases could be ready 'in case of emergency, to start production'.[72]

The CPE also began to study poison gases at about this time. In early May, Ipat'ev sent a request to the GAU to produce phosgene for shells at a factory near Ivanovo-Voznesensk, an industrial city about 130 miles northwest of Moscow. An engineer at this factory had earlier sent a request to Ipat'ev, proposing the preparation of phosgene. This engineer and the factory owner were brought to Petrograd, and arrangements were being made for the production of 600 puds [9.6 tons] of phosgene for shells when news came of the Germans' first gas attack in Poland at the very end of May.[73]

The War Ministry gave responsibility for both offensive and defensive gas warfare to Prince Oldenburg, the Supreme Chief of the Army Sanitary and Evacuation Service (SCASES). Oldenburg was the husband of Tsar Nicholas II's sister and thus influential in military politics. The War Ministry also sent Ipat'ev, as chairman of the CPE, to Oldenburg to discuss the gas attacks. Oldenburg told Ipat'ev that the Germans had released chlorine from cylinders, and asked whether such cylinders could be produced in Russia.[74] Ipat'ev said it would take two factories four to five months. The War Ministry gave the task to the CPE, which sent A. A. Solonin, professor of artillery at the Artillery Academy, and A. E. Makovetskii, an instructor at the Petrograd Technological Institute, to the two factories to supervise.[75] However, it appears that the War Ministry did not have complete faith in the ability of the Russian chemical industry, and so asked the British for supplies of chlorine.[76]

The relationship between the CPE and the SCASES with regard to gas warfare is not clear from published information.[77] The War Ministry realized that the SCASES was not capable of organizing production, and asked the CPE to assume this task in addition to explosives. Ipat'ev apparently at first agreed, but soon decided the to develop a different, but effective gas mask. However, Prince Oldenburg and the CSASES were not able to organize this work efficiently, and work proceeded very slowly during 1915 and early 1916.

Meanwhile, other organizations, such as the WCC and various non-governmental groups, had begun to conduct research on gas masks. In July 1915, soon after the German gas attack, the Medical Commission of the All-Russian Union of Zemstvos and Towns established an Experimental Commission in Moscow under the direction of S. I. Chirvinskii, a professor at Moscow University. V. M. Gorbenko, an instructor at the Moscow Higher Technical School and a member of the Experimental Commission, set up two chambers to test the effectiveness of different substances. At first, the Experimental Commission focused on a wet mask, finding – rather unsurprisingly – that a mask with 35 layers of gauze was superior to the existing one of 20 layers.[92]

Other organizations concentrated on developing a dry mask. As early as June 1915, in a talk given to the Sanitary-technical section of the Russian Technical Society, N. D. Zelinskii (director of the Central Laboratory of the Ministry of Finance) proposed using activated charcoal. Charcoal used to remove impurities from alcohol could also be used in a gas mask. Zelinskii was familiar with charcoal, since the Ministry of Finance supervised the purification of alcohol for tax purposes. Zelinskii promoted the use of activated charcoal for gas masks, while working to find the best methods.[93] He gave talks to the Experimental Commission of the Medical Commission for the All-Russian Union of Zemstvos and Towns (Zemgor), the Central WIC, and the WCC.[94] By August 1915, the Experimental Commission recognised the superiority of Zelinskii's method and asked him to make a systematic study of factors such as the diameter of the filter and the container for the charcoal. Professor A. E. Favorskii conducted experiments in his Petrograd University laboratory that demonstrated the effectiveness of Zelinskii's method. Shortly afterwards, the engineer E. L. Kummant developed a rubber mask with a container to hold the activated charcoal. The combination of Zelinskii's activated charcoal and the Kummant's rubber mask produced the first effective gas mask for Russian troops.[95]

During the summer, while Zelinskii was developing his mask using activated charcoal, professors at the Mining Institute in Petrograd began to develop their own dry mask using as their model an existing mining mask, which used soda lime to trap the poison gases. The Petrograd Section of the Experimental Commission became interested and by late September decided to place large orders for the 'Mining Institute' mask. This mask was effective against chlorine, but was cumbersome and complex. The Experimental Commission in Moscow decided to test the two masks. The results, in late October, showed that the Mining Institute mask was 'completely unsuitable for general use.'[96] Despite this result, the SCASES decided to support the Mining Institute mask instead of the Zelinskii-Kummant mask, perhaps owing to the influence of N. A. Ivanov, head of the Petrograd Section of the Experimental Commission. Ivanov was appointed the head of the gas mask section of the SCASES, where he continued to champion the Mining Institute mask, all the while making changes to overcome its obvious defects, including adding a separate chamber for activated charcoal. Not content to work on its own mask, the SCASES tried to put roadblocks in the path of the Zelinskii-Kummant mask. The SCASES controlled the supply of charcoal, and only small amounts were released to Zelinskii. Moreover, according to Ipat'ev, Zelinskii kept secret his method for activating charcoal. Apparently, Zelinskii and Kummant were interested in obtaining royalties for each gas mask produced.[97]

By the end of 1915, the General Staff was becoming nervous about the slow progress of research on a gas mask. Military leaders believed that the Germans would resume the offensive, using chemical weapons in the spring, and wanted their own troops supplied with gas masks. At the end of November 1915, the General Staff asked Prince Oldenburg to place 'especially urgent orders' for a suitable mask by spring. The General Staff repeated this at the end of December, asking Prince Oldenburg to report the number of masks he had ordered. In early January, Oldenburg's assistant replied that six million Mining Institute type had been ordered, although a different type of mask would be needed to protect against phosgene and cyanide.[98] The General Staff was not satisfied and pressed Oldenburg and the SCASES to develop a more universal mask. At the end of January 1916, the General Staff wrote to Oldenburg, stating that it would be 'undesirable' to place orders only for the Mining Institute mask, and that the Zelinskii mask appeared to meet the criterion of universality. Still, the SCASES placed only a small order for the Zelinskii-Kummant gas masks with the WIC.[99]

By February 1916, the General Staff was complaining about the 'inertness' of the SCASES and asked the War Ministry to take over matters relating to gas masks.[100] In fact, by this time, the General Staff had decided to place orders for one million Zelinskii-Kummant masks with the WIC, even though placing such orders was not part of the General Staff's responsibilities.[101]

It was at this point – in February 1916 – that the General Staff decided to create a War-Chemical Committee to coordinate the production of masks as well as gas warfare and the instruction of troops. The War Minister supported the creation of this new organization and on 16 February 1916, wrote to Tsar Nicholas II stating that it was 'unsuitable' to order the Mining Institute mask when it was clear that the Zelinskii-Kummant mask was superior. Moreover, the minister argued that the army had not yet received a satisfactory gas mask because there was no unifying organization.[102] The Tsar approved the creation of the War-Chemical Committee on 2 March 1916.[103]

Prince Oldenburg did not easily relinquish control. He hurried to the Stavka near the front where the Tsar was directing the war, insisting that he retain control over all 'sanitary' matters, and demanded that the Special Council for Defense, which included representatives from the General Staff, be prohibited from taking action. He further demanded that the Central WIC be prohibited from dealing with gas masks. Oldenburg stated 'that he might not be able to prohibit civilian organizations from preparing these [Zelinskii-Kummant] masks...[and] that they could make as many as they wished, but he would not allow them to be delivered to the army'.[104]

However, Prince Oldenburg was unsuccessful, and on 19 March, the War Minister made a resolution 'to immediately place an order for the Zelinskii type of gas mask in a quantity sufficient to provide them for all of the army, and for the immediate provision of the charcoal for [these masks].'[105] The new War-Chemical Committee of the General Staff soon placed orders for three million Zelinskii-Kummant masks with the Central WIC, and for 800,000 with the All-Russian Union of Zemstvos.[106]

## THE CHEMICAL COMMITTEE

The creation of the War-Chemical Committee by the General Staff came as a complete surprise to the GAU, as the GAU had jurisdiction over some tasks now given to the Committee. However, a few days after the War-Chemical Committee was formed, General A. A. Polivanov was replaced as War Minister by General D. S. Shuvaev. The Chief of the GAU, General A. A. Manikovskii, persuaded the War Minister that his new organization would function better under the GAU. Manikovskii argued that all matters relating to chemistry – including masks, explosives and all raw materials – should be united in one organization. Further, this new organization should come under control of the GAU, since it already had the CPE and CPAG under its authority. Grand Duke Sergei supported Manikovskii's proposal. On 4 April 1916, the War Minister asked Ipat'ev to draw up regulations for this new Chemical Committee, together with a list of staff members. Ipat'ev's draft regulations were approved on 7 April and ratified by the Tsar on 16 April 1916.[107]

However, as often was the case in Russia, the creation of a new Chemical Committee (CC) in the GAU did not mean the abolition of the War-Chemical Committee of the General Staff. The General Staff had created a Council for the War-Chemical Committee, and this continued to function. While the CC had day-to-day control of the production of chemicals and products, the Council for the War-Chemical Committee retained oversight over the CC.[108] The relationship between the two organizations is not clear, although Ipat'ev obviously saw the Council for the War-Chemical Committee as an unwarranted intrusion into his sphere of influence.[109]

Ipat'ev was appointed the first chairman of the CC, reporting directly to General Manikovskii, Chief of the GAU. The chairman of the CC was also to keep in touch with General M. V. Alekseev, Chief of Staff to the Tsar. The CC was composed of five branches: 1) explosives, 2) poison gases, 3) incendiaries and flamethrowers, 4) gas masks, and 5) acids. Each branch held meetings and formed commissions. Decisions were made in general meetings held at least once a week. The permanent members of the CC included representatives from the military, academe, and private non-governmental organizations, including two members of the Academy of Sciences, the Directors of the Scientific-Technical Laboratories of the War and Naval Ministries, representatives from the WICs and Zemgor, a representative from the War-Chemical Committee of the Russian Physical-Chemical Society, the heads of each individual branch of the CC, and representatives from the GAU and the General Staff. The CC could also invite consultants to its meetings and ask them to conduct investigations. According to Ipat'ev, the consultants most often used were professors at the various higher educational institutions in Petrograd, including N. S. Kurnakov, L. A. Chugaev, A. E. Favorskii, A. A. Iakovkin, G. V. Khlopin, V. E. Tishchenko, and several others.[110]

From its very beginning in April 1916, the CC employed the regional bureaus already established by the CPE to coordinate chemical activities, and expanded the regional bureaus in Petrograd, Moscow, Kazan, the Upper Volga region [in the town of Kineshma], Kharkov [for the south], and Baku [for the Caucasus region]. A separate bureau to deal with benzene was set up in Kharkov to supervise factories in the Donets Basin. While these regional bureaus were under the control of the chairman of the CC (and ultimately, under the Special Council of Defense), they were given considerable autonomy. Ipat'ev had great respect for the capabilities of those chosen to head the regional bureaus.[111] Finally, in addition to regional bureaus, the CC had a chemical battalion to instruct soldiers in the fourteen commands at the front.[112]

The Chemical Committee at the GAU began to function even before its official ratification on 16 April 1916. Its most urgent task was to speed up the production of the Zelinskii-Kummant mask. Large orders had already been placed with the Central WIC and Zemgor, but they had been given neither specifications, nor activated charcoal, since Prince Oldenburg had released none to Zelinskii. Accelerating the production of the Zelinskii-Kummant gas mask would prove to be difficult, because Prince Oldenburg had managed to retain control over the production of the Mining Institute mask for the SCASES, forcing a struggle over raw materials. Oldenburg insisted that his orders be filled first. Not only that–Prince Oldenburg and the SCASES secured the right to appoint the head of the gas mask branch (branch IV) of the CC.[113]

On 8 April 1916, Ipat'ev convened in Petrograd a meeting of people involved in different aspects of gas mask production, and called for commissions in Petrograd and Moscow to work out the specifications. The two commissions called for assistance from chemists who were already consultants in other fields supervised by the CC. Zelinskii was sent to Moscow to supervise the production of activated charcoal at several alcohol warehouses. Chemists active in the Petrograd commission included Khlopin, Favorskii, Chugaev, and I. I. Andreev, head of the inorganic section of the Central Scientific-Technical Laboratory of the War Ministry. Ipat'ev noted that these commissions worked hard to determine specifications, and to accelerate production in time for the summer offensive.[114]

Gradually, the consultative committees and the CC increased production and delivery of the Zelinskii-Kummant mask According to official reports, on 8 September 1916, troops on the Western Front had 631,219 Zelinskii-Kummant masks; 300,000 of the improved wet gauze masks; 146,000 of the imported English helmet-type gas masks; and 80,000 of the Mining Institute type of gas mask.[115] On 7 September, the War Ministry issued an order to stop production of the Mining Institute mask, and a few weeks later sent another order to discard the Mining Institute masks and to replace them with the Zelinskii-Kummant masks.[116] On

4 October, the General Staff wrote to the CC that 'unfortunately, up to the present time, over one million of the completely unsuitable Mining type gas masks have been delivered to [the army] while the full provisioning of the army with the [Zelinskii-Kummant] activated charcoal type has been held back until now. While the Zelinskii-Kummant gas masks have significant deficiencies, they do save lives from the deadly effects of asphyxiating and poison gases.'[117]

Increasing the production of the Zelinskii-Kummant mask was not an easy task for the CC, as it had to deal with Prince Oldenburg, as well as to cope with the problems involved with manufacturing in Russia. The CC was forced to accept Prince Oldenburg's choice as the head of the Gas Mask Section. This was N.A. Ivanov, a professor at the Military-Medical Academy who worked at SCASES. Ipat'ev was reluctant to accept this appointment, since Ivanov was a supporter of the Mining Institute mask. Ipat'ev was overruled, but before Ivanov could take up his duties, Prince Oldenburg offended the military command to such an extent that it transferred gas mask orders to the CC.[118]

Although the Zelinskii-Kummant mask was effective against many gases, it had significant flaws. When the activated charcoal was broken up into a powder during movement to the front, it lost much of its effectiveness. The charcoal also had a tendency to fall out of its container, so a stopper was added. Unfortunately, poorly trained troops often forgot to remove this stopper. When they could not breathe, they tore off their masks, which left them exposed.[119] Work by Prince Avalov added an exhaust valve, easing problems with respiration and the inhalation of charcoal dust. Modified masks were sent to the front during the closing months of 1916.[120]

In the summer of 1916, the CC set up its laboratory. In October, G. V. Khlopin was appointed head. There, he worked with I. I. Zhukov, N. T. Prokof'ev, and M. V. Tarle to investigate the absorbing properties and the influence of moisture on charcoal. One of the laboratory's successes was Prokof'ev's development in the autumn of 1916 of an improved type of wet mask. This mask received the name 'Chemical Committee' and was given to troops in late 1916.[121] Scientists at other laboratories continued to share in the work. One such was the laboratory of the Central Scientific-Technical Laboratory of the War Ministry, headed by I. I. Andreev, whose work was conducted mainly during the second half of 1916 and the first half of 1917.[122]

By the second half of 1916, the strain of the war on Russia's social and economic system was acute. The organizations set up to manage the chemical industry struggled valiantly. However, as time went on, problems arose that could not be overcome. One was transportation. Even at the beginning of the war, the railroad system in Russia was not sufficient to meet the large demands placed on it. As rolling stock and personnel were committed to transporting troops and supplies, Russian industry suffered. One of the main problems was that many Russian factories were located far from the sources of their raw materials. For example, sulfuric acid factories were built along the Baltic in order to use imported pyrites. When the imports were cut off, they were forced either to switch to domestic sources or stop production. However, since domestic sources were found only in the distant Urals, the chemical industry was faced with one of two unpalatable choices: either use tremendous amounts of hauling capacity to transport bulky materials to existing factories, or else build new factories closer to the source. The second choice was not optimal because many factories had to be built at the same time, and the construction industry was overextended. In effect, the government decided to transport large quantities of pyrites to existing factories and at the same time to build a smaller number of new factories. This solution worked reasonably well, and sulfuric acid production gradually increased throughout 1916.[123]

Similar problems arose with other chemicals. For example, in early 1915, crude benzene was produced in the Donets Basin region, but there was not a single factory that could fractionate the crude benzene into purified benzene and toluene that could be used for explosives. Thus, the crude benzene had to be shipped to Petrograd, where it was purified. The toluene was shipped to the Samara-Sefiev explosives factory (on the lower Volga River) to produce TNT, while the benzene was shipped to Moscow to make phenol and then back to the Solvay factory to produce picric acid.[124] Eventually, new factories were built in the Donets Basin region that could fractionate the crude benzene, but these of course required considerable resources and time to build. Ipat'ev and the Chemical Committee tried to cope with the transportation problems by having a special 200 soldier military command under their control. The Chemical Committee would assign soldiers to help assure that the supplies actually arrived at the factories and the front. Otherwise, some of these supplies would be commandeered by others needing these supplies and products. Ipat'ev noted that it was particularly difficult to ship supplies to factories in the Moscow region because there were so many other defense factories there needing these same types of supplies that the Chemical Committee's shipments often were diverted.[125] It is interesting to note that many of these same types of problems arose during the intense industrial growth of the First Five-Year Plan (1928–1932) under Stalin.

The experience of Branch III (incendiaries and flamethrowers) of the CC illustrates the problems facing the chemical industry in 1916. As with so many products, incendiaries were not normally manufactured in Russia before the war. So the CC had to find sources of red and yellow phosphorous, barium and strontium nitrates, powdered aluminum, metallic magnesium, antimony sulfide, and many other chemicals. The Russian military did not do much work with these products before the formation of the CC in 1916. And by that time, according to Ipat'ev, it was becoming difficult to set up new factories to produce items not previously made in Russia. The CC managed to set up only a few factories to produce phosphorous and barium nitrate, while factories to produce the other chemicals needed for this Branch were never finished. Some factories, such as a cyanamide factory in Grozny, remained on the drawing board, despite the plans of a special commission headed by Prof. D. P. Konovalov. As flamethrowers had never been used by the Russian army, the CC began to examine different mechanisms. Two types were ready for testing by early 1917, and were introduced to the chemical battalions. However, as with so many other products, Russian factories had no time to manufacture them in quantity before the country withdrew from the war.[126]

## PLANNING FOR THE FUTURE

Perhaps for this reason, by late 1916, some far-sighted Russians began to make plans for the future. Chemists, in particular, looked to what the end of the war might mean for the chemical industry. Their writings were usually pleas that Russia not revert to conditions existing before the war. In 1916, these chemists began to take steps towards a peacetime economy. As Ipat'ev later noted, 'The greatest achievement of the Chemical Committee was its work simultaneously organizing for wartime production as well as preparing for the transition to peacetime conditions.'[127] He was acutely aware that the end of war with Germany would mean a great shift in demand for chemical products. All during his time as head of the CPE and the CC, Ipat'ev attempted to balance wartime needs with post-war expectations. For example, he knew that sulfuric acid would be vital for industry after the war, so he was willing to expand its production (whilst resisting what he considered to be excessive expansion). On the other hand, phosgene had limited uses in a post-war environment, so he was reluctant to build many new factories for its production.

In the autumn of 1916, M. I. Lisovskii, head of the Southern regional bureau of the CC in Kharkov, asked Ipat'ev to plan for a transition to a peacetime economy. Ipat'ev suggested organizing a meeting of all the owners of chemical plants in southern Russia. At the same time, Ipat'ev organized similar meetings for chemical industrialists in Moscow and Petrograd. To these meetings came not only chemical industrialists, but also academic specialists who had taken part in chemical industry during the war. This attempt to maintain contact between academic chemists and industrialists was one of the most important fruits of the war. These meetings led to a special Preparatory Commission, assigned to survey and collect information about chemical manufacturing in Russia. The meeting in Petrograd included representatives from chemical organizations that had been involved in the war, including the War-Chemical Committee of the Russian Physical-Chemical Society and the Society for the Assistance and Development of the Chemical Industry in Russia.

The Moscow meeting also proposed the formation of a new organization to bring together the regional bureaus of the CC, the War-Chemical Committee of the Russian Physical-Chemical Society, and the various WICs, under the direction of the CC. Factory owners supported the idea and collected funds for the new organization. However, Ipat'ev was reluctant to have the CC assume this role, and proposed that the War-Chemical Committee take on this task. Eventually, after much discussion, Ipat'ev agreed to petition the GAU to allow the CC to participate in the demobilization of the chemical industry. The GAU approved his petition and a Preparatory Commission was formed, along with a Central Organization in Moscow. Ipat'ev was elected chairman, with P. I. Walden, who lived in Moscow, taking over day-to-day activities.[128] By August 1917, the Preparatory Commission had gathered a substantial amount of data, but the Bolshevik Revolution of October 1917 changed the situation. In early 1918, the Preparatory Commission and its data were placed under the authority of the newly-organized chemical section of the Supreme Council of the National Economy.[129] A new era had begun.

## CONCLUSION

What can we conclude from the experience of Russia's chemical mobilisation and its wartime agencies? Probably the clearest finding is that Russia's attempts to provision its army were chaotic, haphazard, and disorganized, with many overlapping and competing committees, commissions, and jurisdictions. The gas mask story is perhaps the best example of the range of difficulties besetting the system. Several organizations competed with – and actively hindered – each other, significantly delaying the provision of gas masks to troops at the front. There was no one organization with the authority to coordinate production, even after the creation of the CC. Adding to the chaos, many non-governmental organizations also attempted to contribute to wartime production. For example, in September 1914, the Imperial Russian Technological Society formed a 'Commission for Industry connected to the War'. In June 1915, in the wake of the shell crisis and the retreat along the Eastern front, nineteen societies and social organizations joined together and used this commission as a basis for forming the "Committee for Military-Technical Assistance."[130] This committee had separate sections for gas masks, poison gases, raw materials, and other items. It is not hard to imagine how difficult it was to coordinate so many different combinations of governmental and non-governmental organizations.

The supply of explosives appears to have been the best organized sector of Russia's chemical mobilisation, but even here there were problems. The first major decision faced by the military was whether to produce explosives by expanding domestic production or to rely on imports. Historical practice had been to negotiate contracts abroad for war materials greater than domestic industry could produce. That is, the Russian military was unwilling to expand domestic production above peacetime levels, for fear of having excess capacity after the war. This was the path taken in the first few months of the war. However, it soon became apparent that foreign orders would not arrive in time to meet immediate needs. This situation forced Ipat'ev to organize an increase in domestic production. The organizations he set up gradually increased production, although they never were able to meet the demands of the military. The main problem Ipat'ev faced was not to increase the production of any one component too fast, lest he divert resources from other targets. Thus he resisted dramatic increases in sulfuric acid production, as he knew that a large increase would hurt other sectors of industry. These attempts are significant because 'outsiders' tried to impose their views on the military. Moreover, Ipat'ev, as head of the CPE and later the CC, was forced to defend his actions, with the implication that had he not been persuasive, the military would have agreed to calls for greatly increased production.

The chaotic conditions surrounding chemical mobilisation were a natural consequence of a lack of planning for a long war. Since the Russian military did not expect a long war, they did not think beyond their usual reliance upon foreign orders. When it was necessary to turn to Russian sources, there was no coordinating structure. By the summer of 1915, it was glaringly obvious to impartial observers that foreign orders could not meet demand, although many members of the military high command continued to put their hopes in foreign orders rather than trying to develop domestic production. The crisis with the production of military goods, such as shells, eventually grew so severe that politicians gained considerable influence over industrial policy. Thus, the lack of planning, coupled with the extreme demands, opened up the way for a large number of non-governmental organizations to participate in producing military goods. However, these organizations often had definite political agendas, which complicated the coordination with governmental organizations. The all-too-frequent result was paralysis in production.

Nonetheless, by the time industrial production collapsed in 1917 due to problems with transport, lack of raw materials, etc., many concrete steps had been taken to prepare the future of Russia's chemical industry. Whereas before the war, Russia produced few raw materials for chemical industry, the war forced the country to search for domestic sources of supply. This effort continued throughout the 1920s and 1930s, and resulted in the Soviet Union discovering a wide range of new resources, especially in Siberia. The discovery of these domestic sources of raw materials was a crucial factor that made it feasible for Stalin to turn the Soviet Union towards economic autarky after the late 1920s. Wartime Russia also began building factories to produce many products needed by the Russian chemical industry, instead of relying on imports. This effort would also continue after the war. In addition, the number of workers at chemical plants markedly increased during the war. In 1913, the Russian chemical industry employed 33,000 workers. This number rose to 54,000 in 1915 and peaked at 117,000 in 1917.[131] After the end of the war, these experienced workers were available to be tapped for skilled labor during the Soviet period. Finally, Russian academic chemists began to participate in solving problems for industry. The cooperation of industry and academia continued after the wartime emergency, opening new channels for the organisation of applied science (particularly under the aegis of the Academy of Sciences) in the Soviet Union. In the history of Russia's chemical industry, The Great War witnessed the beginning of a new order rather than the end of an old regime.

### ACKNOWLEDGEMENTS

Research for this essay was supported in part by New Mexico State University. The author wishes to thank Professor Igor S. Dmitriev, Director of the D. I. Mendeleev Archive and Museum at St. Petersburg State University, for assistance with research, as well as for many stimulating conversations about the history of Russian chemistry.

### NOTES

[1] See N. N. Smirnov (ed.), *Rossiia i pervaia mirovaia voina* [*Russia and the First World War*] (St. Petersburg: Izdatel'stvo 'Dmitrii Bulanin', 1999), 9–10 for a discussion of work on the Russia's involvement in the war.
[2] Norman Stone, *The Eastern Front: 1914–1917* (New York: Charles Scribner's Sons, 1975).
[3] L. F. Haber, *The Poisonous Cloud: Chemical Warfare in the First World War* (Oxford: Clarendon Press, 1986).

4 Alexei.Kojevnikov, 'The Great War, the Russian Civil War, and the Invention of Big Science', *Science in Context* XV (2), (2002), 239–275.

5 For example, Alexander Vucinich's massive two-volume *Science in Russian Culture* (Stanford: Stanford University Press, 2 vols., 1963, 1970) devotes only a few pages to science in Russia during the war. Likewise, Loren Graham's *Science in Russia and the Soviet Union* (Cambridge: Cambridge University Press, 1993) almost entirely ignores the war.

6 See Kojevnikov, *op. cit.* note 4; Nathan M. Brooks, 'Chemistry in War, Revolution, and Upheaval: Russia and the Soviet Union, 1900–1929', *Centaurus* 39 (1997), 349–367; David L. Hoffmann and Peter Holquist (eds.), *Cultivating the Masses: The Modern Social State in Russia, 1914–1941* (Ithaca: Cornell University Press, 2002); Peter Holquist, *Making War, Forging Revolution: Russia's Continuum of Crisis, 1914–1921* (Cambridge: Harvard University Press, 2002).

7 For example, see V. S. Lel'chuk, *Sozdanie khimicheskoi promyshlennosti SSSR* [*The Creation of the USSR Chemical Industry*] (Moscow: Nauka, 1964), and A. L. Sidorov, *Ekonomicheskoe polozhenie Rossii v gody pervoi mirovoi voiny* [*The Economic Condition of Russia during the First World War*] (Moscow: Nauka, 1973). These volumes very briefly mention the production of explosives and munitions, or the role of chemists in the chemical industry. But see E. V. Trofimova, 'Proizvodstvo vzryvchatykh beshchestv v gody pervoi mirovoi voiny [The Production of Explosives During the First World War],' *Otechestvennaia istoriia*, No. 2 (2002), 147–152. This short article contains little information not available elsewhere, but does provide references to archival sources, making it an important starting point for further study.

8 For example, see V. I. Kuznetsov and A. M. Maksimenko, *Vladimir Nikolaevich Ipat'ev, 1867–1952* (Moscow: Nauka, 1992).

9 Zemgor was the 'Joint Committee of the Union of Zemstvos and of Towns for the Supply of Military Equipment and Munitions'. It was created in June 1915 as an aid to the military effort after the disastrous defeats suffered by the Russians in Galicia and Poland. For more information, see Paul Gronsky and Nicholas Astrov, *The War and the Russian Government* (New Haven: Yale University Press, 1929), 175.

10 Betweeen 1892–1895, the Imperial Navy engaged Dmitrii Mendeleev to undertake research on different types of smokeless gunpowder. Mendeleev conducted investigations, eventually producing a type that he termed superior to that used by the French and British. However, the Naval Ministry rejected Mendeleev's gunpowder, despite the support it received from younger, lower-ranking officers. See Michael D. Gordin, 'A Modernization of "Peerless Homogeneity:" The Creation of Russian Smokeless Gunpowder', *Technology and Culture*, 44 (2003), 677–702. Also see A. G. Beskrovnyi, *Russkaia armiia i flot v ZIX veke* [*The Russian Army and Navy during the 19th century*] (Moscow: Naula, 1973), 391–5, for statistics on the production of smokeless gunpowder.

11 Sidorov, *op. cit.* note 7, 364.

12 *Ibid.*, 85–86.

13 V. N. Ipat'ev and L. F. Fokin, Khimicheskii komitet pri Glavnom Artilleriiskom Upravlenii i ego deiatelnost' dlia razvitiia otechestvennoi khimicheskoi promyshlennosti [The Chemical Committee of the Main Artillery Administration and its Activities in the Development of the Domestic Chemical Industry] (Petrograd: Nauchnoe Khimiko-tekhnicheskoe Izdatel'stvo, 1921), 5.

14 E. R. Goldstein, 'Military Aspects of Russian Industrialization: The Defense Industries, 1890–1917' (Unpublished Ph.D. dissertation, Case Western Reserve University, 1971), 148. At this time, the official exchange rate was two rubles to the U.S. dollar.

15 For example, see L. F. Haber, *The Chemical Industry, 1900–1930: International Growth and Technological Change* (Oxford: Clarendon Press, 1971), 22–24; and Lel'chuk, *op. cit.* note 7, 15–38, for examples of Russian chemical imports during these years.

16 For an extensive discussion of some of the many blunders in Russian planning, see Stone, *op. cit.* note 2, chapter 1. Stone emphasizes the mal-distribution of types of artillery as one of the root causes of the artillery's poor performance, while downplaying the actual shortage of shells.

17 Reprinted in Frank A. Golder (ed.), *Documents of Russian History, 1914–1917* (Gloucester, Mass.: P. Smith, 1964), 190–2.

[18] V. A. Sukhomlinov, 'Rossiia khochet mira, a gotov k voine' [Russia wants peace, but is prepared for war], *Birzhevye Vedomosti* (27 fevralia 1914 goda), reprinted in Mikhail Lemke, *250 dnei v tsarskoi stavke (25 sentiabria 1915-2 iiulia 1916)* (St. Petersburg, 1920), 6.

[19] Bruce W. Menning, *Bayonets Before Bullets: The Imperial Russian Army, 1861–1914* (Bloomington: Indiana University Press, 1992), 246–247. Note that shortly after war was declared, Tsar Nicholas II decided to change the name of St. Petersburg to Petrograd, as the old name allegedly sounded German (although it was actually derived from the Dutch).

[20] 'Perepiska V. A. Sukhomlinova s N. N. Ianushkevichem' [*The Coreespondence of V. A. Sukhomlinov with N. N. Ianushkevich*], *Krasnyi arkhiv*, No. 1 (1922), 247–8.

[21] A. A. Manikovskii, Boevoe snabzhenie russkoi armii v mirovuiu voinu [The Military Supply of the Russian Army in the World War], (Moscow: Gos. Izd-vo, Otdel voennoi lit-ry,, 2nd ed., 1930), vol. 1, 124.

[22] Goldstein, *op. cit.* note 14,173–180.

[23] Sidorov, *op. cit.* note 7, 27–9.

[24] For a discussion of Russian attempts to purchase supplies in the United States, see Dale C. Rielage, *Russian Supply Efforts in America during the First World War* (Jefferson, North Carolina: McFarland, 2002).

[25] Stone, *op. cit.* note 2, 157; Sidorov, *op. cit.* note 7, 598.

[26] Vladimir N. Ipatieff, *The Life of a Chemist* (Stanford: Stanford University Press, 1946), 195.

[27] Rielage, *op. cit.* note 24, 23–28.

[28] *Ibid.*

[29] Ipatieff, *op. cit.* note 26, 196. Ipat'ev received his training in chemistry at the Artillery Academy and worked after graduation with Baeyer in Munich.

[30] This view permeates Ipat'ev's memoirs. Even though the memoirs were written years after these events, the desire for a domestic chemical industry was shared by many Russian chemists during the war and afterwards. For example, see L. Chugaev, 'O merakh k razvitiiu v Rossii khimicheskoi promyshlennosty [Measures for the development of the chemical industry in Russia]', *Delovaia Rossiia*, No. 14–15, (26 April 1915), 2–3; No. 15, (4 May 1915), 2–3. Also see N. N. Liubavin, 'Voina s Germanii i Russkaie khimiki [*The War with Germany and Russian Chemists*]', *Biulleteny Russkoi khimicheskoi literatury* (vesennyi sem. [spring semester], 1916), 6–8; and 'Peredovaia [Editorial]', *Zhurnal khimicheskoi promyshlennosti*, I (1), (Nov.-Dec. 1924), 3.

[31] Ipatieff, *op. cit.* note 26, 196. Note that Ipatieff is an alternative transliteration of Ipat'ev.

[32] *Ibid.*

[33] Stone, *op. cit.* note 2, 150.

[34] Lewis H. Siegelbaum, *The Politics of Industrial Mobilization in Russia, 1914–17: A Study of the War-Industries Committees* (New York: St. Martin's Press, 1983), 37–39. The Special Council is important because it brought industrialist into the process for the first time. However, at first only the largest industrialists—mainly the traditional suppliers located in Petrograd—were allowed to participate. This provoked smaller industrialists—in Moscow and elsewhere—to set up their own institution: the War-Industries Committee.

[35] Ipatieff, *op. cit.* note 26, 197.

[36] Ipat'ev and Fokin, *op. cit.* note 13, 16–17. A number of different Commissions were set up at about this time to oversee the production of military goods.

[37] Ipatieff, *op. cit.* note 26, 198.

[38] *Ibid.*, 199.

[39] *Ibid.*, 202 for figures on production.

[40] Ipat'ev and Fokin, *op. cit.* note 13, 21.

[41] *Ibid.*, 21.

[42] Bernard Pares, *The Fall of the Russian Empire*, (London: J. Cape, 1939), 230.

[43] Stone, *op. cit.* note 2, chapter 9, claims that the shell crisis of 1915 was exaggerated by politicians in order to obtain a greater role in the conduct of the war.

[44] Lel'chuk, *op. cit.* note 7, 20. Haber, *op. cit.* note 15, 171, gives Russian production of sulfuric acid in 1913 as 165,000 metric tons.

⁴⁵ L. F. Fokin, *Obzor khimicheskoi promyshlennosti v Rossii* [Survey of the Chemical Industry in Russia] (Petrograd: Nauchnoe Khimiko-tekhnicheskoe izdatel'stvo, 1920), part 1, 16–17.
⁴⁶ *Ibid.*, 9, 16.
⁴⁷ *Ibid.*, 18.
⁴⁸ Trofimova, *op. cit.* note 7, 148.
⁴⁹ *Ibid.*
⁵⁰ Ipat'ev and Fokin, *op. cit.* note 13, 32.
⁵¹ *Ibid.*, 32–33.
⁵² *Ibid.*
⁵³ *Ibid.*, 33
⁵⁴ Ipatieff, *op. cit.* note 26, 203–204.
⁵⁵ Trofimova, *op. cit.*, note 7, 150.
⁵⁶ *Ibid.*, 150.
⁵⁷ Ipat'ev and Fokin, *op. cit.* note 13, 33–34.
⁵⁸ *Ibid.*, 34. Note that this is according to Ipat'ev and Fokin. This conclusion needs to be verified using archival documents.
⁵⁹ For information about KEPS, see A. V. Kol'tsov, *Sozdanie i deiatel'nost Komissii po izucheniiu Estestvennykh Proizvoditelnykh Sil Rossii, 1915–1930* [The creation and activities of the Commission for the Study of Natural Resources of Russia, 1915–1930], (St. Petersburg: Nauka, 1999); and Kojevnikov, *op. cit.*, note 4.
⁶⁰ Ipat'ev and Fokin, *op. cit.*, note 13, 34–36. For more details about the construction of this plant, see: A. K. Kolosov, 'Vozniknovenie v Rossii proizvodstva azotnoi kisloty okisleniem ammiaka po sposobu I. I. Andreeva' [The origin in Russia of the production of nitric acid by the oxidation of ammonia according to the process of I. I. Andreev], *Materialy po istorii otechenstvennoi khimii* (Moscow: Akademiia Nauk SSSR, 1953), 196–209.
⁶¹ For a detailed description of the WICs, see Siegelbaum, *op. cit.* note 34. The War Ministry tended to rely on a small group of large industrialists in Petrograd to provide the bulk of military orders. By the spring of 1915, the problems with supplying needed military goods were clear to all observers. This opened the door to a wider range of producers.
⁶² Ipat'ev and Fokin, *op. cit.*, note 13, 28–9.
⁶³ *Ibid.*, 29.
⁶⁴ See Zelinskii's report about this work: *Zhurnal Russkogo Fiziko-Khimicheskogo Obshchestva*, 47, (1915), 1807. Zelinskii continued this work after the revolution, and this process became an important one for the Soviet chemical industry. Y. Yuriev and R. Levina, *Life and Work of Academician Nikolai Zelinsky* (Moscow: Foreign Languages Publishing House, 1958), 89–94.
⁶⁵ Ipat'ev and Fokin, *op. cit.* note 13, 29–30.
⁶⁶ *Ibid.*, 30–31.
⁶⁷ Sidorov, *op. cit.* note 7, 118. During the entire course of the war, Russia produced only 40,000 of these very heavy shells; more than twice this amount came from imports. During 1916, the GAU recognized the need for more of these shells, but claimed that it was impossible to increase production due to problems with the lack of metal, labor, and transport. The Russians had sufficient quantities of all types of heavy shells to help produce a breakthrough of the Austro-Hungarian lines in the summer of 1916, although the Russian generals pleaded for more shells. *Ibid.*, 120.
⁶⁸ *Ibid.*, 119.
⁶⁹ A. A. Manikovskii, *Boevoe snabzhenie Russkoi armii v voinu 1914–1918 gg.* [The military supply of the Russian army during the war 1914–1918], part 3 (Moscow: n. p., 1920), 560.
⁷⁰ Haber, *op. cit.* note 3, 36–39.
⁷¹ M. V. Krasil'nikov, 'K istorii snabzheniia russkoi armii protivogazom N. D. Zelinskogo', in N. A. Figurovskii, *Ocherk vozniknoveniia i razvitiia ugol'nogo protivogaza N. D. Zelinskogo* [A sketch of the Origin and Development of the Charcoal Gas Mask of N. D. Zelinskii] (Moscow: Akademiia Nauk SSSR, 1952), 193, quoting archival documents. The exact numbers according to these archival documents are 8,932 gas casualties, 1,101 gas fatalities.

[72] Quoted in Kojevnikov, *op. cit.* note 4, 245. In January 1915, the Germans had attacked Russian troops with about 1,800 shells filled with tear gas. However, the gas was not effective, probably due to the cold temperatures, and the Russians suffered no casualties from this attack. See Ia. M. Fishman, *Voenno-khimicheskoe* delo (Moscow, 1929), p. 16. This attack perhaps stimulated the Russian military to begin discussing the possibility of gas warfare.

[73] Ipat'ev and Fokin, *op. cit.* note 13, 36–37. Ipat'ev claims that he also thought of the use of phosgene independently of this engineer.

[74] According to Haber, the Germans probably used a mixture of chlorine and phosgene, expecting the Russians not to notice the addition of phosgene. Haber, *op. cit.*, note 3, 37.

[75] Ipat'ev and Fokin, *op. cit.* note 13, 37.

[76] Joachim Krause and Charles K. Mallory, *Chemical Weapons in Soviet Military Doctrine: Military and Historical Experiences, 1915–1991* (Boulder: Westview Press, 1992), 20. This must be used with caution as it contains many errors.

[77] The main sources were all written by Ipat'ev, so archival research will be necessary to confirm his accounts.

[78] Ipat'ev and Fokin, *op. cit.* note 13, 37.

[79] *Ibid.*, 37–38.

[80] Augustin M. Prentiss and George J. B. Fisher, *Chemicals in War; A Treatise on Chemical Warfare* (New York: McGraw-Hill, 1937), 658, 661. However, Kojevnikov, *op. cit.* note 4, 246, citing archival documents, claims that no more than two tons of all types of gases were ever delivered to the front by the time Russia stopped fighting in 1917. The reason for this discrepancy is not clear.

[81] Ipat'ev and Fokin, *op. cit.* note 13, 39.

[82] Ipat'ev and Fokin do not mention the exact method used for the simpler production of chlorine at these plants. Ibid., 38.

[83] *Ibid.*, 38.

[84] *Ibid.*, 39.

[85] *Ibid.*, 40.

[86] B. G. Klimov, ed., *Opytnyi zavod Gosudarstvennogo Instituta Prikladnoi Khimii, 1916–1926* [The Experimental Factory of the State Institute of Applied Chemistry, 1916–1926], (Leningrad, 1927), 26.

[87] Ipat'ev and Fokin, *op. cit.*, note 13, 40.

[88] See Brooks, *op. cit.* note 6.

[89] *Trudy Voenno-khimicheskogo komiteta*, No. 2 (1918), 66.

[90] *Ibid.*, 8–10.

[91] For information about the founding of the Institute of Applied Chemistry and the early work performed there, see the Russian State Archive of the Economy [RGAE], f. 3429, op. 7, d. 1083, ll. 1–111. Also see: Iu. I. Solov'ev, *Nikolai Semenovich Kurnakov 1860–1941* (Moscow: Nauka, 1986), 139–146; and Klimov, ed., *op. cit.* note 86.

[92] M. M. Dubinin, 'Predislovie [Foreword]', in N. A. Figurovskii, *Ocherk vozniknoveniia i razvitiia ugol'nogo protivogaza N. D. Zelinskogo* [A Sketch of the Origin and Development of the Charcoal Gas Mask of N. D. Zelinskii] (Moscow: Akademiia Nauk SSSR, 1952)), 5. Also see M. M. Dubinin, 'O razvitii protivogazovogo dela v Sovetskom Soiuze [The develoment of gas masks in the Soviet Union]', *Materialy po istorii otechestvennoi khimii* (Moscow: Akademiia Nauk SSSR, 1953), 163–172.

[93] N. D. Zelinskii and V. S. Sadikov, 'Ugol' kak sredstvo bor'by s udushaiushchimi i iadovitymi gazami. Eksperimental'noye issledovannye *1915–1916 gg.* [Charcoal as an agent in the fight with asphyxiating and poisonous gases. Experimental work of 1915–1916]', in N. D. Zelinskii, *Sobranie trudov* [Collected works], (Moscow: Akademiia Nauk SSSR, 1960), vol 4, 59–145. Also see N. A. Figurovskii, *Ocherk vozniknoveniia i razvitiia ugol'nogo protivogaza N. D. Zelinskogo* [A sketch of the origin and development of the charcoal gas mask of N. D. Zelinskii] (Moscow: Akademiia Nauk SSSR, 1952).

[94] Krasil'nikov, *op. cit.* note 71, 194–195.

[95] Figurovskii, *op. cit.* note 93, 112. For a brief description in English of Zelinskii's gas mask and its development, see Ipatieff, *op. cit.* note 26, 218–225.

[96] Quoted in Figurovskii, *op. cit.* note 93, 95.

97 Ipatieff, *op. cit.* note 26, 218. However, Figurovskii categorically denies that Zelinskii tried to keep his method of activating charcoal a secret, pointing to numerous letters Zelinskii sent to various people. Moreover, Figurovskii claims that Zelinskii never wanted royalties. Unfortunately, Figurovskii does not quote any document to support his claims, nor does he cite any references. Figurovskii, *op. cit.* note 93, 113–114. It appears likely that there was conflict or jealousy between Ipat'ev and Zelinskii, and probably not only concerning gas masks. For example, by the summer-autumn of 1916, after Ipat'ev had taken control of all gas mask production, he excluded Zelinskii from further research on improving methods to activate charcoal. *Ibid.*, 126–127. This time Figurovskii does quote from letters from Zelinskii.

98 Krasil'nikov, *op. cit.* note 71, 196.
99 *Ibid.*
100 *Ibid.*, 197.
101 Ipat'ev and Fokin, *op. cit.* note 13, 45.
102 Krasil'nikov, *op. cit.* note 71, 197.
103 *Ibid.*
104 *Ibid.*, 198.
105 *Ibid.*
106 *Ibid.*
107 Ipat'ev and Fokin, *op. cit.* note 13, 46.
108 *Ibid.*, 49.
109 Ipatieff, *op. cit.* note 26, 221. It also is not clear how long the Council for the War-Chemical Committee continued to function.
110 Ipat'ev and Fokin, *op. cit.* note 13, 47.
111 Ipatieff, *op. cit.* note 26, 220–221.
112 Ipat'ev and Fokin, *op. cit.* note 13, 47. For information on the chemical battalions, see N. N. Ushakova, *Nikolai Aleksandrovich Shilov* (Moscow: Nauka, 1966), 33–47. Shilov, a professor at Moscow Higher Technical School, became the commander of one of the three main chemical battalions.
113 *Ibid.*, 48.
114 *Ibid.*, 49.
115 Krasil'nikov, *op. cit.* note 71, 200, citing archival sources.
116 *Ibid.*, 201.
117 *Ibid.*
118 Ipat'ev and Fokin, *op. cit.*note 13, 52.
119 *Ibid.*, 54.
120 Krasil'nikov, *op. cit.* note 71, 200–201.
121 Dubinin, *op. cit.* note 92, 10–11; Ipat'ev and Fokin, *op. cit.* note 13, 49.
122 Dubinin, *op. cit.* note 92, 11.
123 Ia. M. Bukhshpan, *Voenno-khoziastvennaia politika* [*Military-Economic Policies*] (Moscow-Leningrad: Ranion – Institut Economiki, 1929), 365.
124 Ipat'ev and Fokin, *op. cit.* note 13, 26.
125 *Ibid.*, 65–66.
126 *Ibid.*, 70–71.
127 *Ibid.*, 72.
128 Walden was a member of the Imperial Academy of Sciences, elected in 1910.
129 Ipat'ev and Fokin, *op. cit.* note 13, 72-3.
130 L. V. Darda and N. A. Figurovskii, 'Organizatsiia promyshlennosti khimicheskikh reaktivov v Rossii (1900–1917) [*The Organization of the Chemical Reagent Industry in Russia (1900–1917)*],' *Istoriia i metodologiia estestvennykh nauk*, XXVIII (1982), 13–19.
131 S. Strumilin, 'Khimicheskaia promyshlennost' SSSR', *Problemy ekonomiki*, no. 2 (1935), 118.

# TECHNICAL EXPERTISE AND U.S. MOBILIZATION, 1917–18: HIGH EXPLOSIVES AND WAR GASES

## INTRODUCTION

During the First World War, most of the organisation and technology behind the production of high explosives in the United States lay in private hands. When President Woodrow Wilson and Congress brought the country into the war in April 1917, the scope of the Federal government increased dramatically, but the Armistice arrived long before the government could supplant private industry as the principal source of technical expertise. Nevertheless, the government played an important role in diffusing expertise across the industrial landscape. By 1917, the military organisation responsible for procuring explosives for most of American consumption — the Ordnance Department — became an important agent in the dissemination of industrial expertise in synthetic organic explosives production.[1]

In contrast to explosives, Americans in April 1917 possessed very little technical expertise in war gases, whether in the military or private sphere, and the organisation of production consequently developed very differently. Military officials followed tactical chemical warfare developments in Europe without obtaining productive know-how; nor did American manufacturers show any inclination to produce war gases for the European market. They understood the extremely hazardous nature of war gas production—more dangerous than explosives—and few manufacturers found the potential rewards worth the risks, even after April 1917. Initially, the generation of new expertise came from neither industry nor the military, but from non-military, scientific units of government and from university chemists. Military officials also had strategic reasons to organise production differently. The war generated few new kinds of high explosives unknown before the war, but combatants in chemical warfare continually sought new gases to surprise and outmaneuver the enemy. Secrecy mattered more. The pervasive lack of expertise, the hazards of gas production, and military strategy all contributed to the creation of a more centralized, government-controlled, government-owned mobilization apparatus, in which the loci of technical expertise lay outside private firms.[2]

This essay considers the role of technical expertise, as the U.S. government mobilized the resources of the nation for the production of synthetic organic high explosives and war gases. Differing levels of expertise contributed to the creation of significantly different organisational structures to mobilize production in the two categories of weapons.

## THE EXPLOSIVES INDUSTRY MOBILIZES

The American economy was drawn into the European conflict thirty months before the United States officially entered the war.[3] Although the British blockade hurt certain sectors of U.S. transatlantic trade, it also created new opportunities for entrepreneurs. Moreover, unprecedented Allied spending greatly stimulated relevant sectors in the American economy. In the first year of the war, Allied purchasing produced a chaotic bliss for American suppliers. Allied governments sent armies of representatives to the United States in search of everything from mules, food, and leather to steel, artillery, and explosives. France, Britain, Russia, and Italy established offices to supervise purchasing and inspections.[4]

High explosives, particularly TNT (trinitrotoluene) and picric acid (trinitrophenol), were among the most sought after products in the American market, and established explosives firms like DuPont and Hercules grew quickly to accommodate millions of dollars in Allied contracts. The rush to make these synthetic organic explosives was complicated by shortages in related commercial chemicals. In the four decades before 1914, Germany dominated the production of synthetic dyes and pharmaceuticals, and the war left American markets starved for them. The shortages prompted entrepreneurs to try their hand at synthetic organic chemicals, an experiment that resulted in success for simpler chemicals; in severe difficulty, if not failure for complex chemicals; and in not a few accidental plant explosions and chemical poisonings. The chemical similarities among TNT, picric acid, dyes, and certain pharmaceuticals such as aspirin meant competition for scarce technical expertise, but also continued employment for such expertise after the war. The prospect of an enduring domestic industry encouraged young chemists to specialize in synthetic organic chemistry and lured existing chemists from other fields to retool. In addition, the military and commercial chemicals shared the same raw materials (benzene, toluene, and phenol) and similar processes (nitration and sulphonation). This 'dual use' allowed proponents of a domestic synthetic organic chemicals industry to exaggerate the chemical relationship when they lobbied Congress for protective legislation. They argued that the peacetime existence of dyes plants and of a pool of skilled chemists contributed to national security through their potential conversion to military uses.[5]

Almost immediately at the outbreak of war, the US synthetic organic chemicals market felt the nearly simultaneous shock of three different pressures—shortages, protectionism, and Allied purchasing. Owing to the Allied demand for high explosives, American production of these simpler synthetic organics increased rapidly. Prices for chemicals such as toluene and benzene raced upwards as the British and French governments, and private speculators, competed for scarce raw materials. Toluene went from 30c per gallon to $5 per gallon, and benzene soared from 25c to $1.75.[6] The chemicals trade press expressed astonishment at the sheer size of the contracts. The amounts of money that changed hands often remained private, but occasionally the public might learn that the French had signed a contract for picric acid in May 1915 worth nearly $1 million, rumoured to produce profits close to 20 per cent.[7] By the end of the war, $1 million would not seem like a large contract in explosives.

For shell explosives, the Allies purchased TNT and picric acid. By 1915, the British favoured TNT (trinitrotoluene), while the French preference throughout the war remained picric acid (trinitrophenol), which yielded a more potent blast, but which also carried a greater risk of instability and premature detonation. In general, the two kinds of explosives were made by different firms, and indeed the State–industry relationship developed variations based on the two groupings. Neither explosive was new to American industry, but there was little reason before the war to produce either on a large scale. At the outbreak of war, manufacturers, whether experienced or not, signed contracts with the Allies to produce them in unprecedented quantities.[8]

A list of the largest manufacturers of TNT includes well-known names, most having corporate connections with DuPont (E.I. du Pont de Nemours & Company). Based in Wilmington, Delaware, since 1802, DuPont began by making gunpowder, and took up high explosives late in the nineteenth century. In 1902, the firm established its first research laboratory and increasingly broadened its product line. Before August 1914, DuPont was the only company in the United States producing TNT; although it manufactured small quantities, it had the skills to scale up production. The war made DuPont into one of the nation's largest firms, as it became a top manufacturer of smokeless powder (nitrocellulose) as well. Following antitrust violations, a Federal court in 1912 created from DuPont two new firms, the Atlas Powder Company and the Hercules Powder Company. During the war, Atlas became the largest producer of ammonium nitrate (the most important high explosive that was not a synthetic organic chemical), and Hercules became a large producer of TNT.[9] In May 1915, Hercules received its first war contract for TNT from the Russian government; this it manufactured in California, and later sold to the British as well.[10] In 1913, a former DuPont executive set up Aetna Explosives Company, which manufactured TNT, picric acid, and other military explosives. However, while DuPont and Hercules charged customers the cost of plant construction, Aetna borrowed heavily to expand, and collapsed into receivership during the last two years of the war.[11]

Americans had even less prewar experience with picric acid than with TNT; it had been made only experimentally in the United States. Burgeoning French demand lured some American makers into picric acid contracts, but these firms tended to be smaller than those that made TNT. Indeed, the ranks of picric acid manufacturers comprised such unfamiliar names as Butterworth Judson, Goodrich-Lockhart, Everly M. Davis, and the Nitro Chemical Company.[12] DuPont made negligible amounts, and these were primarily used as raw material for ammonium picrate; Hercules and Atlas made none. The Semet-Solvay Company, a builder of by-product coking ovens since the 1890s, moved into the production of synthetic chemicals, including high explosives, during the war. Only Aetna and Semet-Solvay held contracts to make significant amounts of both TNT and picric acid.[13]

The chemistry of picric acid posed few scientific difficulties, but developing a safe and practical process proved challenging. In 1915, for example, Aetna delivered its first French contracts for TNT and picric acid five months late. Such delays obliged the French government to send technical advice as well as capital.[14] By the end of 1917, Aetna was supplying two-thirds of the TNT and one-half of the dry picric acid that the French government purchased in the United States.[15] When chaotic market conditions overwhelmed Allied purchasers, they turned to J.P. Morgan & Company, a company that had made its fortune partly by directing European investment into American railroads and other assets. In January 1915, J.P. Morgan agreed to coordinate and supervise purchases for the British, and a similar agreement was concluded with France four months later. To oversee these enormous contracts, the bankers hired Edward R. Stettinius, Sr., a respected executive of Diamond Match, to run a new 'Export Department'. For the next two years, Stettinius was the mastermind of a complex organisation that handled more than $3 billion in Allied contracts. Contracts to explosives firms accounted for at least a fifth of this amount.[16]

At his Export Department, Stettinius devised strategies to obtain reliable deliveries of explosives and other products at better prices. He not only guided J.P. Morgan, but also significantly influenced US military purchasing guidelines after April 1917. First, he stopped publicizing the products the Europeans demanded, a process which had only alerted the market to raise prices. Instead, he sought contracts privately and quietly, settling on generous but not outrageous prices.[17] Second, as he valued reliability and dependability, he approached large, established firms. In high explosives, that meant DuPont, Hercules, Aetna, and Semet-Solvay. This policy encouraged further concentration in the economy. Finally, in fields where too few or no experienced suppliers existed, he identified possible candidates. For example, he encouraged the steel industry to accelerate its conversion from beehive to by-product coke ovens, the primary source of raw materials for high explosives. Meanwhile, French demand led J.P. Morgan to cultivate new manufacturers, such as Butterworth-Judson, which became one of the largest picric acid contractors. To run his Department, Stettinius chose knowledgeable people to oversee industries they already knew. Eventually, his office had a staff of 175.[18]

The supply of *matériel* brought unprecedented prosperity to many makers of high explosives. The demands of the war also brought about a limited, private form of mobilization. President Wilson's official commitment to neutrality, and public resistance against joining a foreign war, meant that, in April 1917, the United States was militarily unprepared. The war-related economy, however, was much better prepared. The asymmetry soon became very apparent.[19]

## THE STATE MOBILIZES EXPLOSIVES

Historians have noted the speed of US mobilization relative to that of the Allies, from whom the Americans borrowed many policies and procedures.[20] However, even a relatively rapid transition could not transform the American economy quickly enough to integrate smoothly into an ongoing war.[21] Generous estimates give April or May 1918—a full year after the US entry into the war—as the time when American mobilization achieved effective coordination, although one could delay that date by many months in some sectors.

For high explosives, the declaration of war made the Federal government immediately relevant to production. Still, the havoc in Washington, DC, tested Allied patience. No longer the Allies' representative in the United States, Stettinius noted to a colleague that 'we are repeating all of the mistakes that were made by England during the first months of the war'. The 'autocratic control' developed in Britain to coordinate production appeared to be as 'offensive to our democratic ideals' as it had initially been to the British.[22] For almost a year, until March 1918, the army, navy, and rapidly mutating civilian mobilization boards struggled among themselves, and with their enormous tasks, showing little of the efficiency so valued by the 'Progressives' of the early 20th century.

Even under government direction, mobilization relied heavily upon the private sector. That private firms possessed the necessary technical know-how was the underlying assumption. No one in the United States seriously contemplated a military takeover of the economy, which was not only politically impossible, but logistically irrational. Still, the Federal government alone possessed authority to coordinate national efforts. Legislation in 1916 gave government power to commandeer plants, but this typically resulted in cooperation rather than takeover. The logic of mobilization depended upon creating partnerships — between firms with know-how, and the federal government with legal and financial resources.[23]

The task of coordinating high explosives production fell to the mobilization boards—first the Council of National Defense (CND), set up in 1916, and then the War Industries Board (WIB) in 1917. Initially a sub-agency within the CND, the WIB became the primary mobilization board, particularly after its reorganisation in 1918. These industrial boards had one thing in common: they brought businessmen and government officials together. While it pained Wilson and other 'Progressives' to give more power to business, in the year before April 1917, most political, military, and business leaders came to see this as necessary. Wilson therefore aimed to use private expertise, while retaining overall control.[24]

Both the CND and WIB attempted to coordinate the economy by economic planning, but they lacked authority to enforce decisions. After President Wilson strengthened the WIB in March 1918, coordination occurred more smoothly, although the WIB still coddled as much as coerced business into cooperation. The WIB created 'cooperative committees' to work with individual industries. In most cases, concentrated industries with few firms—such as the explosives industry—proved much easier for the WIB than the management of more competitive industries. The military branches continued to make their own purchases, but both the Navy and the War Departments increasingly cooperated with the WIB in setting priorities.[25] Within the WIB, there was a chemicals division with about twenty sections. One section dealt specifically with explosives, and related sections with supplies of raw materials (such as toluene and nitric acid), synthetic dyes, and intermediates. The chemicals division worked with firms through an association of manufacturers, and also maintained a file of statistical information.[26]

Before the United States entered the war, scientists and engineers contributed to 'preparedness' through two quasi-governmental organisations. In July 1915, the US Navy called upon Thomas A. Edison to head its new Naval Consulting Board, which drew on the major engineering societies for technical advice. More important, the National Academy of Science created the National Research Council (NRC) in 1916 to mobilize research for national defence.[27] The NRC identified technical problems, which they assigned to scientists and engineers in universities and government agencies. Marston Bogert, a Columbia University chemist, chaired the NRC's chemistry committee. In July 1918, the NRC established an explosives committee, which did not formally convene until late August. While the explosives committee garnered praise for collecting and organizing information, the NRC's impact in the field of synthetic organic explosives was relatively slight.[28]

Even after the United States entered the war, private firms sat at the centre of American mobilization. When these firms came into contact with the government, they typically dealt with the War Department. Historian Paul Koistinen blames the War Department for many failures, noting that it never underwent the reorganisation in procurement that the Navy experienced at the turn of the 20th century, and that wartime demands overwhelmed its bureaucracy. Indeed, Secretary of War Newton D. Baker and his military leadership vigorously resisted attempts to subordinate army purchasing to civilian-run mobilization boards, and only after a year of chaos, culminating in a serious crisis in the winter of 1917–1918, did the War Department begin to cooperate with the War Industries Board.[29] Charged with purchasing munitions for the War Department, the Ordnance Department thereafter worked closely with the explosives manufacturers.

## EXPLOSIVES, TECHNICAL EXPERTISE, AND THE ORDNANCE DEPARTMENT

The performance of the Ordnance Department mirrored the larger pattern of American mobilization. Unprepared in April 1917, Ordnance officials faced a steep learning curve, and attained a respected level of competence only a year later. In high explosives, the Department inherited a relatively effective munitions industry. DuPont and Hercules, among others, had expanded rapidly, adjusted quickly, and by 1917 had little need for outside technological input. Ordnance officials built on their experience, and had the war continued into 1919, the Ordnance Department's contribution to the dissemination of 'best practice' would have been even greater.

In the spring and summer of 1917, the Ordnance Department gathered information and advice, especially from Edward Stettinius. If Stettinius was annoyed to receive duplicate inquiries from officers of different ranks and offices, he hid it gracefully. Brigadier General William Crozier, Chief of Ordnance, tried to recruit Stettinius into his department, and although Stettinius turned down the offer, many of his Export Department staff found positions in Ordnance and other agencies.[30] Sharing lists of manufacturers, Stettinius advised the Ordnance Department on contracts, prices and cancellation clauses. The Ordnance Department adopted almost entirely his advice to stay with large, established firms whenever possible.[31] After the war, Ordnance officers endorsed his policy, arguing that it would have been 'suicidal' to grant contracts to 'companies who were not entirely familiar in every respect' with high explosives. The Department was unable to avoid inexperienced firms completely, but it preferred and encouraged the 'companies with the best technical knowledge and organisation'.[32]

The War Department also approached the Allies for information. Ordnance officers acknowledged the relationship between Allied purchasers and American suppliers, and strove to avoid interrupting a working system. Unlike J.P. Morgan, the government could bring the force of law to negotiations with American manufacturers, and therefore compelled more favourable prices than Allied governments had paid.[33] What struck the trade press about picric acid contracts in October 1917 was the disparity in prices paid by the Allies, ranging from $1 to $1.25 per pound, as contrasted with the prices of 60 ¢ and 63 ¢ per pound paid by the US government.[34] Consequently, when existing contracts expired, the Allies placed subsequent orders through the US government to obtain the better prices. This was not always simple. When, for example, the French realized that the US government had contracted for Aetna's output of TNT and picric acid — which they had expected for the duration of the war — they sent harsh letters to the Department. In this case, the plant's output continued to go to France, but through American contracts.[35]

After April 1917, Allied technical advice typically flowed through the US military, rather than directly to firms. Attachés stationed in London and Paris stayed abreast of new technical developments and became conduits for information passing in each direction. For example, in the spring of 1918, the British offered the United States technical information on a continuous process for the manufacture of TNT. Ordnance officials shared the British information with several firms, soliciting manufacturers' assessments of the method's applicability to the United States. In this case, the American manufacturers rejected the process as more dangerous and no more efficient than their current practice.[36] In early 1918, the National Research Council posted scientific attachés to London, Paris, and Rome, but the NRC believed that the War Department had undercut them by insisting on using military attachés. A British plan for the NRC to provide thirty-five American scientists to help and learn from the British never received funding.[37]

The most significant transfer of technology from the Allies was the British process for making 'amatol', a mixture of TNT and ammonium nitrate. When, in 1917, the demand for TNT outstripped supply, the British extended their store of TNT by mixing it with the more plentiful ammonium nitrate—first in proportions of 50–50, and then, 20–80. Ordnance officers expected the United States would encounter similar shortages, and made inquiries of their own arsenals, of DuPont, and of Britain.[38] In October 1917, two British military experts travelled to the United States, bearing stacks of confidential reports that encapsulated the British state of the art.[39] In inorganic explosives as well, the Americans also borrowed British technology: a huge Atlas plant at Perryville, Maryland, which reproduced a Brunner Mond facility to make ammonium nitrate, began manufacturing in July 1918.[40]

However much Ordnance officers learned from J.P. Morgan and the Allies, American firms were the primary suppliers of technical knowledge. Then, Ordnance officers became distributors of the firms' expertise. The voracious market for high explosives stretched the American pool of expertise, as when employees from established firms struck out on their own, aiming to land the rich Allied or Ordnance contract that would give their 'start-up' a share of the market.[41] An Ordnance officer later argued that the expansion of the industry had left the 'experienced talent... consequently very thinly distributed over the existing production'.[42] For example, two former Butterworth-Judson employees established a new firm, and obtained an Ordnance contract to make picric acid; but when the technically competent partner died, the firm was left scrambling to find a replacement.[43] Before prohibitive legislation in August 1918, firms also poached employees from rivals (or from the Bureau of Mines).[44] With the demand for expertise, Ordnance officers believed that increased production depended upon their ability to expand and distribute technical knowledge.

Throughout the war, DuPont not only dominated the production of explosives, but also set the standard by which Ordnance officials judged the performance of other firms. After the Department invited explosives makers to Washington to learn each firm's capacity and potential, Ordnance officials tried to set prices for TNT based on DuPont's cost structure. If DuPont could sell TNT for 50 ¢ per pound, then the others should, too. However, when the other firms hesitated, the Department had to increase its offer, not least because other firms knew that DuPont already had most of the necessary raw materials locked up in contracts. To obtain as much TNT as it wanted, the Department had to offer 52 ¢ or 55 ¢ per pound to a few of the other manufacturers.[45]

To manage production, statistics were vital. To compare firms, the Department gathered data on raw materials, yields, and rate of output. To a degree that made some executives unhappy, Ordnance officers shared their information. When manufacturers signed contracts with the government, they agreed to maintain accounting records and to make these available on demand to Ordnance officials. By mid-1918, Ordnance officials were visiting under-performing plants, and suggesting changes in equipment and procedure, based on their observations of better performing firms.[46]

DuPont rarely needed (or heeded) advice from Ordnance officials. Officials recommended practices they most likely learned from DuPont, and for the most part, the Department seemed satisfied with DuPont's cooperation. On at least one occasion, however, the Delaware firm protested. In the summer of 1918, Ordnance officials asked for plans showing plant layouts, which most firms supplied. DuPont's executive committee, however, refused to turn over the plans of its highly productive TNT plant in Barksdale, Wisconsin. In the end, DuPont's executives relented, because their engineers had inspected the plans that were submitted by other firms, and this Ordnance officials used as leverage to demand a fair return. No doubt other firms gained more than DuPont from such exchanges.[47]

Smooth delivery and increased output drew Ordnance officers into issues of safety and health. Manufacturing high explosives was dangerous, even more so in the hands of inexperienced producers, and disastrous explosions interfered with output. Two debilitating explosions levelled TNT plants in the United States, and another two factories working under US contracts in Canada.[48] Picric acid, made in the traditional 'pot process', released so many dangerous fumes that Ordnance officials pushed firms to adopt newer, safer nitrating units. In choosing contractors, officials evaluated the risks associated with each plant. Sometimes, in periods when demand neared total plant capacity, officials felt obliged to give contracts to manufacturers whose plants frightened even the inspectors. Although officials fought to keep a dangerous plant operating if they needed the output, they also pushed firms to make their plants safer. This produced a course of action that neither endorsed dangerous practices nor effectively eliminated them. In general, Ordnance officials believed, usually correctly, that the more experienced producers like DuPont and Hercules were the safest. To reduce the risks, the more established manufacturers evolved standard procedures in everything from plant layout and equipment selection to daily sanitary and cleaning habits.[49]

Explosives firms occasionally asked the Department for help with other branches of government. Especially troublesome was the general shortage of labour, aggravated by the loss of skilled men to the military. Several firms submitted to the Department lists of essential employees, only to be told that Ordnance officers could not interfere with the rationing of labour, which was handled by a different agency.[50] Firms also faced restrictions from State governments, which regulated labour safety and health, and they were also subject to local town and neighbour complaints. In those cases, Ordnance officers vouched for the importance of a firm, but a state government could and sometimes did close down an exceptionally dangerous plant unless modifications were made.[51]

In the end, the Department's managerial superstructure contributed significantly to the transfer of technology across the US explosives industry. Statistics indicated where to find best practice, and the Department used their information to improve the performance of firms with less expertise. Information gathered from the plants of DuPont and Hercules ended up in the hands of smaller firms in South Bend, Indiana, or Lansing, Michigan. When the war drew to a close, the Ordnance officers sought to codify and preserve the technical information they had acquired from firms and the Allies. They systematically gathered process descriptions and wrote technical reports in the hope that the information would be useful in any future conflict, and cut time required for any future mobilisation.[52]

## MOBILIZING FOR WAR GAS PRODUCTION: AN ALTERNATIVE MODEL

Gas warfare on the western front began in April 1915, when the Germans released a cloud of chlorine.[53] Very quickly, the Allies adapted to the new weaponry, and each side engaged in an escalating scientific process of challenge and response.

Seeking ever more potent gases, as well as more reliable gas masks, chemists in all the belligerent countries joined in intensive, collaborative efforts. Although chlorine remained an integral ingredient of most gases, most subsequent gases were synthetic organic chemicals, aliphatics (chain molecule) rather than the aromatics (ring molecule) found in most high explosives. In the summer of 1917, the Germans introduced the most notorious of the war gases, mustard gas, which became a key challenge to the Allied gas programmes. In the twenty months of their participation in the war, Americans mobilized an impressively large gas warfare organisation, but it contributed relatively little to wartime operations, and the most tangible products would have arrived only in 1919. The structure of American gas mobilization, however, offers an intriguing comparison to the structure of high explosives mobilization.

Unlike high explosives production, the United States possessed little prewar capacity in war gases entering the war in April 1917, and some Americans later believed they were slow to recognize 'the great importance of gas warfare.'[54] American gas mobilization involved not only the coordination of resources and existing expertise, as in explosives, but also the generation of an enormous amount of new expertise. The loci of gas weapons expertise lay outside U.S. chemical firms, and outside the military as well. Instead, leadership came from non-military government agencies and the academic chemical community, both of which worked closely with their Allied counterparts. As the war progressed, however, the military grew increasingly central to gas development and production. By the Armistice, the military had integrated most centres of gas expertise into its command, both by taking over previously autonomous, non-military government programmes and by expanding units within the military.

In the wartime relationships among government, industry, universities, and Allies, the history of American gas warfare differed considerably from the history of high explosives. Although high explosives and war gases both centred on synthetic organic chemicals, gases posed much greater production hazards than explosives. More significantly, gases also demanded a greater level of technical expertise. The American experience of war gases reflected the relative dearth of expertise and the urgent need to acquire it.

Initially, the primary impetus behind the American gas programme came not from the military or from the chemical manufacturers, but from a non-military government agency and the academic chemical community. The Bureau of Mines, familiar with gases and gas masks from their work on American mining, led the development of the American gas programme under director Van H. Manning. As Manning offered his Bureau's help to the military in February 1918, the NRC almost simultaneously asked the Bureau's personnel to direct gas research. This began a fruitful collaboration between the Bureau and the NRC. Acting with extraordinary bureaucratic entrepreneurship, Manning orchestrated a research programme that peaked at 1900 people, including 1200 chemists and other technical staff, concentrated in Washington, DC, in new laboratories. Whereas the NRC's Committee on Explosives was almost an afterthought, its Committee on Noxious Gases played a vital role in identifying significant research problems, coordinating academic chemists, and communicating with the Allies. It performed as the 'clearinghouse' of scientific information that the NRC's founders had envisaged, and helped to guide the Bureau's research programme. The Bureau also received advice and guidance through its advisory committee, comprising military officers, chemical manufacturers, and government and academic chemists.[55]

Advice also came from abroad. The relationship with the Allies developed differently in war gases, because the American gas programme needed Allied expertise in a way the US high explosives industry simply did not. When Americans contemplated the organisation of their gas programme, they took the British model into consideration. Americans also knew that an 'understanding' existed with the British government, which was willing to 'exchange experts whenever necessary.'[56] In fact, the director of gas research, George Burrell, later recalled that the Bureau designed its research strategy only after 'literature...from France and England had been digested.'[57] Although the Allied gas programmes communicated rather badly with one another, the Americans developed regular channels with the Allies, particularly the British.[58] Through the NRC and the military, American emissaries were in close touch with the British gas efforts, learning from British successes and failures. As research progressed, the British shared their reports in meetings or even via cable, and Americans developed a system of bi-weekly reports that communicated research results not only internally but also to the Allies.[59]

In early April 1917, the NRC committee convened a series of meetings that proved instrumental in structuring the American gas effort. Manning marched the committee through a series of basic questions: Who should do the U.S. research? Should it be under the military? Where should it be located? Should the military (in addition to the NRC) send chemists to Europe? What organisational structure should the gas research take?[60]

Among other things, the Bureau made the development of an adequate gas mask its first priority. Establishing gas 'defence' centred research on charcoal and other materials. 'Offensive' research, which started more slowly—getting underway in July 1917—by early 1918 attracted some of the best chemists in the country.[61] In May, the Bureau sent representatives around the eastern United States to recruit chemists to gas research, and several universities became 'branch laboratories' in the Bureau's gas programme. Increasingly, however, the programme's leadership attempted to consolidate the bulk of the research in Washington, DC. Already in April 1917, the American University had offered its facilities, and in September the Bureau centered its research on the campus, which grew from two buildings to 'over sixty' by the Armistice. Important work still occurred in the branch laboratories, but at its American University Experiment Station the Bureau pioneered focused interdisciplinary research involving groups of scientists. Over 1200 chemists worked at the Experiment Station by the summer of 1918, and the facility employed a total staff of 1700 at the Armistice.[62]

The gas researchers strove to understand the gases already in use in Europe, including chlorpicrin, phosgene, and, after July 1918, mustard gas. Lacking previous experience with the gases, the chemists needed to devise workable processes in the laboratory before they could guide full-scale production. By 1918, the chemists also sought new gases. The most promising was Lewisite, a chlorine and arsenic compound, which was identified in Washington, DC and secretly manufactured outside Cleveland. A shipment was on the way to Europe when the Armistice arrived.[63]

While the Bureau of Mines and the NRC concentrated on replicating existing war gases and designing new ones, the production of gases fell under the military's jurisdiction. As with the War Department more broadly, the gas programme underwent several reorganisations during 1917 and 1918. Initially lodged in the Trench Warfare Section of the Ordnance Department, the gas programme ultimately achieved the status of an independent military unit within the War Department with the creation of the Chemical Warfare Service (CWS) at the end of June 1918. The CWS consolidated gas-related programmes scattered elsewhere in the government, including that of the Bureau of Mines. These changes partly reflected the increasingly central and complex role of the military in gas production.[64]

In October 1917, the Trench Warfare Section of the Ordnance Department established the Gunpowder Neck Reservation, a piece of the Aberdeen Proving Ground in rural Maryland, as its centre for gas development. While expecting to rely upon private firms to make the gases, the military planned to build at Gunpowder Neck a large plant in which to fill shells with the gases. The scope of the reservation expanded over the next several months as the military added manufacturing and laboratory facilities. In March 1918, the army removed the Gunpowder Neck Reservation from the jurisdiction of the Trench Warfare Section and gave it autonomous standing within the Ordnance Department, renaming it the Edgewood Arsenal in early May.[65]

The formation of the Chemical Warfare Service, and the prior arrival of General William Sibert to take command of the military's war gas efforts, brought another significant transition, particularly in the research programme of the Bureau of Mines. Known even by the Secretary of War as one of the 'most stubborn' of military men, Sibert was convinced that the war gas projects needed consolidation and should be placed entirely under his command. Even before Sibert's arrival, William Walker used his position as the commander of the Edgewood Arsenal to expand the range of military investigations into chemical warfare research, which already had the Bureau of Mines leadership wary of competing jurisdictions.[66] Dismayed at the possibility that he would lose his project, Van Manning protested against military control and attempted to retain autonomy for the Bureau's research programme. His external advisory committee came to his defence as the military attempted to take over gas research. Manning's operation, they argued, was 'complex and delicate but well articulated and working with an efficiency and enthusiasm which has greatly impressed us.' Many of them had the model of corporate research in mind and believed that 'best practice...place[d] the research work under separate control,' protected and away from the immediate demands of the producers.[67] To Sibert, however, the analogy fell short. In war, the only demands were immediate demands, and he believed that the chemists were too isolated and autonomous. He wanted them to respond more directly to the military's needs. His view prevailed, and President Wilson signed the Executive Order transferring the Bureau of Mines research programme to the new Chemical Warfare Service on 26 June 1918.[68]

The protests by the Bureau of Mines obscured the reality that a research chemist would often follow his project from the laboratory stage through development and production. Several personnel on the military side began the war with the research programme, including such high-ranking officers as William Walker and William McPherson.[69] In most respects, however, the Bureau of Mines and the military retained a division of labour until the creation of the Chemical Warfare Service. Large-scale production of gases belonged to the military.

As with the gas programme more generally, gas manufacturing evolved and mutated over the course of the war. From semi-scale production at small firms cooperating with the Bureau of Mines, manufacturing increasingly became more militarised. From the beginning, full-scale production fell under the jurisdiction of the Ordnance Department, which contracted with several firms to make war gases in 1918. For strategic, legal, and technical reasons, however, the military began to consolidate production into government-owned and -operated facilities, also onto a single, giant compound in rural Maryland. While the relationship between the military and industry was problematic in both high explosives and war gases, the balance of power was very different because of the chemical characteristics of the war gases and because of the greater lack of experience with them.

Some of the earliest plant production of war gases occurred under the Bureau of Mines as they scaled up their research. Typically, when the researchers were ready for semi-scale plant production of a war chemical, the Bureau and NRC personnel contacted willing commercial manufacturers to undertake production. Quite often, personal or professional ties determined the location of the semi-scale plants. For example, prior to April 1917, two MIT chemical engineers consulted at the American Synthetic Color Company in Stamford, Connecticut, where they advised on production of picric acid. Those engineers provided the link to the NRC, which encouraged the firm to undertake semi-scale production of chlorpicrin. After experiments in the late summer and autumn 1917, the firm received a full-scale contract from the Ordnance Department in December.[70] The same month, the Oldbury Electro Chemical Company, following its work on phosgene for dyes in 1916, received a contract from the Ordnance Department after doing research in cooperation with the Bureau of Mines and the NRC.[71]

These two companies—the American Synthetic Color Company and the Oldbury Electro Chemical Company—exhibited many characteristics typical of the firms receiving war gas contracts. Few firms were large, well-known producers; most had in-house or academic consulting chemists with close contacts to either the NRC or Bureau of Mines; and many were also venturing for the first time into synthetic dyes production. Rarely did experience and expertise run deeply, but these had what small amounts existed. That Oldbury began experimenting with phosgene for dyes in 1916 gave them a year's headstart on most firms in accumulating relevant expertise.

By the end of 1917, despite the early contracts for chlorpicrin and phosgene, the Ordnance Department personnel realized they faced steep obstacles in procuring war gases on the market. Very few manufacturers expressed a willingness to make the gases. From the industrialists' point of view, war gases offered few rewards and many risks. First, few of the war gases would have civilian applications, which meant investment would be of little value after the war. Second, manufacturers recognized the extreme hazards of war gases to their firm, whether through employee casualties or through the liability they carried should an accident endanger a local community. The railroads also hesitated to carry such materials, adding to costs and risks. To protect their firms, manufacturers insisted that the Federal government assume liability, and all of the 'private' contracts were consequently written to give ownership to the government. Thus, the manufacturers came under military supervision.[72]

The military officers also soon realized that, even among manufacturers willing to cooperate, the broad lack of expertise exacerbated the hazards of war gas manufacture. Even a relatively established (if still small) manufacturer, Dow Chemical, suffered a tragic accident in their experimental mustard gas plant in June 1918. Three employees died and several more were hurt after exposure during equipment cleaning and maintenance. The military shut down Dow's operation (partly because other plants also made mustard gas).[73] The American Synthetic Color Company's chlorpicrin plant never lived up to expectations, and experienced such sloppy management that military inspectors saw it as a potential disaster for the city of Stamford.[74] Beyond these two stark examples, the other half dozen firms engaged in gas manufacture frustrated military planning with their delays, failures, accidents, and low outputs. Most troubles lay in a combination of inexperience and risk. Even when a unit of the more experienced National Aniline firm undertook research and production of mustard gas, the military officers had already begun to focus their attention on government plants.

Construction of gas manufacturing facilities at the Gunpowder Neck Reservation had been underway for four months by April 1918 when the organisation received responsibilities for the 'outside plants,' as the privately-run firms were known. They remained organisationally a part of the Edgewood Arsenal for the remainder of the war, because, despite their relatively poor performance, the 'outside firms' remained relevant strategically. The Ordnance officers could imagine an accident at the main Edgewood Arsenal facility that would shut down the entire complex, in which case the outside firms would need to compensate for the lost production. As construction accelerated at the Edgewood Arsenal, however, the outside firms became sideshows. The Ordnance officials built chemical plants for phosgene, chlorpicrin, and mustard gas, as well as chlorine plants, shell-filling plants, and the infrastructure to support the entire facility. In June 1918, Edgewood gases began shipping to Europe, although output in 1918 was slowed by shortages of shells for the gases. As in explosives, the planned capacity for 1919 was even greater.[75]

At the Armistice, the Edgewood Arsenal was an imposing complex, poised for yet further expansion in 1919. Because of the hazards of gas and Americans' technical inexperience with it, government played the central role in the mobilization of the gas programme. From the Bureau of Mines and the Ordnance Department, the Chemical Warfare Service inherited and oversaw the immense research and production efforts, and private firms remained secondary, if not peripheral. With a few exceptions, the Federal government increasingly centralized the administration and physical location of the research and production wings of the gas programme. As it shepherded a new and more powerful gas, Lewisite, from research through development and production at a top secret facility near Cleveland, the CWS provided a glimpse of its tighter control over the entire process. Chemists in Washington, DC worked together to develop a process that was scaled up in Cleveland behind barbed wire and entirely with military personnel.[76]

## CONCLUSION

The government-industry relationship in the US production of synthetic organic chemicals during First World War developed differently among high explosives and war gases. The vast expansion of high explosives production prior to American entry to the war gave the private explosives firms experience and expertise that exceeded the military's for the duration of the war. The challenge for the Ordnance Department lay in distributing and expanding privately held technical knowledge to accommodate the rising demand for high explosives. Increasingly through 1917 and 1918, the Ordnance officials used their power to share the best practice techniques of top firms like DuPont and Hercules with newer and less experienced firms. In the case of high explosives, the firms generated any necessary new knowledge, and the government authorities participated minimally in the production of knowledge. Both geographically and administratively, the high explosives mobilization remained decentralized, despite the transfers of technology among firms by Ordnance.

The reluctance of private manufacturers to undertake production of war gases, the greater need for secrecy, and the sheer hazards of the gases all contributed to increasing centralization of production in gases. Under the guidance of the Bureau of Mines and National Research Council, the gas research programme drew hundreds of university and other chemists into the government war work. Gas chemistry was demanding, and while not a dominant factor in the overall Allied effort, it engaged and required American chemists far more extensively than did the production of explosives. The production programme at the Edgewood Arsenal put that knowledge into practice under the close supervision of the Ordnance Department and then the Chemical Warfare Service. With private firms incapable of supplying much expertise in either research or production, the gas mobilization effort faced a more daunting uphill challenge than the high explosives mobilization.

The Federal government consequently dominated gas mobilization in a way that simply was unnecessary in high explosives, where firms possessed the research and production expertise.

In both sectors, the military and civilian mobilization authorities planned for even more enormous outputs for 1919; and had the war continued for another year or two, the two administrative approaches may very well have converged to some degree. The government would have owned an increasing number of explosives plants, and Americans rapidly gained scientific and production expertise in war gases. However, the distinction between the two approaches would no doubt have remained. Despite government ownership, explosives plants very clearly would still have been operated by private firms, and they would have been even more geographically decentralized with new Hercules plants in California. In the gas programme, except for continued small use of private firms, no trend away from the centralization at the Edgewood Arsenal existed in future plans. In addition, 'progress' in gas development meant only more deadly gases that would require even greater administrative control, tighter security, and enhanced secrecy.

## ACKNOWLEDGEMENTS

The author would like to thank the organizers, the editors, and the participants of the conference. Part of the funding for this research came from National Science Foundation grants SBE 9212816 and SES 0080496.

## NOTES

[1] The Ordnance Department belonged to the War Department. The Department of the Navy had its own ordnance unit, but it was smaller and less influential in shaping the production of synthetic organic explosives and, especially, war gases. Over the years, historians have developed a sizeable literature on explosives and U.S. mobilization. Particularly central are Paul A.C. Koistinen, *Mobilizing for Modern War, 1865–1919* (Lawrence: University Press of Kansas, 1997), upon which I have drawn extensively; and also Robert D. Cuff, *The War Industries Board* (Baltimore: Johns Hopkins University Press, 1973). An earlier but useful reference is Arthur Pine van Gelder and Hugo Schlatter, *History of the Explosives Industry in America* (New York: Columbia University Press, 1927; reprint 1998). There are also useful firm histories, particularly David A. Hounshell and John K. Smith, Jr., *Science and Corporate Strategy: Du Pont R&D, 1902–1980* (New York: Cambridge University Press, 1988), and Davis Dyer and David B. Sicilia, *Labors of a Modern Hercules* (Cambridge, MA: Harvard Business School Press, 1990). Benedict Crowell, an Assistant Secretary of War and Director of Munitions, assembled an official military history, which is a good starting point, although it glosses over problems. Crowell, *America's Munitions, 1917–1918* (Washington, DC: Government Printing Office, 1919).

[2] The standard account of chemical warfare is L.F. Haber, *The Poisonous Cloud: Chemical Warfare in the First World War* (Oxford: Clarendon Press, 1986). Daniel P. Jones has written extensively on the American gas programme, and readers of his work will recognize his influence in this essay. See Jones, 'The Role of Chemists in Research on War Gases in the United States during World War I' (Unpublished PhD dissertation, University of Wisconsin, 1969), and Jones, 'Chemical Warfare Research during World War I: A Model of Cooperative Research', in John Parascandola and James C. Whorton (eds.), *Chemistry and Modern Society* (Washington, DC: American Chemical Society, 1983), 165–185. Crowell, *op. cit.* note 1, also discusses war gases.

[3] Koistinen, *op. cit.* note 1, 105ff.

4   National Archives and Records Administration (afterwards, NARA; Washington, DC/College Park, MD), RG 107, Office of the Secretary of War, Entry 191, General File, File 12, 'Memorandum' from J.P. Morgan & Company to Wilson administration (probably Colonel E. House), 19 May 1917, enclosed in Stettinius to Wainwright, Assistant Secretary of War, 17 April 1922.
5   Kathryn Steen, 'Wartime Catalyst and Postwar Reaction: The Making of the US Synthetic Organic Chemicals Industry, 1910–1930' (Unpublished PhD dissertation, University of Delaware, 1995).
6   F. Carrington Weems, *America and Munitions: The Work of Messrs. J.P. Morgan & Co. in the World War* (New York: privately printed, 1923), 23–26, 99–103. The Morgan Library, New York City, owns a copy of this rare book and has a microfilm version.
7   'Million Dollar Deal in Picric Acid', *Weekly Drug Markets*, 1 (21 July 1915), 10.
8   Van Gelder and Schlatter, *op.cit.* note 1, 940–948.
9   Hounshell and Smith, Jr., *op. cit.* note 1; Alfred D. Chandler, Jr., and Stephen Salsbury, *Pierre S. du Pont and the Making of the Modern Corporation* (New York: Harper and Row, 1971), 291–300; and Van Gelder and Schlatter, *op. cit.* note 1, 944–948. Atlas submitted a proposal to make TNT in July 1917, but there were no supplies of toluene, and the Ordnance officers postponed a contract until supplies were available. NARA, RG 156, Records of the Chief of Ordnance, Entry 764, Box 141, Burns to Atlas Powder Company, 7 August 1917, 471.868/46. Had the war gone into 1919, Atlas would have presided over a large government-owned TNT plant in Perryville, Maryland. See NARA, RG 156, Entry 528, Box 2, ex. 18, A.W. Hixson and Major C.F. Backus, 'History of Explosives Section; Production Division; Explosives, Chemicals, and Loading Division', part II, vol. I, 1 December 1918, Appendix.
10  DuPont took up nitroglycerine and dynamite in 1880. Van Gelder and Schlatter, *op. cit.* note 1, 563; Dyer and Sicilia, *op.cit.* note 1, 94–95.
11  *Ibid.*, 145–150; Van Gelder and Schlatter, *op. cit.* note 1, 541–544.
12  Van Gelder and Schlatter, *op. cit.* note 9, 940–944; NARA, RG 156, Entry 524, Box 13, Pl. 50, 'List of Powder, Explosives, & Loading Plants, Government Owned and Privately Owned, Operated on War Department Contracts', 10 November 1918, 7–9.
13  In both cases, explosions sharply reduced TNT output. Hixson and Backus, *op. cit.* note 9, 15.
14  University of Virginia Special Collections (Charlottesville, Virginia), Papers of Edward Reilly Stettinius, Sr., Acc. 2723 (afterwards, Stettinius, Sr., Papers), Box 44, Memorandum on Aetna Explosives Company contracts with France, January 1918; NARA, RG 156, Entry 764, Box 138, 471.862 Aetna, 15 November 1917.
15  NARA, RG 156, Entry 764, Box 138, Lacombs to Hoffer, 15 November 1917, 471.862/40 Aetna, with multiple enclosures.
16  Koistinen, *op. cit.* note 1, 121–126; Kathleen Burke, *Britain, America and the Sinews of War, 1914–1918* (Boston: George Allen and Unwin, 1985), 23; Weems, *op. cit.* note 6, *passim*.
17  'Memorandum Regarding Activities of J.P. Morgan & Co. as Purchasing Agents for the British and French Governments', enclosed in Stettinius to Wainwright, *op. cit.* note 4.
18  *Ibid.*; Koistinen makes the point about purchasing policies encouraging economic concentration, Koistinen, *op. cit.* note 1, 121–126; Burke, *op. cit.* note 16, 23; William B. Williams, *History of the Manufacture of Explosives for the World War, 1917–1918* (Chicago: University of Chicago Press, 1920), 39–59.
19  Koistinen, *op. cit.* note 1, 139–165.
20  Daniel T. Rodgers, *Atlantic Crossings: Social Politics in a Progressive Age* (Cambridge, MA: Belknap Press of Harvard University, 1998), 285–286.
21  Koistinen, *op. cit.* note 1. The following paragraphs draw on chs. 8–10, *passim*.
22  Stettinius, Sr., Papers, Box 21, folder 376, Stettinius to Hardy, Chicago, 27 October 1917.
23  Koistinen, *op. cit.* note 1, 152, 160–163, 177, 181.
24  *Ibid.*, 139–140, 166–182, 198–216. The Council of National Defense was championed by Hollis Godfrey, an engineer (rather than businessman) who was familiar with prewar attempts in both Britain and the United States to establish a mechanism for efficiently managing a wartime economy. Koistinen, *op. cit.* note 1, 156. The mobilization boards are also discussed in Cuff, *op. cit.* note 1. See also Bernard M. Baruch, *American Industry in War, A Report of the War Industries Board* (New York: Prentice-Hall, 1941; includes reprint of March 1921 report).

[25] Koistinen, *op. cit.* note 1, 230–244.
[26] Williams Haynes, *American Chemical Industry* (New York: D. Van Nostrand, 1945), vol. 2, 45–52, 355; Baruch, *op. cit.* note 24, 165–207; and Charles H. McDowell, 'The Work of the Chemical Section of the War Industries Board', *Journal of Industrial and Engineering Chemistry*, 10 (October 1918), 780–783.
[27] Daniel J. Kevles, *The Physicists: The History of a Scientific Community in Modern America* (Cambridge, MA.: Harvard University Press, 1971, 1987 edn.), 102–118, 137–138.
[28] National Academy of Sciences (Washington, DC), NAS-NRC Central File, 1919–1923, Divisions of the NRC, Chemistry and Chemical Technology, Activities, 'War Activities of the Division of Chemistry and Chemical Technology'; NAS-NRC Central File, 1914–1918, Divisions of the NRC, Chemistry and Chemical Technology, Committee on Explosives Investigations, Munroe to Hale, 21 March 1919.
[29] Koistinen, *op. cit.* note 1, 182–187.
[30] Stettinius, Sr., Papers, Box 82, Stettinius to Dunn, 30 March 1917; Stettinius to Burns, 26 October 1917; Crozier to Stettinius, 4 February 1917. Leland L. Summers was a former Export Department employee who became one of Bernard Baruch's chief assistants in the War Industries Board, where he had special responsibility for chemicals. Koistinen, *op. cit.* note 1, 137, 175–176.
[31] As Koistinen points out, this policy reinforced the trend toward large integrated firms, a tendency already underway four decades before the war, against which many Progressive reformers protested. Koistinen, *op. cit.* note 1, 163 *et passim*.
[32] Hixson and Backus, *op. cit.* note 9, 19.
[33] An example of Ordnance officers' concern to avoid interrupting flows is found in Stettinius, Sr., Papers, Box 82, Hoffer to Dunn, 26 March 1917, enclosed in Dunn to Stettinius, 29 March 1917. When Ordnance officers met with firms, they explicitly asked about current and future obligations under Allied contracts. In August and September 1917, officers made several inquiries about available capacity. NARA, RG 156, Entry 764, Box 137, 471.862. When J.P. Morgan and the British ended their purchasing agreement, Stettinius pointed out that the US government's ability to set prices was a principal reason the British should work through the government rather than through J.P. Morgan. Stettinius, Sr., Papers, Box 84, J.P. Morgan & Company to Spring-Rice, 25 May 1917. On at least one occasion, Ordnance officers threatened to commandeer a plant when its managers balked at the official price for picric acid. See NARA, RG 156, Entry 764, Box 139, Major J.H. Burns, memorandum, 'Picric Acid: New England Chemical Company', 6 October 1917, 471.862/2, New England Chemical Company.
[34] 'Contracts for $7,830,000 in Picric Acid', *Drug and Chemical Markets*, 4 (10 October 1917), 9.
[35] Lacombs to Hoffer, *op. cit.* note 15.
[36] NARA, RG 156, Entry 764, Box 141, 471.868/144, Military Attaché, London, to Chief of Ordnance, 10 December 1917, and 471.868/152, Backus to Production Division, 7 March 1918.
[37] Roy MacLeod, 'Secrets among Friends: The Research Information Service and the 'Special Relationship' in Allied Scientific Information and Intelligence, 1916–18', *Minerva*, 37 (4), (1999), 201–233; NAS-NRC Central File, 1914–1918, Policy Files, Administration, Divisions of the NRC, General Relations, Research Information Service, Appointments — Scientific Attachés — State Department, Gilby to Welles, 17 January 1918; NAS-NRC Central File, 1914–1918, Administration, Public Relations, Inauguration (Speeches), George Ellery Hale, 'War Activities of the National Research Council', address given in the Engineering Societies Building, New York, 28 May 1918, 919–923; NAS-NRC Central File, Policy Files, Administration, Executive Board, Projects, Scientists for Service in Allied Laboratories (Proposed), Edgar to Bumstead, 25 April 1918; Manning to Hale, 13 May 1918; Hale to Bumstead, 13 June 1918.
[38] National Archives, RG 156, Entry 36, Box 2682, Burns to Picatinny Arsenal, 11 June 1917, 471.86/1901 Miscellaneous; Burns to E.I. DuPont de Nemours and Company, 25 June 1917, 471.86/1924 Miscellaneous; Crozier to Gordon, 1 August 1917, 471.86/1994 Miscellaneous.
[39] Hixson and Backus, *op. cit.* note 9, 10, 15.
[40] NARA, RG 156, Entry 764, Box 135, Lt Egbert Moxham, 'Propellants and Explosives in the War', 17 December 1918, 471.86/880; Van Gelder and Schlatter, *op. cit.* note 1, 954–956. So as not to leave the Russians out: Ordnance officers consulted with the Russian government in their purchases of TNA

(trinitroaniline) in the United States. NARA, RG 156, Entry 36, Box 2682, Shinkle to Tschekeloff, 25 April 1917, 471.86/1042 Miscellaneous.

[41] For example, W.W. Edwards left the Aetna research labs and helped build the General Explosives Company in Missouri. NARA, RG 156, Entry 36, Box 2682, Edwards to Peters, 6 June 1918, 471.86/2525.

[42] Moxham Report, *op. cit.* note 40.

[43] NARA, RG 156, Entry 764, Box 138, Backus to New York District Office, 7 May 1918, 471.862/4 Everly M. Davis.

[44] NARA, RG 156, Entry 764, Box 138, Goodrich Lockhart Company to Peters, 17 August 1918, 471.862/34 Goodrich Lockhart; Box 137, Morey to Backus, 20 June 1920, 471.862/266, 3; Entry 36, Box 2682, W.C. Cope, Eastern Laboratory, DuPont, 'Investigation of TNA', 1 May 1917, 471.86/1123 Miscellaneous.

[45] NARA, RG 156, Entry 764, Box 141, J.H. Burns (?), memorandum, 'Method of Purchasing Refined TNT', 2 July 1917, 471.868/25; J.H. Burns, memorandum, 'TNT Federal Dyestuff & Chemical Corporation', 20 July 1917, 471.868/36. TNT prices would sink to about 30 ¢ per pound at the end of the war. Hixson and Backus, *op. cit.* note 9, 20.

[46] NARA, RG 156, Entry 764, Box 141, Burns to Semet-Solvay Company, 18 October 1917, 471.868/64; Box 137, Morey to Backus, 20 June 1918, 471.862/266; Backus to New York District Ordnance Office, 2 August 1918, 471.862/309; Hixson and Backus, *op. cit.* note 9, 18–20; Box 143, Backus to French, Federal Dyestuffs & Chemical Corporation, 19 June 1918, 471.868/7 Federal Dyestuff.

[47] NARA, RG 156, Entry 764, Box 143, Williams to E.I. DuPont de Nemours Powder Company, 28 September 1918, 471.868/243 DuPont; Buckner to Backus, 7 August 1918, 471.868/180 DuPont; Williams to DuPont, 1 October 1918, 471.868/249 DuPont. Once DuPont finally sent the Barksdale plans, Ordnance immediately sent a copy to Atlas. Backus to Cook, Atlas Powder Company, 27 September 1918, 471.868/242 DuPont.

[48] Hixson and Backus, *op. cit.* note 9, 15.

[49] NARA, RG 156, Entry 764, Box 137, Backus to New York District Ordnance Office, 2 August 1918, 471.862/309; Hixson and Backus, *op. cit.* note 9, 18–20.

[50] NARA, RG 156, Entry 764, Box 137, Morey to Backus, 'Picric Acid, Goodrich-Lockhart Plant, Labor Shortage' report, 2 August 1918, 471.862/309; Box 139, O'Brien, O'Brien Synthetic Dye Company, 6 November 1918, 471.862 O'Brien Synthetic Dye Company (50–100); Badgley to Backus, 12 November 1918, 471.862 O'Brien Synthetic Dye Company (50–100); Box 141, Holloway to Burns, 23 July 1917, 471.868/39; Burns to Holloway, 1 August 1917, 481.868/40.

[51] NARA, RG 156, Entry 764, Box 142, Wollenberg to Backus, 21 March 1918, 471.868 Aetna (1–50); Wollenberg to Palmer, Department of Labor and Industry of Pennsylvania, 21 March 1918, 471.868 Aetna (1–50); Palmer to War Industries Board, 23 May 1918, 471.868/45.

[52] In relation to TNT, officers believed their reports could be 'extremely valuable to any manufacturer just beginning', but suspected that picric acid processes still needed improvements they expected to come in 1919. See Hixson and Backus, *op. cit.* note 9, 16–18.

[53] Haber, *op. cit.* note 2, provides the most complete overview of the war's gas story.

[54] George A. Burrell, 'The Research Division, Chemical Warfare Service, USA,' *Journal of Industrial and Engineering Chemistry* 11 (February 1919), 93–104, on 93.

[55] Jones, 'The Role of Chemists', chapters 1–3, and 'Chemical Warfare Research', 169–170, 173–174, *op. cit.* note 2; NARA, RG 70, Entry 80, Box 2, National Research Council, Committee on Noxious Gases, 'Meeting of April 21, 1917'; RG 70, Entry 46, Box 110, Manning, 'Historical Report to the Secretary of the Interior on the Origin and Development of the Research Work of the Bureau of Mines on Gases Used in Warfare, February 1, 1917 to March 1, 1918'.

[56] NARA, RG 70, Entry 46, Box 110, National Research Council, Committee on Noxious Gases, meeting transcript, 7 April 1917, p. 3.

[57] Burrell, *op. cit.* note 54, 94.

[58] Haber, *op.cit.* note 2, 131. Haber considers that the British and French communicated badly with one another.

[59] Burrell, *op. cit.* note 54, 101; NARA, RG 70, Entry 80, Box 3, Manning, 'Report to the Secretary of the Interior on the Research Work of the Bureau of Mines on War Gas Investigations, July 1, 1917 to May 15, 1918', 3–4.
[60] Jones, 'Chemical Warfare Research', *op.cit.* note 2, 170; NRC, Committee on Noxious Gases, 7 April 1917, *op. cit.* note 56; 'Meeting of April 21, 1917', *op. cit.* note 55.
[61] Burrell, *op. cit.* note 54, 94–97, 102.
[62] Manning, 'Historical Report...February 1, 1917, to March 1, 1918,' *op. cit.* note 55; Jones, 'The Role of Chemists,' *op. cit.* note 2, 115–118, 122; Jones, 'Chemical Warfare Research', *op. cit.* note 2, 173–177.
[63] Jones, 'The Role of Chemists', *op. cit.* note 2, 146–150; US Army Military History Institute (Carlisle, PA), Harry M. St. John, 'Standard Methods for the Manufacture of New G-34 Development Division', Nela Park, Cleveland, Ohio, Chemical Warfare Service, 29 March 1919.
[64] Crowell, *op. cit.* note 1, 396–399; Charles H. Herty, 'Gas Offense in the United States, A Record of Achievement,' *Journal of Industrial and Engineering Chemistry* 11 (January 1919), 5–12.
[65] Other gas-related programmes included the development of gas masks under the direction of the Surgeon General (head of the Medical Department), and gas shell under the Artillery Ammunition Section in the Department of Ordnance. *The Edgewood Arsenal*, published as an issue in *The Chemical Warfare*, 1 (5), (March 1919), 7–9; see also 'Chemical Warfare Service,' *Journal of Industrial and Engineering Chemistry*, 10 (September 1918), 675–684.
[66] *op. cit.* notes 64 and 65; NARA, RG 70, Entry 80, Box 4, 'Memorandum Regarding Conference Held in the Office of the Secretary of War, from 3:00 p.m. to 4:45 p.m., May 25, 1918, regarding the Proposed Transfer of the War Gas Investigations of the Bureau of Mines to the War Department, Under Major General Sibert, Chief of the Gas Service', quotation on p. 9.
[67] NARA, RG 70, Entry 46, Box 110, William H. Nichols, *et al.*, to Franklin K. Lane, Secretary of the Interior, 16 May 1918.
[68] NARA, RG 70, Entry 80, Box 4, Woodrow Wilson, 'Executive Order', 26 June 1918; 'Memorandum', 25 May 1918, *op. cit.* note 66.
[69] Burrell, *op. cit.* note 54, 99; Jones, 'The Role of Chemists', 141–142.
[70] NARA, RG 175, Entry 8, Box 14, 'Brief Historical Sketch Concerning the Inception and Operation of the Stamford Plant of Edgewood Arsenal,' enclosed in William McPherson, 'Report of the Director of Outside Plants, Historical Sketch', n.d. [early 1919].
[71] 'Historical Sketch of Experimental Plant Leased by the Government from Oldbury Electro Chemical Co.' [fragment] enclosed in McPherson, *op. cit.* note 70; NARA, RG 175, Entry 8, Box 14, Engineering Bureau to Control Bureau, 'Experimental Plant of Oldbury Electro-Chemical Co., Niagara Falls, NY, for Temporary and Immediate Manufacture of L-3', 20 February 1918.
[72] *op. cit.* notes 64 and 65.
[73] NARA, RG 175, Entry 5, Box 2, untitled report beginning, 'On Wednesday, June 26, 1918' [c. July 1918].
[74] NARA, RG 175, Entry 8, Box 8, E.E. Free, 'Report on Conditions at the American Synthetic Color Co., Stamford, Conn.', 4 April 1918.
[75] Herty, *op. cit.* note 64, 5–12; 'The Edgewood Arsenal,' *The Chemical Warfare*, 1 (March 1919), 9–12.
[76] St. John *op. cit.* note 63.

# OPERATING ON SEVERAL FRONTS: THE TRANS-NATIONAL ACTIVITIES OF ROYAL DUTCH/SHELL, 1914-1918

INTRODUCTION

During the First World War, The Netherlands stayed neutral. This was no 'accident', as if the belligerents had forgotten to occupy this small country, nor was it a sign of passivity. On the contrary, keeping neutral status was hard work, both for the Dutch government and for the belligerents. Indeed, well before the war, the German government had come to the conclusion that Dutch neutrality could be benificial to the German war effort. Dutch trade could act as a 'windpipe', as it were, through which strategic products could be imported. As also the British found it unnecessary to occupy Dutch territory, The Netherlands remained officially neutral to the end of the war.[1]

Despite its neutral status, however, the Dutch army and navy were mobilised during the first days of August 1914, and remained at the ready for the rest of the war. During the first two years of the war, the economy prospered, but after the summer of 1916, the situation deteriorated. The British blockade was felt more and more, and the beginnings of unrestricted submarine warfare by Germany had serious consequences for Dutch shipping.

The means available to the Dutch to supply their fleet and troops with munitions were modest. The country possessed one state-owned factory for the production of shells and cartridges, which since the 17th century had been located in the town of Delft, but which between 1895 and 1900 was transferred to a locality called 'Hembrug', west of Amsterdam. There was only one explosives manufacture, called the 'Gezamenlijke Buskruidmakers van Noord-Holland, Utrecht, and Zeeland', which also dated from the 17th century. During the Dutch Republic of the 17th and 18th centuries, there were black powder mills in almost all of the seven United Provinces. In 1807, thirteen powder mills still existed, all privately owned. In the next twenty years, however, several of these mills closed down, and the remaining seven merged in 1844 to form the privately-owned 'Gezamenlijke Buskruidmakers'. After the merger, all but two plants were scrapped. The two remaining mills were located at Muiden, east of Amsterdam, and at Ouderkerk a/d Amstel, south of Amsterdam. Between 1885 and 1895 they were modernised, with the help of German, and later French, know-how. The mill at Muiden became the centralised facility for the production of smokeless powder, and the mill in Ouderkerk a/d Amstel was converted into a plant for the production of nitroglycerine, gun cotton, and concentrated nitric acid.[2]

Until 1914, it was common practice for the Dutch to import large amounts of explosives and ammunitions from Germany, Britain, and other countries. During the war, imports from the United States continued for a while, but rising prices and strategic considerations made it a matter of crucial importance to increase domestic production. The number of workers at the goverment ammunitions factory at Hembrug then grew from 1578 at the end of 1914, and 3285 in 1915, to more than 8500 in 1917. The explosives plants at Muiden and Ouderkerk expanded as well. By 1916, production had grown to over four times pre-war figures.[3]

In the course of 1915, it became increasingly clear to the Dutch military that modern warfare, as illustrated by the trenches in Belgium and France, demanded huge amounts of high explosives. Unfortunately, Dutch companies had no experience whatsoever with the production of these modern war-chemicals. It was in these circumstances, that the Anglo-Dutch oil company Royal Dutch/Shell entered the scene – a company which, to the great surprise of anyone but insiders, was already occupied with the chemical technology of dyestuffs and explosives.

The activities of Shell took place both in neutral Holland, and in Britain and, to a smaller extent, in Germany, from 1906. Then, a year before their merger, both the Royal Dutch Petroleum Company and Shell began research and development on so-called aromatic 'intermediates' (i.e., simple chemicals which can be converted into more complex substances, such as explosives and dyes) from petroleum. Three years later, a production unit for nitrobenzene and nitrotoluene(s) came into operation at a site near Düsseldorf, Germany. During the war, research in the Netherlands intensified and extended to the conversion of nitrocompounds into dyestuffs and explosives. TNT factories were erected in Britain and The Netherlands, and serious attempts were made to establish a synthetic dye business. After the war, however, work on these aromatic chemicals appeared to be of no commercial value, and in 1919, Shell's board of directors stopped research and production in these fields. Therefore, at first glance, it seems that these war time activities were without significance for the future of the company.

Taking a longer view, however, the picture changes. I will argue that the war effort of Royal Dutch/Shell in the field of (petro)chemicals contributed decisively to the development of organizational capabilities that played a crucial role when, in the late 1920s, the company was facing the competition of Standard Oil and IG Farben. The earliest chemical activities of the Anglo-Dutch oil company stimulated the emergence and extension of the company's first research laboratory, in Amsterdam, and contributed to the formation of a staff with a sound knowledge of chemistry and chemical engineering.

To understand the chemical activities of Royal Dutch/Shell helps to explain why this company was among the first oil companies to succesfully diversify into petrochemicals. It also reveals an interesting case of a company which had part of its operations in a neutral country, and another parts in both belligerent countries.

## DIVERSIFICATION INTO PETROCHEMICALS

In the literature on Royal Dutch/Shell, and on the petrochemical industry in general, little attention has been paid to the fact that between 1906 and 1919, this oil company was actively engaged in research on, and production of, aromatic intermediates, explosives and synthetic dyes. Most authors emphasize that the Anglo-Dutch oil company entered the world of petrochemicals only in 1927, as a result of an alliance between the powerful German IG Farben trust and Shell's most important competitor, Standard Oil.[4]

From the point of view of Royal Dutch/Shell, this alliance between the world's largest chemical company and the world's largest oil company formed a major threat to its position in the world market. During the first months of 1927, the Anglo-Dutch company decided to embark upon organic chemical and catalytic research in the hope that this would strengthen its bargaining position. A few months later, on 18 August 1927, during difficult negotiations between IG Farben, Standard Oil and Royal Dutch/Shell in London over hydrogenation technology for coal and heavy oil fractions, Carl Bosch, one of the bosses of IG Farben, suddenly embarked upon a monologue in German. He prophesized that IG Farben, with its gasoline from lignite, would kill the world oil industry just as it had killed Indian indigo cultivation and the Chilean nitrate business.[5]

Sir Henry Deterding (1866–1939), the undisputed leader of Royal Dutch/Shell, was furious, as was his junior colleague Guus Kessler (1888–1972). Within a few days, Kessler and his staff had decided that they should try to damage IG Farben in one of its core businesses: synthetic ammonia. The idea was that Royal Dutch/Shell would erect a nitrogen fertiliser plant close to the coke ovens of the Dutch Hoogovens company (now part of Corus), where Guus's brother Dolf Kessler (1884–1945) was one the general managers. They would then erect ammonia plants all over the world, based on natural gas. For the oil company, this would be a first step in the creation of a large chemical industry. It would be 'a new departure'. As Guus Kessler wrote to his fellow managing director, Frits de Kok (1882–1940):

> We should have confidence, energy and courage enough to develop this new chemical part of our business even if it does not give profits to start with.... We have the money, the best raw material, and the best geographical position, and if we can work ourselves out of the routine to which we have been accustomed up to now in our more or less simple oil business as we have been developing hitherto, we should certainly be in a position to develop our industry into a chemical industry on a very complete scale.[6]

The tone of the letter shows the self-confidence of the managers of Royal Dutch/Shell and their awareness that they were taking very crucial decisions. Indeed, there have been few years in the history of Royal Dutch/Shell more important than 1927. Within a year, no less than six important steps had been taken by the Anglo-Dutch Group, which together drove the company forward in the direction of petrochemicals. These were: (1) the enlargement of the Group's research laboratories in Amsterdam, and the formation of a new department for chemical research;

(2) the founding of a large chemical engineering research laboratory in Emeryville, California; (3) the erection of a synthetic ammonia plant near the Hoogovens steel works in the Netherlands, mentioned above; (4) the erection of a synthetic ammonia plant in California, based on natural gas; (5) the building of petrochemical plant in California for the production of solvents; and (6) the appointment of a number of university professors of chemistry and chemical engineering as scientific advisors to the Group.[7]

Although these decisions were a watershed in the history of the company, there was also continuity. Apart from Sir Henry Deterding, who was not involved in the technicalities, there were four main figures responsible for 'the new departure': Daniel Pijzel (1877–1972), Guus Kessler, Frits de Kok, and Willem Knoops (1879–1953). With the exception of Kessler, who had studied mechanical and electrical engineering, all were chemical engineers from the Technical University (then Technische Hogeschool) at Delft. Pijzel, De Kok and Knoops not only had an education in chemistry and chemical engineering; they had all been involved in the chemical activities of the Group before and during the war.[8]

Willem Knoops was the founder of the first research laboratory of Royal Dutch in 1906, and three years later, became the manager of the nitration plant that Royal Dutch/Shell set up in Germany. After the war, he was made the technical supervisor of all the Group's activities in Europe, and in 1927, was appointed head of its newly-founded chemical department. His colleague, Daniel Pijzel, was the chief chemical engineer of Royal Dutch/Shell. In 1906, he founded the first 'technological department', or engineering office, of Royal Dutch at its headquarters in The Hague, and later became the head of all technical operations of Shell Oil in the USA. The third chemical engineer was Frits de Kok. He was Knoops's successor as head of the Royal Dutch research laboratory in Rotterdam until its transfer to Amsterdam in 1914. He succeeded Pijzel as manager of the engineering department in The Hague, in which role he coordinated all work on explosives and dyes during the war, both in Holland and Britain. After the war, he became one of the managing directors of the Group, until he succeeded Deterding as chairman of the Board ('Directeur-Generaal') in 1937.

These short biographies show that Knoops, Pijzel and De Kok were the leading figures in the chemical operations of Royal Dutch/Shell between 1906 and 1927. It was because of their previous experience that they could act quickly and decisively when the circumstances in 1927 made such action necessary. In the remainder of this essay, I will argue that the company's chemical activities between 1906 and 1919, gave it a powerful incentive to develop the technology and commerce of petrochemicals. As we shall see, the start of research was largely the product of youthful ambition, and not the result of a particular strategic vision of general management.

## THE PROBLEM OF BORNEO OIL

The story of the chemical activity at Royal Dutch begins in 1905, when Henry Deterding, the general manager of the Asiatic Petroleum Company – the sales cartel of the oil producers in the Dutch Indies, which included the British Shell company – wrote the participating companies, informing them that the market for Borneo oil had collapsed. The large world market for Borneo oil-based kerosene had been outpaced by American kerosene: it was too heavy, it became brown during storage, and it did not give a nice flame in standard lamps. Petrol from Borneo oil had marketing problems as well. The situation was a disaster for Shell and Royal Dutch, which both possessed very large oil reserves in Borneo. Deterding urged both companies to resolve this problem.[9]

The technical capabilities of both companies were remarkably modest. At the time, Royal Dutch employed only three men with a university or Technische Hoogeschool training in chemistry. These included Willem Sluyterman van Loo (1871–1946), the manager of the petrol refinery of Royal Dutch at Rotterdam; Adriaan Coster van Voorhout (1875–1945), chemist at the laboratory of the Royal Dutch paraffin plant at Pangkalan Brandan, Sumatra; and Dr. Paul Schreckenberger, the technical manager of the petrol refinery of the Benzinwerke Rhenania, a German subsidiary of Royal Dutch, at Reisholz near Düsseldorf. Shell, in its turn, employed only a few Russian chemical technologists at Balik Papan on Borneo, as well as a young Cambridge chemist, Robert Waley Cohen (1877–1952), in London. The joint trading organization, the Asiatic, employed one Dutch chemist, Dr. John Waterloo van Geuns (1874–1927), in an analytical laboratory near Singapore.[10]

All these experts were involved in day-to-day activities, and had no time to pay serious attention to the Borneo oil problem. Moreover, a joint effort by Royal Dutch and Shell was out of the question, because in 1905 – despite commercial cooperation within the Asiatic – there was still rivalry between the companies. Therefore, independently – and knowing nothing about the other's activities – Royal Dutch and Shell decided to consult academic science.

At the end of 1905, Waley Cohen contacted a young organic chemist from his *alma mater* Cambridge, Dr. Humphry Owen Jones (1878–1912), to invite him to investigate samples of Borneo oil products. Jones analysed the samples, according to the procedures of academic organic chemistry, which differed from the usual commercial petroleum testing methods. In the course of 1906, he discovered that Borneo oil – and the petrol fraction in particular – contained high amounts of aromatic substances, which were then produced only from coal tar. It also appeared that these aromatic substances were causing problems in the use of Borneo oil fractions. American oil, and oil from Sumatra, contained only tiny traces of aromatic substances. Borneo oil appeared to be an exception. By this unexpected result, the commercially unattractive Borneo oil suddenly became interesting to organic chemistry. Waley Cohen decided to build a pilot plant near London, and appointed the chemist R.E. Robinson to develop a process for the separation of the aromatic substances from Borneo oil.[11]

In the meantime, the manager the Rotterdam petrol redistillation plant of Royal Dutch, Sluyterman van Loo, had in December 1905 asked his former teacher, professor Sebastiaan Hoogewerff (1847–1934) of the Technische Hoogeschool at Delft, to help him with solving the Borneo oil problem. Hoogewerff put his assistant Willem Knoops on the case. Knoops' approach differed from the scientific method of Jones. He began to manipulate the Borneo oil technically. With help of a catalyst and hydrogen, he developed a so-called hydrogenation process which early in 1906 resulted in a perfectly clear and lighter Borneo oil, which did not have the disadvantages of the original product.

The general managers of Royal Dutch – Deterding and Hugo Loudon (1860–1941) – were delighted with this result. They offered Knoops a position within Royal Dutch on his own terms, a very unusual step. Knoops asked for his own laboratory and got it. In a few years, this laboratory, became the first central research laboratory of Royal Dutch/Shell. In his laboratory Knoops first worked on the technical development of his hydrogenation process. In the course of that research he also discovered, in late 1906, just like Jones, that the presence of aromatic substances accounted for the disadvantageous properties of Borneo oil.[12]

After the amalgamation of Royal Dutch and Shell in 1907, it was decided to concentrate scientific and technical work in Holland. A pilot plant for the removal of aromatic substances from Borneo oil was built at Pernis, near Rotterdam. There, Sluyterman, Knoops, and an assistant Dutch chemist were employed on the new process. At the same time, technical work in Britain was stopped and Robinson was sent to Holland to join the others. Between 1906 and 1908, Knoops and his team developed a commercial process for the extraction of aromatic substances, mainly toluene, from Borneo petrol. Two new plants were built between 1907 and 1909: a petrol rectification plant at the Royal Dutch petrol refinery near Rotterdam, and a nitration plant at the petrol refinery of the Rhenania at Reisholz. After these came into operation, Royal Dutch/Shell was able to produce kerosene and petrol fractions that could compete with the American products and, as a by-product, aromatic intermediates for the chemical and explosives industries.[13]

## THE REISHOLZ NITRATION PLANT

After the two new plants had come on stream, the processing of Borneo oil was as follows. First, at Balik Papan on Borneo, crude oil was separated into seven fractions, one of which was Borneo 'straight run' petrol (i.e. the petrol fraction which comes over during the first distillation of crude oil), and another was kerosene. Kerosene was freed from aromatic substances at Balik Papan, but the proportion of aromatic chemicals was too small to justify their recovery. Therefore, 'straight run' petrol, which was transported by tanker to Rotterdam, was in practice the only oil-based source of aromatics.[14]

The second step in the processing of Borneo petrol was carried out at Rotterdam, where a new distillation and rectification plant for producing 50,000 to 60,000 metric tons a year came on stream early in 1908. In this plant two consecutive 'sharp' fractional distillations were carried out. The product of the first fractional distillation, which contained approximately 40% of toluene, was called 'plus minus toluol petrol'. When that product was rectified again a product called 'toluol petrol'

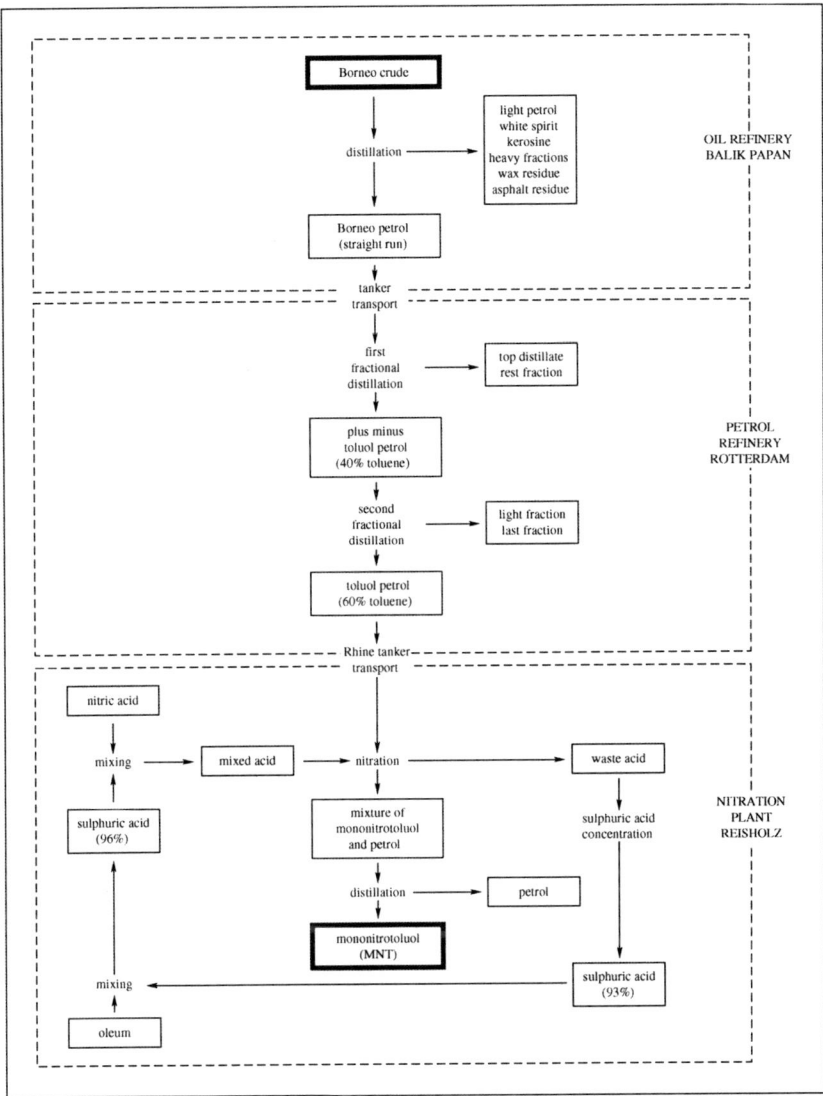

Figure 1. The process of extracting aromatic intermediates from Borneo oil for the dyestuffs and explosives industries. (Source: Company Archives, Shell International, The Hague)

resulted, which contained about 60% of toluene. This 'toluol petrol' was shipped to the Rhenania petrol redistilling plant at Reisholz for further treatment.[15]

The third step was carried out at Reisholz. There, Knoops built a nitration plant which, after great difficulties, came on stream at the end of 1909. Knoops was appointed manager of the plant.[16] At the Reisholz nitration plant the aromatic substances were separated from the 'toluol petrol' in a two-step process, which consisted of nitration carried out with a mixture of concentrated nitric acid and concentrated sulphuric acid, followed by a distillation. As an end result, mono nitro toluenes (MNT) were produced, and petrol freed from aromatic substances was distilled off. The initial design capacity of the plant was 2,000 metric tons of MNT a year. Three years later, Knoops succeeded in re-designing the plant to a capacity of 5,000 tons MNT a year. He also built a sulphuric acid concentration plant, to concentrate the waste acid of the nitration to 93% sulphuric acid which, after mixing with oleum, was ready for re-use.[17]

The question of whether to build the nitration plant in Rotterdam, in Reisholz, or at Thameshaven near London, had been a matter of intense debate between Deterding, Loudon and Waley Cohen. In the end, two circumstances influenced the decision in favour of a German location. The first was the far lower price of nitric acid on the German market. The second was the fact that the extensive German dye industry formed the largest single market for the endproducts of the process: ortho- and para-MNT.

After extensive negotiations with German dye producers and with the explosives manufacturer Allendorf at Schönebeck a/d Elbe, Royal Dutch/Shell decided in June 1910 to sell their total production of MNT to the German 'Anilin Syndikat', from which it could get nitric acid in return on favourable conditions. This aniline cartel included the major German producers of synthetic dyes, as well as a few smaller firms.

During its first years, the Reisholz factory was a technical and commercial disaster. Royal Dutch/Shell lacked any practical experience with large scale chemical processes. Numerous start-up problems in the nitration plant, and even greater technical troubles in the sulphuric acid concentration unit, meant that Knoops, supported by Pijzel's engineering department in The Hague, was constantly occupied with repairs, and with research on process improvement. By May 1912, total building costs had risen to Dfl. 900,000 (= £75,000), almost three times the initial budget. In 1910 and 1911, there was a net exploitation deficit of Dfl. 175,000 (= £14,600), or even Dfl. 370,000 (= £30,900) after the customary writing-off.[18]

By mid-1912, when the plant was at last running smoothly, the market for the other important product of the plant, petrol freed from aromatics, had changed dramatically. Whereas in Europe petrol had first been used in industry – for instance, in dry cleaning and paints – motor cars had suddenly become the most important market. More significant, car manufacturers discovered that the addition of aromatic substances, such as benzene, improved the performance of combustion engines. So Borneo petrol now seemed to be an almost perfect product, for which no tedious separation processes were necessary. With respect to petrol, the disadvantages of 1905 gave way to the advantages of 1912. Had these facts been known earlier, Deterding and Loudon would never have ordered the building of the Reisholz plant, because for them the market for oil products, not the market for chemicals, was the core business of Royal Dutch/Shell. It was only because by 1912 Knoops had succeeded in turning MNT into a profitable product, that the Reisholz nitration plant remained in operation.

Halfway through 1913, Knoops handed over the day-to-day management of the plant to his newly-appointed assistant, Michel Bastet (1889–1963), giving him time to direct his attention to the expansion of other German businesses of Royal Dutch/Shell. Then, in August 1914, came the war. Bastet was called back to Holland, where he served in the army, but Knoops remained in Germany as technical manager of the Mineralölwerke Rhenania (another subsidiary of Royal Dutch) until after the Armistice in 1918.[19]

## ROYAL DUTCH/SHELL IN FRANCE AND BRITAIN

When the war began, Royal Dutch/Shell was in a difficult position. As said before, part of its General Management was located in London; another part was in The Hague, in a neutral country; and at least four of the company's refineries and plants were in Germany. In London, Deterding coordinated the international relations of the Group, in particular those involving Britain and its allies. In The Hague, Hendrik Colijn (1869–1944) – later the Dutch Prime Minister, who succeeded Hugo Loudon as general manager of Royal Dutch in 1913 – tried to make the Dutch subsidiaries of the Group more independent of London, in accordance with Dutch neutrality. At the same time, however, he coordinated the interests of the Group inside Germany and the other Central powers.[20]

With the outbreak of hostilities, the German nitration business came to a halt, but petrol deliveries to Germany continued for a while, via Holland. When, after a few months, the British blockade and the policies of the Dutch government made these deliveries impossible, Royal Dutch continued to serve the German market until the end of the war with products from its Rumanian oil fields and refineries.[21]

The war had equally significant consequences for Borneo oil, and suddenly put great weight on the group's pre-war experience of producing of (nitro)toluene. In December 1914, faced by a shortage of toluene for TNT production, Deterding received requests from the French – and a little later, from the British government – to supply toluene-rich Borneo oil. After initial contacts, things went rapidly. Frits de Kok – who had been assistant to Sluyterman van Loo at the Borneo petrol refinery at Rotterdam, before his appointment as head of the research laboratory of Royal Dutch – was sent to France to construct a fractional distillation plant for the production of 'toluol petrol' (with 60% toluene), according to the design that had been in successful operation in Rotterdam since 1908.[22]

In January 1915, Deterding and the other leaders decided to dismantle the entire Borneo petrol refinery at Rotterdam and ship it to Britain. Hendrik Colijn – who, as a former Minister of War, had excellent relations with Dutch government circles – succeeded in persuading the Minister for Foreign Affairs, Hugo's younger brother John Loudon (1866–1955), to give permission for the export of the technical equipment to Britain. John Loudon, who more than anyone carefully guarded Dutch neutrality, was not aware that the Rotterdam plant would be important for the

British war industry. Colijn had told him only that, because of the British blockade, the plant was standing idle for lack of feedstock from Borneo. By promising the government that Royal Dutch would immediately order an identical refinery from Dutch machinery and equipment manufacturers, so that a new refinery would be ready at the end of the war, Colijn obtained the necessary permit. But when this permission arrived, the refinery had already been shipped to Britain. Within little more than two weeks after Deterding's decision, Sluyterman van Loo and his men had succeeded in dismantling all the equipment – which, ironically, had been built in 1907 by a German firm (Heckmann, of Berlin). During the night of 30 January 1915, a convoy of two Royal Navy vessels trans-shipped all installations to Brentford, near London, whence they were taken by train to Portishead on the Severn estuary, where the factory was rebuilt and started on 3 April – after only nine weeks. Sluyterman van Loo became the works manager of this crucial fractional distillation plant, and remained in Britain until the end of the war.[23]

There was no experience in Britain of the production of TNT from 'toluol petrol'. Accordingly, De Kok, head of the Royal Dutch research laboratory and just back from France, went to Britain to assist in the manufacture of TNT from Borneo oil. In London, he joined the American explosives expert Kenneth B. Quinan (1878–1948), who had been brought from South Africa to begin TNT production. De Kok erected a MNT plant at Oldbury, on the basis of experience that Royal Dutch had gained in the Reisholz nitration plant. Whereas the Reisholz factory never produced more than 100 metric tons MTN a week, the Oldbury plant was designed for 450 tons a week. In a second plant, built according to a comparable design at Sandycroft by De Kok and his Dutch assistant V. Henny, an even more remarkable 700 tons a week were produced.[24]

The MNT produced by the Oldbury plant was transported to an adjacent TNT factory, which was erected by Quinan. With the experience thus gained, other refineries and plants were built. After the Sandycroft MNT plant, Dutch chemical engineers assisted in the construction of a second Borneo petrol refinery at Barrow-in-Furness, near Lancaster, and a very large TNT plant at Queensferry, near Chester. Several Dutch chemists and chemical engineers were sent to Britain to assist in the erection and management of these plants (including Sluyterman, De Kok, Henny, I.J.F. Reydon, J.U. de Kempenaer, and S.H. Bertram). This happened partly because of the expertise they had acquired in Amsterdam and Germany, and partly because of the great shortage of chemists in Britain. Nevertheless, it remains a remarkable fact that these citizens of a neutral country participated in the British war effort. Obviously, their role as Shell employees overruled their citizenship.[25]

## ROYAL DUTCH/SHELL AND EXPLOSIVES IN THE NETHERLANDS

The transfer of know-how did not, however, pass one-way only. In the summer of 1915, the Dutch War Ministry decided there was a serious risk that Holland would become involved in the war. If this should happen, high explosives such as

TNT would be indispensible. But the one explosives manufacturer in the country, the 'Gezamenlijke Buskruidmakers van Noord-Holland, Utrecht en Zeeland', made only smokeless powder, nitrocellulose and nitroglycerine.[26] Buying TNT in America was possible but very expensive, so the Ministry decided that a TNT plant should be erected. After negotiations with several companies, including the 'Gezamenlijke Buskruidmakers', the Dutch government requested Royal Dutch/Shell to build and run a TNT plant. The Dutch government would cover all expenses.[27]

Accordingly, De Kok, who had been appointed head of the engineering department of Royal Dutch in September 1915, was called back from England, and together with the chemical engineer Bastet, who had assisted Knoops in Germany, designed the TNT plant and coordinated the construction process. The production of TNT was a two-stage process. First, MNT was produced from 'toluol petrol'; for this stage, Royal Dutch could build on previous experience. Bastet had directed the Reisholz plant for more than a year, and De Kok had coordinated the construction of at least two such plants in Britain. The second stage was the production of TNT from MNT. Royal Dutch had no direct experience of this part of the process, but De Kok had cooperated very closely with Quinan and we may assume that he made use of this know-how in the Dutch TNT factory. Moreover, Bastet, who was to become the technical manager of the new plant, was sent to the laboratory of the Technische Hoogeschool at Delft, in order to investigate the production of TNT from MNT experimentally. At the same time, in Amsterdam, Dr. Willem de Leeuw (1881–1964), De Kok's successor at the Royal Dutch research laboratory, started an extensive research programme on intermediates for explosives and dyes. Several chemists who later were employed in the Royal Dutch TNT plant got their first training in industrial aromatic chemistry in this research group.[28]

Although the government requested a very modest minimum capacity of 10 metric tons of TNT a week, the careful preparation by Royal Dutch/Shell, as well as wartime shortages of feedstocks, equipment and building materials, delayed the construction of the plant – at a site near the government ammunition works at Hembrug – until April 1916, and it did not come on stream until May 1917. Between then and April 1918, when production was stopped, Bastet produced a little over 500 tons of TNT. So, the plant and its production were typical for a neutral country. After the war, Bastet wrote that the entire wartime production of his plant was equal to the amount of TNT expended by the belligerents in one day. In 1917, British TNT plants produced about 1,500 tons a week, more that 150 times the total Dutch capacity.

After the war, the plant was handed over to the Dutch government and dismantled. The 'Gezamenlijk Buskruidmakers' were unable to buy the plant from the goverment, and Royal Dutch/Shell decided to discontinue their activities in this field. For the oil company, the war had been good for business. The Dutch government had invested about 1.5 million Dfl. ($= £125,000$) in the plant and its product, while on the Royal Dutch/Shell side of the balance sheet there was a net profit of about half a million Dfl. ($= £41,700$). This was a very nice profit, which more than compensated the company for the loss of Dfl. 370,000 ($= £30,900$) on the Reisholz nitration plant during 1910 and 1911. Investment in organic chemistry had paid off.[29]

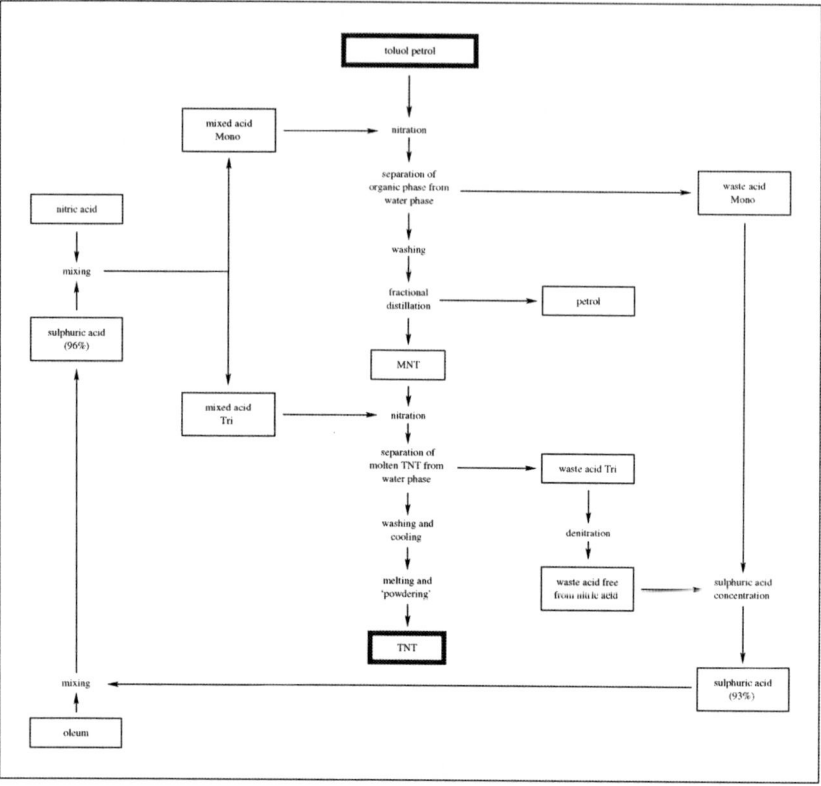

Figure 2. The production of TNT from 'toluol petrol' by a two-step nitration process, followed by fractional distillation (in the case of 'toluol petrol') and purification, as practiced in Royal Dutch' 'Chemische Fabriek' at Hembrug near Amsterdam. (Source: Company Archives, Shell International, The Hague)

## ANGLO-DUTCH DYESTUFFS

In parallel with its activities in explosives, Royal Dutch/Shell also directed its attention to synthetic dyes. For this, there were two reasons. In Britain, there was lively public debate on the need for a large national dyestuffs industry that would make the country self-reliant, and the possibility that British firms would be able to compete with the powerful German dye industry after the war. Robert Waley Cohen, as a member of the Shell board, was strongly attracted by this view, and thought that Royal Dutch/Shell could play an important role on the basis of its Borneo feedstock. In Holland, on the other hand, it was not the feedstock which played a primary role, but the excellent financial position of Royal Dutch, as well as its superior scientific and technical abilities. During the war, there was a shortage of dyes in Holland, and several small companies that wanted to start their own production tried to involve the wealthy Royal Dutch/Shell, as one of the few Dutch firms with practical experience in aromatic organic chemistry.[30]

During the first half of 1915, Waley Cohen contacted an organic chemistry professor at Cambridge, William (later, Sir William) Pope (1870–1939), and gave financial support for a research programme concerning the production of dyes from Borneo oil. Waley Cohen also intended to erect pilot plant facilities at Cambridge, in which Dutch chemists from the Amsterdam laboratory could do development work on dyes, under the supervision of De Kok and other experienced company chemists, with Pope as advisor. On 9 July 1915, Waley Cohen expressed his view clearly in a letter to the Royal Dutch headquarters in The Hague:

> I think there is no doubt whatever that after the war we shall find ourselves entering into the dye business, as the extraordinary purity of the benzol and toluol which can be obtained from our Borneo products would give us a very great advantage.[31]

Royal Dutch/Shell had assets, which the other oil companies did not possess. But to exploit this option fully, it would be necessary to have more chemists in the employ of Royal Dutch/Shell. Waley Cohen therefore asked the managing directors in The Hague to recruit new chemists. When there was not enough to do in the Amsterdam research laboratory, he wrote, they could be sent to Britain to work in:

> the various laboratories which we are putting up in this country, in connection with works we have already erected here, and where they could become familiarised with the work under Mr. de Kok, Mr. de Kempenaer, and, we hope also Mr. Reydon.[32]

It would be a distortion of the truth to state that the general managers of Royal Dutch – Hendrik Colijn and Cornelis Pleyte (1865–1951) – were pleased with this letter. Four days earlier, when De Kok arrived in Holland bearing Waley Cohen's order that he should dismantle the Rotterdam pilot plant of the Group, erected in 1907 by Sluyterman van Loo and Knoops, and take it to Cambridge in order to start technical research on synthetic dyes, Pleyte reacted furiously. He immediately sent a telegram to London in which he stated that he strongly objected to this 'first step to remove the chemical centre of the Group from Holland to England'. After the merger of 1907, it had been decided that London would be the commercial headquarter of the Group, and The Hague and Amsterdam would be its technical and scientific nerve-centres. From the Dutch point of view, Shell's war effort in Britain was undermining a carefully negotiated balance of power, that reflected different national interests.

During July and August 1915, there were frequent exchanges of letters and telegrams between London and The Hague. Although Deterding at first agreed with Waley Cohen, it was decided in the end that Cambridge and Amsterdam would both be active in research on dyes and their intermediates. Pope would do research on xylenes as a starting material, and De Leeuw and his group would investigate the chemistry of benzene and toluene derivatives. So, the general managers of Royal Dutch succeeded in upholding the strong position of the Amsterdam laboratory within the Group. Between early 1916 and 1919 a large part of the Amsterdam research effort was directed to aromatic intermediates for dyes and explosives. After Pope's early promising results with his xylene derivatives, Waley Cohen and Colijn agreed in May 1916 that Cambridge and Amsterdam should cooperate more closely. Pope would do the scientific part of the research, and Amsterdam the technical experiments on a larger scale. In July 1916 De Kok, who then worked in The Hague, was appointed technical coordinator of all work done in the field of dyestuffs and intermediates, both in Holland and in Britain.

In the meantime, Waley Cohen sought to speed up the Group's work in the actual production of synthetic dyes, by acquiring an existing company. Between late 1916 and May 1917, he negotiated with the managing directors of the Clayton Aniline Company of Manchester, in which he tried to obtain a controlling interest. After many months, Shell and Clayton Aniline failed to reach an agreement. Nevertheless, Waley Cohen and other London based managers of Royal Dutch/Shell still thought in 1917 that their company would have a shining future in the dye business. In January 1917, Deterding proposed investing Dfl. 50,000 to Dfl. 100,000, then in Holland a very large sum of money (£4200–8300), in research on dyes alone. Colijn reacted far from enthusiastically, and replied that he thought Dfl. 12,000 (£1000) would be enough for the time being.[33]

This more down-to-earth approach of the Dutch-based managers of the Group did not prevent the leaders of Royal Dutch going a step further than their British colleagues. They reacted positively to a request by the 'Chemische Fabriek Naarden' to invest in a new national synthetic dye company that would make aniline and naphtol dyes, based on intermediates produced by 'Naarden'. Writing to Deterding on 21 February 1918, Colijn expressed his 'absolutely firm conviction' –

> that the dye industry proper is not in our line. The manufacture of these products on a commercial scale involves very special requirements technically, for which we would have to create a separate staff of technical men, whilst the selling organisation also would have to answer requirements quite different from those of our own selling organisation.
>
> 'On the other hand', he added, 'I have always favoured the idea of supporting other Concerns in their efforts to make themselves independent from the German dye monopoly, the more so as it has been proved that for the manufacture of various products, petroleum derivates can be used as starting material.[34]

Given this reasoning, Colijn decided to support the 'Naarden' initiative. De Kok was appointed to the Advisory Boards of both 'Chemische Fabriek Naarden' and the new dye company, 'Nederlandsche Kleurstoffen-Fabriek'. At the Amsterdam research laboratory, De Leeuw and the team did some research on behalf of 'Naarden' and the 'Kleurstoffen-Fabriek', without asking for compensation. The general feeling – particularly during the war, and especially in the case of such strongly patriotic people as Colijn – was that the Dutch chemical and oil industry should cooperate in setting up a strong national synthetic organic chemical industry. In fact, following the war, the 'Kleurstoffen-Fabriek' turned out to be a commercial disaster. Following years of heavy losses, the company was eventually sold to I.G. Farben in 1931.[35]

In 1917, with the Amsterdam laboratory and the 'Naarden' initiative in mind, Colijn and De Kok looked to the role that Royal Dutch/Shell might play in the dye business after the war. Accordingly, they asked De Leeuw to report on the future of the international dye industry, and to evaluate the position of the Group. De Leeuw was a genuine scholar, a bibliophile with philosophical aspirations, who had a far-wider perspective than the average laboratory chemist. He took the request seriously. Working intermittently for more than a year, in January 1918, De Leeuw produced a masterpiece. His 300-page book on the *Kleurstoffen uit petroleum. Algemeene beschouwingen over de industrie der synthetische kleurstoffen* (*Dyes from petroleum. General reflections on the synthetic dye industry*), was a penetrating analysis of the entire German, British, French, and American industry. The report also contained a study of the market opportunities for the Borneo products.[36]

De Leeuw's conclusions must have come as a disappointment to those within Royal Dutch/Shell who thought that the Anglo-Dutch oil company might become a leading chemical company, comparable even to BASF, if it continued to pursue its research and acquisitions. On the contrary, De Leeuw argued that the company had a strong position only in the toluene market. Its position in the benzene and xylenes markets was far less rosy, and other strategic raw materials, such as naphtalene and anthracene could only be produced from coal tar. On the basis of its own feedstocks, Royal Dutch/Shell could therefore never realise the same economies of scope as the German dye industry. De Leeuw agreed with De Kok that the company should try to build up its position in the dyestuffs business very gradually, starting with a few intermediates and dyes, based on its own feedstock, and then expanding step by step. This process would take years.

After the war, the prospects of starting a dyestuffs business declined rapidly. In 1918 and 1919, dyes at Royal Dutch/Shell were still in the development stage, and British, Dutch and German textile manufacturers soon made clear that, just as before the war, they would start buying the superior German dyes as soon as they were on the market again. Moreover, as Deterding had prophesied in March 1918, nearly all countries implemented protectionist measures of some kind, in order to support their own national dye industries. The chances for dye manufacturers, especially newcomers, from small countries such as Holland were slight, and the fate of the 'Nederlandsche Kleurstoffen-Fabriek' proved these fears wellfounded. In January 1919, Deterding and Waley Cohen therefore decided to discontinue all research and development work on dyes, explosives, and their intermediates. Royal Dutch kept its financial stake in the 'Kleurstoffen-Fabriek', but the original strategic importance of this investment vanished completely.[37]

## CONCLUSION

When we reflect on the role of Royal Dutch/Shell in the Dutch war effort, a few conclusions stand out. First, it is clear that, far from being unaffected, the science-based industries of neutral countries, such as The Netherlands, were strongly touched by the war. That the war had an important effect on the Dutch economy is common knowledge. What is less widely known, is the impact of the war on military technologies. In high explosives, although in quantitative terms Dutch activities were relatively small, in qualitative terms the impact was important. Dutch chemists and chemical engineers together succeeded in developing new processes, just as did their colleagues abroad, and their abilities were highly regarded in Britain.

Second, it is clear that the critical role of innovation in this field was played not by traditional actors, such as the 'Gezamenlijke Buskruidmakers', or the military, but by a relative newcomer – Royal Dutch/Shell. The route taken by the company towards the production of aromatic intermediates for explosives and dyes was haphazard, and largely determined by factors outside the control of its general managers. These included the peculiar composition of Borneo oil, changing views about combustion engines, and the circumstances of the war. Between 1906 and 1914, the company had set up a well-organized research department. When war broke out, this department was well prepared to embark on research on intermediates for both dyestuffs and explosives. As we have seen, the Amsterdam laboratory worked not in isolation, but in close cooperation with chemists in Cambridge and elsewhere in Britain.

This brings us to a third conclusion – i.e. the role played by Anglo-Dutch cooperation and rivalry. Archival research in the Central Company Archives of Shell International shows an astonishing number of letters and telegrammes between Shell headquarters in London and The Hague during the war. Despite the British blockade and German submarine warfare, there was a permanent interaction. And despite rivalries between these centres of power, it is striking to note that there was considerable support for the British war effort on the part of Dutch chemists, engineers, and the managers of Royal Dutch. As compared with this, the contribution of the British part of the Group to the needs of The Netherlands was rather modest. The Dutch goverment, which tried to remain strictly neutral, appears ignorant of what was going on within the Anglo-Dutch oil company.

It appears, fourthly, that the continuing business of Royal Dutch/Shell in Germany played little part in our story. Although there were extensive exchanges of technology and chemicals between Germany and The Netherlands during the war, especially during 1914–1916, there is no evidence that these played a role in the explosives and dyestuffs business of Royal Dutch/Shell.[38] Nor is there evidence that the German munitions effort was aided by the company. The Group seems to have limited its Central European activities to supplying the German economy, as well as, probably, the German military, with oil products. The German nitro factory of the Group was closed down in 1914, and Bastet, its operational manager, was called back to Holland. There he built upon his pre-war German experiences, but there is no evidence that during the war there was a similar exchange between Bastet and chemists of the Group in Germany (esp. Knoops), as took place between Bastet and the Group's chemists in Britain. Nor were there other strong German influences on the Dutch oil industry during the war. It seems that Colijn, who was pro-German, managed to keep the Central European business of the Group strictly separate from operations in The Netherlands and the Dutch-Indies. By so doing, Colijn – just like Deterding – took a long-term stategic view of the interests of the Group. He carefully avoided making decisions that could put the post-war business of Royal Dutch/Shell at a risk. It is clear that, in the midst of the war, Deterding and Colijn knew that there would be a post-war situation which they would have to take into account.

In the immediate post-war years, Royal Dutch/Shell discontinued its activities in the field of aromatic organic chemicals. It would be wrong, however, to conclude that these activities had been useless. Important experience had been gained. At the beginning of 1906, Royal Dutch and Shell together employed only about six or seven chemists and chemical engineers, whereas in 1919, their numbers had risen to dozens. In Amsterdam, the Group developed a central research laboratory that employed about thirty chemists, analysts, and technicians. Under the successive leadership of Knoops, Van Tienen, De Kok, and De Leeuw, the laboratory acquired skills in research management, and built up a bank of knowledge on the oil industry, and even of chemical industry. At the managerial level, De Kok, Pijzel and Knoops were intensively involved in chemical processes, and so gained experience that proved extreemly useful when Royal Dutch/Shell again came into the orbit of the chemical industry in 1927.

When circumstances led to the decisions of 1927, the expertise gained before 1919 helped Shell evaluate its options. At the scientific and technical level, know-how was diffused into other activities within the Group. Some chemists of the Amsterdam laboratory, including De Leeuw, went to the USA, where they assisted in the development of cracking technologies. For this, their previous experience in the chemical field were of crucial importance. Other war staff left Royal Dutch to work with counterpart Dutch companies in explosives and dyes, including 'Naarden', the 'Nederlandsche Kleurstoffen-Fabriek', and the 'Gezamenlijke Buskruidmakers'. It is no exaggeration, therefore, to state that the wartime research of Royal Dutch/Shell in explosives and dyes had significant implicatons for the Dutch chemical industry as a whole. Borneo oil and the war together produced an unexpected result – an innovative, scientific-technical organisation which would put its stamp on the Dutch economy in the decades to come.[39]

ACKNOWLEDGMENT

The author would like to thank Ashgate Publishers for granting permission to reproduce parts of this chapter from the authors work Explosives from Oil: The Transformation of Royal Dutch/Shell during World War I from Oil to Petrochemical Company" (pp. 385–407), in: Brenda J. Buchanan (ed.), *Gunpowder, Explosives and the State: A Technological History*

## NOTES

[1] For the history of The Netherlands during the war, see M. Frey, *Der Erste Weltkrieg und die Niederlande. Ein neutrales Land im politischen und wirtschaftlichen Kalkül der Kriegsgegner* (Berlin: Akademie Verlag 1998), esp. 362-ff, H.P. van Tuyll van Serooskerken, *The Netherlands and World War I: Espionage, Diplomacy and Survival* (Leiden: Brill 2001), esp. 331–350, Maartje Abbenhuis, 'Rustig te midden van woedende golven; de moeilijke verdediging van de neutraliteit', in Hans Andriessen, Martin Ros and Perry Pierik (eds.), *De Grote Oorlog, kroniek 1914–1918*, (Soesterberg: Aspect 2002), vol. 1, 303–324, and M.M.H. Timmermans, 'De inbeslagname van de Nederlandse koopvaardijvloot door de geallieerden in 1918; de schepenrequisitie in het licht van de verhouding tussen overheid en bedrijfsleven', in Andriessen, Ros and Pierik (eds.), *idem.*, 177–197.

[2] On the history of Dutch explosives manufacture, see [G. de Bruin], *Het geslacht Bredius: honderdvijftig jaren buskruitfabricatie in Nederland, 1772–1922* (Amsterdam: Nederlandsche Springstoffenfabriek [1947]), G. de Bruin, *Buscruytmaeckers. Ervaringen en lotgevallen van een merkwaardig bedrijf in Holland* (Amsterdam: Nederlandsche Springstoffenfabriek 1952), H. Nägele and D. Schaap, *Geen oorlog, geen munitie. De geschiedenis van 300 jaar militaire produktie* (Haarlem: Fibula-Van Dishoeck 1979), J. Lenselink, *Vuurwerk door de eeuwen heen* (Amsterdam: De Bataafsche Leeuw 1991), esp. 28–31, 62–64, H.A.M. Snelders, 'De scheikundige laboratoria van de Delftse Artillerie-Inrichtingen', in H.L. Houtzager *et al.* (eds.), *Kruit en krijg. Delft als bakermat van het Prins Maurits Laboratorium TNO* (Amsterdam: Rodopi 1988), 207–217, C.W. de Jong, 'Uit Delft en weer terug', *idem.*, 219–228.

[3] Nägele and Schaap, *op. cit.* note 2, 79, [De Bruin], *Het geslacht Bredius, op. cit.* note 2, 25.

[4] English language histories of Royal Dutch/Shell are: F.C. Gerretson, *History of the Royal Dutch*, 4 Vols. (Leiden: E.J. Brill 1958), *Royal Dutch Comp. Diamond Jubilee Book* (The Hague 1950), K. Beaton, *Enterprise in Oil: A History of Shell in the United States* (New York: Appleton-Century-Crofts 1957), and R.J. Forbes and D.R. O'Beirne, *The Technical Development of Royal Dutch/Shell* (Leiden: E.J. Brill 1957). The recent book by Stephen Howard, *A Century in Oil. The 'Shell' Transport and Trading Company, 1897–1997* (London: Weidenfeld & Nicholson 1997), reproduced, unfortunately, several of the many errors present in R. Henriques, *Samuel Marcus, First Viscount Bearsted and Founder of The 'shell' Transport and Trading Company, 1853–1927* (London: Barrie and Rockliff 1960), and in R. Henriques, *Sir Robert Waley Cohen, 1877–1952: A Biography* (London: Secker and Warburg 1966). For the history of petrochemicals, see P.H. Spitz, *Petrochemicals: The Rise of an Industry* (New York: John Wiley 1988).

[5] For a detailed discussion of this episode, see E. Homburg, J.S. Small and P.F.G. Vincken, 'Van carbo- naar petrochemie, 1910–1940', in J.W. Schot *et al.* (eds.), *Techniek in Nederland in de twintigste eeuw*, Vol. 2 (Zutphen: Walburg Pers 2000), 332–357, esp. 351–352. See also: Beaton, *op. cit.* note 4, 502–519, Forbes and O'Beirne, *op. cit.* note 4, 455–469, 479, Spitz, *op. cit.* note 4, 82–83, Joh. de Vries, *Hoogovens IJmuiden: ontstaan en groei van een basisindustrie* (IJmuiden: Koninklijke Nederlandsche Hoogovens en Staalfabrieken 1968), 320–328, H. Gabriëls, *Koninklijke Olie: de eerste honderd jaar, 1890–1990* ('s-Gravenhage: Shell Internationale Petroleum Maatschappij 1990), 102–103.

[6] Shell International Central Archives, The Hague (hereafter SICA): Algemeen Technisch Archief 1916–1945 (49) - Researchwerk-Algemeen (180), Kessler to De Kok, 25 August 1927. See also Forbes and O'Beirne, *op. cit.* note 4, 456–457, Gabriëls, *op. cit.* note 5, 102.

[7] Homburg, Small and Vincken, *op. cit.* note 5, 351–353, Spitz, *op. cit.* note 4, 83–84, Forbes and O'Beirne, *op. cit.* note 4, 464–467, 472, Henriques, *op. cit.* note 4, 289–290, J.H. Schweppe (ed.), *Research aan het IJ. LBPMA 1914 - KSLA 1989. De geschiedenis van het 'Lab Amsterdam'* (Amsterdam: Shell Research 1989), 48–51, 59–64.

[8] These, and other, biographical details are all based on the study of the personnel files of these individuals, present in SICA (note 6). See also: E. Homburg, A. Rip and J.S. Small, 'Chemici, hun kennis en de industrie', in Schot *et al.* (eds.), *op. cit.* note 5, 298–315, esp. 301.

[9] SICA (note 6): Correspondentieboeken (3) - Deterding (105), Deterding to Capadose, 23 August 1905, Gerretson, *op. cit.* note 4, Vol. 2, 320–322.

[10] Cf. Homburg, Rip and Small, *op. cit.* note 8, 300–302.

[11] SICA (note 6): Correspondentieboeken (3) - Benzine Installatie Rotterdam (37), Lab. B.P.M. Schiedam (200) and R. Waley Cohen (365), and Kopieboeken (5) - Bataafsche Petroleum Mij. (6), Koninklijke Algemene Zaken, Conf. (727) and Directeur-Generaal (900), correspondence on/with Jones and Robinson 1906–1909. See also: Gerretson, *op.cit.* note 4, vol. 4, 22–25; H.O. Jones and H.A. Wootton, 'The Chemical Composition of Petroleum from Borneo', *Journal of the Chemical Society. Transactions*, 91/2 (1907), 1146–1149, J. Shorter, 'Humphrey Owen Jones, F.R.S. (1878–1912), Chemist and Moutaineer', *Notes and Records of the Royal Society*, 33 (1979), 261–277. I thank John Shorter for giving me access to his notes on Jones. Robert Henriques erroneously dated Jones' discovery about 1901. See Henriques, *Samuel Marcus op.cit.* note 4, 457–459, 553, 597, and Henriques, *Sir Robert Waley Cohen, op. cit.* note 4.

[12] SICA (note 6): Correspondentieboeken (3) - Deterding (105), Loudon to Deterding, 28 June 1905, and Correspondentieboeken (3) - Benzine Installatie Rotterdam (36), Deterding (107), Lab. B.P.M. Schiedam (200), and Kopieboeken (5) - Koninklijke Algemene Zaken, Conf. (723–724), Directeur-Generaal (899–900), and Historisch Archief (190C) - Confidentieele Correspondentie, correspondence on/with Knoops. Cf. Gerretson, *op. cit.* note 4, vol. 2, 320–322, vol. 3, 10–11, vol. 4, 15–25.

[13] SICA (note 6): Correspondentieboeken (3) - Benzine Installatie Rotterdam (37), Lab. B.P.M. Schiedam (200–201), R. Waley Cohen (365), and Kopieboeken (5) - Bataafsche Petroleum Mij. (6), and Koninklijke Algemene Zaken, Conf. (727 and 728), correspondence on/with Knoops, Groeneveld and Robinson, 15 November 1906-July 1909.

[14] Also the other fractions were processed at Balik Papan and, to a large extent, sold on East Asian markets. The removal of aromatic substances from kerosene was first tried along the lines of Knoops' hydrogenation process, then with the help of concentrated sulfuric acid, but finally, just before the war, Royal Dutch/Shell decided to build a so-called Edeleanu plant in Borneo. This plant was named after the Rumanian chemist, Lazar Edeleanu (1861–1941), who had sold his process to the Anglo-Dutch company in 1911 after extensive research by De Kok and Pijzel had convinced the managing directors that the Edeleanu scheme was superior to the other two processes. See Gerretson, *op. cit.* note 4, vol. 3, 118, vol. 4, 23–30, 170, Forbes and O'Beirne, *op. cit.* note 4, 366–374, and SICA (note 6), Cattaneo to M.D.A.A. Koenig, 1 June 1938.

[15] *Van Rotterdam Charlois naar Rotterdam Pernis, 1902–1977* (Rotterdam 1977), 2, 15, 17, 23, Gerretson, *op. cit.* note 4, vol. 2,. 253–268, 275–286, 319, Vol. 4, 22, 168, Forbes and O'Beirne, *op. cit.* note 4, 37, 305, S. Hoogewerff, 'Een en ander in verband met petroleum', *18e Jaarverslag 1908–1909 van het Technologisch Gezelschap te Delft* (Delft 1909), 15–43. Also see SICA (note 6): Correspondentieboeken (3) - Benzine Installatie Rotterdam (36–40), letters from/to Sluyterman van Loo, 1902–1910, Kopieboeken (5), letter Loudon to Waley Cohen, 1 March 1907, Historisch Archief (190C), letter Bastet to Gerretson, 17 December 1951.

[16] In June 1909, he was succeeded as head of the Rotterdam research laboratory by Willem van Tienen (1877–1966), who, in January 1911, was replaced by De Kok.

[17] For this, and following paragraphs, see: SICA (note 6): Correspondentieboeken (3) - Benzinwerke Rhenania Düsseldorf-Reisholz (33), Benzine Installatie Rotterdam (40), Knoops-Koenigs (194), Lab. B.P.M. Schiedam (200–202), H. Loudon (213–214), and R. Waley Cohen (365), letters 1906–1908, and Knoops' plans and budget of the Reisholz nitration plant, and Kopieboeken (5), Loudon to Waley Cohen, 1 March and 16 May 1907, Loudon to Rudeloff, 13, 18 and 22 April 1907, Loudon to Späth, 30 April and 3 May 1907, Rudeloff to Loudon, 8 October 1908, Waley Cohen to Späth, 26 November 1909, meeting H. Loudon, W. Rudeloff, O. Allendorff and H. Späth, The Hague, 13–14 November 1907. See also H. Flieger, *Unter der gelben Muschel. Die Geschichte der Deutschen Shell* (Düsseldorf 1961), 52–93, Gerretson, *op. cit.* note 4, vol. 2, 285, 322, vol. 4, 23–24.

[18] SICA (note 6).Correspondentieboeken (3), and Kopieboeken (5), letters to/from Beltman, Coenen, Knoops, Rudeloff and Späth, 1908–1912, report H. Späth and W.C. Knoops, 'Bericht über Nitrotoluol', 1 May 1912, Agreement on nitrocompounds with Anilin Syndikat, 20 July 1910/30 September 1910.

[19] Flieger, *op. cit.* note 17, 58–59, 64–66, Forbes and O'Beirne, *op. cit.* note 4, 356. See also SICA (note 6): Historisch Archief (190C), Bastet to Gerretson, 17 December 1951, Correspondentieboeken (3) and Kopieboeken (5): letters Loudon to Rudeloff, 6 May, 4 June and 9 September 1912, Rudeloff

to Loudon, 8 May 1912, Späth to Rudeloff, 9 and 11 May 1912, Rudeloff to Loudon and Capadose, 13 May 1914, Knoops to BPM head office, The Hague, 26 June 1914.

[20] For the Dutch chemical industry during the war, see E. Homburg, 'De Eerste Wereldoorlog: samenwerking en concentratie binnen de Nederlandse chemische industrie', in Schot *et al.* (eds.), *op. cit.* note 5, 316–331. For the role of Royal Dutch/Shell, see P.G.A. Smith, *The 'Shell' that hit Germany hardest* (London, about 1920); F.C. Gerretson, *Geschiedenis der 'Koninklijke'*, vols. 4 and 5 (Baarn: Bosch & Keuning 1973) (unlike the first three volumes of Gerretson's book – which became four volumes in English translation – these two volumes have not been translated into English). On Colijn and Royal Dutch, see H.-J. Langeveld, *Dit leven van krachtig handelen. Hendrikus Colijn, 1869–1944*, Vol. 1 (Amsterdam: Balans 1998), esp. 179–212, and P.W. Klein, 'Colijn en de "Koninklijke"', in J. de Bruijn and H.J. Langeveld (eds.), *Colijn: bouwstenen voor een biografie* (Kampen: Kok 1994), 105–128.

[21] Gerretson, *op. cit.* note 20, Vol. 4, 21, 49–50, 64–69, 87, Langeveld, *op. cit.* note 20, 186, 198–201, Klein, *op. cit.* note 20, 109–110. See also SICA (note 6): Späth to Colijn, 6 October and 2 December 1914, Colijn to Benzine-Installatie (Rotterdam), 12 October 1914, Sluyterman van Loo to BPM, The Hague, 21 November 1914.

[22] Gerretson, *op. cit.* note 20, Vol. 4, 89–91, Smith, *op. cit.* note 20, 11–14. SICA (note 6): Deterding to Pleyte, 22 July 1915.

[23] The story of the transfer of the Rotterdam refinery to Portishead is based on Smith, *op. cit.* note 20, and on two books by Robert Henriques, who mentions an internal Shell-memorandum, dated 1918, on the 'Origin, Production, and Ultimate Sale to the British Government of Borneo Toluol', which I have not seen. As Henriques' books are often unrealiable, I have used archival sources wherever possible. Cf. Henriques, *Samuel Marcus, op. cit.* note 4, 597–601, and Henriques, *Waley Cohen, op. cit.* note 4, 201–205. See also *Van Rotterdam Charlois, op. cit.* note 15, 17, 24, Forbes and O'Beirne, *op. cit.* note 4, 455, Gerretson, *op. cit.* note 20, vol. 4, 92–93, Gabriëls, *op. cit.* note 5, 61. And SICA (note 6): Waley Cohen to Colijn, 21 January 1915, Colijn to Minister of Foreign Affairs, 25 January 1915, Fijenoord to KNPM, 25 February 1915, Fijenoord to BPM, Technical Department, 6 March 1915 and 26 May 1915.

[24] C.S. Robinson, 'Kenneth Bingham Quinan, C.H., M.I.Chem.E. 1878–1948', *The Chemical Engineer*, 206 (1966), 290–297, R.M. MacLeod, 'Chemistry for King and Kaiser: Revisiting Chemical Enterprise and the European War', in A.S. Travis *et al.*, (eds.), *Determinants in the Evolution of the European Chemical Industry, 1900–1939* (Dordrecht: Kluwer Academic 1998), 25–49, esp. 28–34.

[25] *Van Rotterdam Charlois, op. cit.* note 15, 17, 24, 31; Henriques, *Waley Cohen, op.cit.* note 4, 202, 205–207, 228, 290, 293; and Henriques, *Marcus Samuel, op.cit.* note 4, 601–602; Robinson, *op. cit.* note 24, 291;R.M. MacLeod, 'The "Arsenal" in the Strand: Australian Chemists and the British Munitions Effort', *Annals of Science*, 46 (1989), 45–67, esp. 57–59; P. Hendrix, *Henri Deterding: de Koninklijke, de Shell en de Rothschilds* (The Hague: SDU 1996), 171–172, 271. SICA (note 6): Benzine-Installatie to BPM, 18 March 1915, and Historisch Archief (190C) – Chemie, petroleumdestillaat voor het maken van schoensmeer (60), Deterding to C.M. Pleyte, 2 July 1915, Reydon and Roos to BPM, 8 July 1915, Pleyte to Deterding, 12 July 1915, Bastet to Gerretson, 17 December 1951.

[26] On Dutch explosives manufacture and research around 1914–1918, see De Bruin, *Buscruytmaeckers, op. cit.* note 2; Nägele and Schaap, *op. cit.* note 2; De Jong, *op. cit.* note 2; P.W. Scharroo, 'Opstellen over springstoffen', *Technisch Tijdschrift*, 1 (1914), 75–77, 106–107, 144–146; P.W. Scharroo, *Springstoffen: eigenschappen, vervaardiging, werking en toepassingen, voor militaire vernielingen, in den mijnbouw en in den landbouw* (Utrecht, 1914); C.F. van Duin en K. Brackmann, 'Het "waarom" der moderne brisante nitrospringstoffen', *Chemisch Weekblad*, 16 (1919), 501–509; A. Steger, 'Springstoffen en munitie in den modernen oorlog te land en ter zee', *24e Jaarverslag 1914–1915 van het Technologisch Gezelschap te Delft* (Delft 1915), 53–100; P. van Romburgh, 1915, 'Chemie en Onafhankelijkheid', *Verslag van het Verhandelde in de Algemeene Vergadering van het Provinciaal Utrechtsch Genootschap van Kunsten en Wetenschappen* (1915), 9–37, esp. 16–23.

[27] 'Mobilisatie van chemici voor chemische landsverdediging', *Chemische Weekblad*, 12 (1915), 565–566; *75 jaar Vondelingenplaat, speciale uitgave van In geuren en kleuren* (February 1976), 18; Gerretson, *op. cit.* note 20, vol. 4, 93–94. See also SICA (note 6): Historisch Archief (190B) – Springstoffen en - persfabriek aan de Hembrug (39A), Van Royen to BPM, 21 August 1915, Pleyte

to Munitiebureau, 23 September 1915, Minutes of general meetings, Munitiebureau 2 July 1915 - 21 January 1916.

[28] In addition to Bastet, the following chemical engineers were employed in the TNT plant of Royal Dutch at Hembrug, near Amsterdam: F.E. van Haeften, J. ter Horst, M.L.A.A. König, S.H. Bertram, J.G. de Voogt and L.A. van der Ent. For their research activities at the BPM-laboratory at Amsterdam, see Duinmayer's summary of KSLA-reports (about 1957), present in the so-called Abbott-room of the Shell Research and Technology Center at Amsterdam. Gerretson, *op. cit.* note 20, vol. 4, 87–88, 93–94. The history of the TNT-plant presented here, is based on document in SICA (note 6): esp. [1] the archive 'Chemische Fabriek Hembrug' (Nijman-series 4, box 803), [2] Historisch Archief (190B) - Springstoffen en - persfabriek aan de Hembrug (39A), minutes of meetings of the Munitiebureau, August 1915-September 1916, and [3] Kopieboeken (5) - Chemische Fabriek Hembrug 1916–1918 (266–272), Pleyte to Pijzel, Coenen and De Kok, 9 November and 11 December 1915, Van Royen to BPM, 11 September 1915, 20 and 30 March 1916, and 19 May 1916, Pleyte to Munitiebureau, 23 September 1915, Gezamenlijke Buskruidmakers to BPM, 5 October 1915, BPM to Buskruidmakers, 7 October 1915, BPM to Benzine-Installatie, 8 October 1915, Munitiebureau to BPM, 12, 15, 23 and 25 February, Colijn to Munitiebureau, 17 February 1916, De Kok to Van Royen, 3 March 1916, Bastet, 'Globale begroting van een trinitrotoluol (trotyl) fabriek', 6 October 1915, 'Contract tussen de Staat der Nederlanden en de Bataafsche Petroleum Maatschappij', 23 December 1916, Bastet to Gerretson, 17 December 1951.

[29] SICA (note 6): Van Royen to BPM, 23 February and 25 March 1916, Nederhorst to BPM, 25 March 1916, Munitiebureau to BPM, 8 May 1916, De Kok to Munitiebureau, 12 September 1916, Nederhorst to Chemische Fabrieken a/d Hembrug, Report on control of stocks by Woltman and Rebholz, 2 May 1917, W.B. Oort to BPM, afd. T.L., 4 September 1917, Chemische Fabriek Hembrug to BPM, 24 September 1917, W.B. Oort, 'Administratief-Verslag van de Chemische-Fabriek der Bataafsche Petroleum Mij.', December 1917, W.B. Oort, 'Chemische Fabriek BPM, kostprijsberekening trinitrotoluol,' January 1919, Bastet to Gerretson, 17 December 1951. See also: Nägele and Schaap, *op. cit.* note 2, 79–82.

[30] M. R. Fox, *Dye-makers of Great Britain 1856–1976: A History of Chemists, Companies, Products and Changes* (Manchester: ICI 1987), 85–86; MacLeod, *op. cit.* note 24, 35–38; 'De strijd tegen de Duitsche kleurstof-nijverheid', *Holdert's Polytechnische Weekblad*, 9 (1915), 138–140. Shell Research and Technology Center, Amsterdam: abstracts of letters from Waley Cohen to Colijn, 31 January and 30 May 1916, in J. Duinmayer and C. Groeneveld, 'Geschiedenis van het Koninklijke/Shell-Laboratorium, Amsterdam (v/h Laboratorium der Bataafsche Petroleum Maatschappij' (Amsterdam 1957, unpublished typoscript).

[31] SICA (note 6): Correspondentieboeken (3) - Brieven aan Mr. Colijn (83A). See also: SICA (note 6): Algemeen Archief 1890–1945 (8) - Nederlandsche Kleurstoffenfabriek (1800). Shell Amsterdam: Duinmayer and Groeneveld, *op. cit.* note 30, correspondence and reports, 1916–1917, Henriques, *Waley Cohen*, *op.cit.* note 4, 247–248; Fox, *op. cit.* note 30, 127–128.

[32] For this and the paragraph below, see: SICA (note 6): Correspondentieboeken (3) - Brieven aan Mr. Colijn (83A), letters to Colijn and Pleyte, 1915–1916, Deterding (134–135), letters to and from Deterding. Shell Amsterdam: Duinmayer and Groeneveld, *op. cit.* note 30, reports J. ter Horst, 1915.

[33] Robert Henriques picture of Waley Cohen is too positive, relative to his negative picture of the other Royal Dutch/Shell managers. See Henriques, *Waley Cohen*, *op. cit.* note 4, 247–248; Fox, *op. cit.* note 30, 127–128. SICA (note 6): Waley Cohen to Colijn, 25 July 1916, 5 March and 11 May 1917, Colijn to Waley Cohen, 8 August 1916, Colijn to Deterding, 21 February 1918.

[34] SICA (note 6): Algemeen Archief 1890–1945 (8), Colijn to Deterding, 21 February 1918. See also: SICA (note 6): J.E.F. de Kok, 'Kleurstoffen industrie Holland', report 22 August 1917, Algemeen Archief 1890–1945 (8) - Chemische Fabriek 'Naarden' (1596), Nederlandsche Kleurstoffenfabriek (1800), Deterding to Colijn, telegramme 4 March 1918, Deterding to Colijn, 5 March. 1918, Colijn to Chemische Fabriek Naarden, 8 March 1918, Colijn to Deterding, 8 March 1918. Relevant documents on the erection of a synthetic dye plant in The Netherlands, as well as on the cooperation with BPM, are also in the company archives at Quest, Naarden, The Netherlands.

[35] On the history of the 'Nederlandsche Kleurstoffen-Fabriek', see Homburg, *op. cit.* note 20, esp. 322–329; A.M.C. Lemmens and G.P.J. Verbong, 'Natuurlijke en synthetische kleurstoffen in Nederland in de negentiende eeuw', *Jaarboek voor de Geschiedenis van Bedrijf en Techniek*, 1 (1984), 256–275,

esp. 269–273; H. Schröter, 'Cartels as a Form of Concentration in Industry: The Example of the International Dyestuffs Cartel from 1927 to 1939', *German Yearbook on Business History 1988* (Berlin, 1990), 114–144; and documents in the archives of Shell (The Hague), Philips (Eindhoven), and Quest (Naarden), as well as correspondence between I.G. Farben and his father, which were in the possession of (the late) Dr. A.M. ter Horst (Vierhouten, The Netherlands).

[36] W.C. de Leeuw (ed.), *Kleurstoffen uit petroleum. Algemeene beschouwingen over de industrie der synthetische kleurstoffen. Rapport in opdracht der directie der Bataafsche Petroleum-Maatschappij* (Amsterdam: BPM 1918), present in SICA (note 6), The Hague, and Shell Research and Technology Center, Amsterdam.

[37] SICA (note 6): Algemeen Archief 1890–1945 (8) - Nederlandsche Kleurstoffenfabriek (1800), letters Deterding to Colijn, 5 March 1918, Waley Cohen to Colijn, 2 December 1918 and 30 January 1919, De Leeuw to BPM, 18 December 1918 and 24 February 1919; Shell Amsterdam: Duinmayer and Groeneveld, *op. cit.* note 30.

[38] Homburg, *op. cit.* note 20.

[39] Cf. B.P.A. Gales and K.E. Sluyterman. 'Outward bound: The Rise of Dutch multinationals', in G. Jones and H.G. Schröter (eds.), *The Rise of Multinationals in Continental Europe* (Aldershot 1993), 65–98.

# KUHLMANN AT WAR, 1914–1924

INTRODUCTION

On 16 May 1917, an extraordinary court case opened in Paris. This involved two chemical companies, each contesting the right to insert the term 'colouring material' (matières colorantes) in its trading name. The Société Anonyme des Matières Colorantes et Produits Chimiques de Saint-Denis (also known as the Société de Saint-Denis), founded in 1881 by two entrepreneurs called Alcide Poirrier and G. Dalsace, brought the case against a company, founded in November 1916, under the name of Syndicat National des Matières Colorantes (SNMC), before it became the Compagnie Nationale des Matières Colorantes (CNMC) in January 1917.

However, the case was not simply a legal issue between two private companies. The SNMC owed its existence to the State, and, specifically, to the Minister of Trade, Etienne Clémentel, who wished, under the circumstances of total war, to unite French producers and consumers of industrial dyes within a single structure.[1] This case came in response to the wishes of other ministers, notably the Ministre de l'Armement, Albert Thomas,[2] and the military establishment, who were eager to establish a powerful French company to compete in an essential industry hitherto dominated by Germany.

The argument over a name — which the Société de Saint-Denis was to lose— hid a struggle between two models of economic organization – between a private company and a syndicate given preferential treatment by the State.[3] The court case also concealed a political struggle between Alcide Poirrier, director of the Société de Saint-Denis, and the heads of the SNMC, including its managing director, Georges Patart, Ingenieur en chef of the Service des Poudres. Until his death in 1916, Poirrier was a socialist Senator for the Seine Département, while his successor and son-in-law, André Cotelle, was a member of the State Council. Significantly, Patart also managed the Service des Poudres. In the opinion of the Société de Saint-Denis, this represented a conflict between private and public interest. Nonetheless, it was a fait accompli, and one that for France had a direct bearing on the successful prosecution of the war. To understand its development, is to understand the character of French industrial mobilization.

As it happens, the wartime SNMC and CNMC were directed not by the French government, but by French private enterprise. That enterprise was in the hands of one of the most important companies in the history of French chemical industry – the Établissements Kuhlmann. Therefore, in order to understand the CNMC, one needs first to understand the role played by this company during the First World War.

In October 1914, Établissements Kuhlmann faced the prospect of total ruin when its factories in northern France were overrun by invading German forces. Only four years later, however, the company was a picture of strength. In the interim, it had undergone a radical transformation. First, to complement its traditional expertise in inorganic chemistry, it acquired a mastery of organic chemical manufacture. Second, it extended its reach throughout France, through the acquisition of factories in Paris, Marseilles and Rouen. Third, it expanded prodigiously, with its capital increasing ten-fold between 1914 and 1918, and thirty-fold (from FF 6 million to FF 180 million) between 1914 and 1924.[4]

By 1924, Kuhlmann had absorbed the CNMC, and was the leading manufacturer of dyestuffs in France. In 1919, it took advantage of a provision in the Treaty of Versailles stipulating that France was to benefit from German patents in chemical engineering. In July 1924, the Treaty of London guaranteed the company's right to operate under these favourable conditions until 1928. Thus, the war served, in many respects, to accelerate Kuhlmann's development, and helped underpin a period of rapid expansion. The company was also at the heart of links between the State and corporate France, as new models for economic activity were devised and 'temporary civil servants' from the private sector (such as Ernest Cuvelette) took up positions in government.

This essay will review Kuhlmann's inheritance at the beginning of the twentieth century, and its importance to wartime production. It will also examine the company's role in setting up the CNMC, a move that enabled its directors to maintain their domination of the French industrial dye market well after the war.

## FRÉDERIC KUHLMANN'S INHERITANCE

To understand the significance of Établissements Kuhlmann, one must consider its founding father, and the technical and economic situation facing the firm on the eve of the Great War. Charles-Frédéric Kuhlmann (known as Frédéric Kuhlman) was a distinguished scientist and a local 'notable' in the Département du Nord. Born in 1803 in Colmar, south of Alsace, he emigrated to Nancy, where he studied at the Lycée and maybe at the university and also in Vauquelin's laboratory in Paris.[5]. Like many chemists in Wurtz's school Kuhlmann was Protestant. One of the most prolific French applied chemists of the nineteenth century, he became a Member of the Academie des Sciences (one of the five sections of the Institut de France) in 1824, and registered almost fifty patents.[6]

In 1825, Kuhlmann began his first chemical company in Loos-lez-Lille, in the suburbs of Lille. In 1870, this became the Société Anonyme des Manufactures de Produits Chimiques du Nord. In the following years, he also established or bought several chemical factories in the Département du Nord – including sites at Madeleine-lez-Lille in 1842, and at Saint-André-lez-Lille in 1852. According to André Thépot, Kuhlmann's strategy focused on the cultivation of regional markets (e.g., chlorine for the textile industry, sulfuric acid for the iron industry, and a salt mine in Villefranque near Bayonne). Between 1861–1865, he attempted to contrive a merger with the formidable conglomerate of Saint-Gobain. When failed, he arranged instead a cartel agreement with the huge firm, which opened the acid market of the North of France to his ambitions.⁷ At his death in 1881, Kuhlmann was the head of the third largest chemical company in France, and left 10 million francs to his children.

Frédéric Kuhlmann was also a 'notable' figure in the Département du Nord, serving as a member of the Chambre de Commerce de Lille in 1832, and its president from 1848 to 1869. Director of the Monnaie de Lille, he was also on the board of the Compagnie du Nord (a railway company), and a founder of the Société des Crédits Industriels et de Dépôt du Nord, which later became the Crédit du Nord.⁸ In 1831, he married Marie Woussen, the daughter of a spinner in Douai,⁹ and, he married his daughters well within the chemical and financial establishment. Pauline Kuhlmann, for example, married Theodore Kiener, head of the Crédit du Nord and a member of Kuhlmann's board.¹⁰ Frédérique Kuhlmann married Joseph-Emile Colson, a général de brigade, while Romaine Kuhlmann married Auguste Lamy, a professor of chemistry, first at the Faculté de Lille and later at the Ecole Centrale in Paris. Lucie Kuhlmann married Edouard-Donat Agache, the head of two textiles factories.¹¹ All Kuhlmann's sons-in-law became members of the board of Établissements Kuhlmann.

The second and the third generation of the family followed the founder's path. Donat Agache-Kuhlmann, son of Edouard Agache-Kuhlmann, was in the textile industry, and on the boards of the Compagnie des Mines d'Anzin and the Compagnie des Chemins de Fer du Nord. Joining the board of Établissements Kuhlmann in 1914, he became président in 1920. Théodore Barrois-Kiener,¹² son of Théodore Kiener and grand-son of Charles-Frédéric Kiener, became vice-président of the board of Établissements Kuhlmann in 1905, and président of the board of Mines de Lens and of Établissements Kuhlmann in 1919.¹³ Ernest Cuvelette, polytechnician and vice-director of Mines de Lens, married a grand-daughter of Frédéric Kuhlmann in 1909.¹⁴ Thus, the Kuhlmann family built connections with both textiles and coal. Their interests coincided in Mines de Lens, the third most important producer of coal in France in 1914, and the only French company capable of distilling benzene. Before the war, Kuhlmann planed to 'add value' to its production of coal, by building a dyestuffs factory as well.

Despite these impressive achievements, from a technical point of view, Kuhlmann in 1914 looked rather outdated. Of course, it was quite profitable (in 1912, the company recorded a profit of FF 2.2 million), but its product range had changed little since the 1880s.¹⁵ Its only new products were superphosphates, for use as fertilizers. Although Kuhlmann ranked amongst the top producers of sulphuric acid and chlorine, its processes were old-fashioned. In 1914, it was still using the Leblanc (and not the Solvay) process for producing soda, and the Deacon and Weldon processes for producing chlorine.¹⁶ Moreover, it was not involved in the developing fields of organic chemistry that were enriching German industry. Its only novel feature was the degree to which it attracted professional staff. In 1913, to manage its six plants, the board appointed Directeur Général, Paulin Grandel a graduate of the Ecole Polytechnique, who succeeded Paul Stahl graduated of the Ecole Polytechnique as well. They also recruited a number of plant engineers from the grandes ecoles, including four from the Ecole Polytechnique (versus six by Saint-Gobain, and none by the Société de Saint-Denis), six from the Ecole Centrale (versus twelve by Saint-Gobain and none by the Société de Saint-Denis).[17]

At the outbreak of war, therefore, the company was powerful and expanding. Kuhlmann's political presence lent the firm an influence larger than the few factories it owned. All this was threatened by disaster when, on 12 October 1914, German forces seized Lille, and, therefore, most of Kuhlmann's factories. Only a small plant in Amiens (producing only 5 per cent of the Établissements production) remained in French hands. Some 180,000 tons of material, worth FF 8.5 million, were taken by the Germans.[18] Établissements Kuhlmann was threatened with obliteration.

## WARTIME PRODUCTION OF ACID AND CHLORINE

Of course, not only Kuhlmann was touched by the war. The whole French economy was destabilized by the German occupation of northern France.

As Gerard Hardach has shown, by comparison with 1913, France suffered the loss of 49 percent of its coal production, 64 percent of its pig iron, and 58 percent of its steel. In 1915, production fell to very low levels, and only then began to rise.[19] [see Table 1]

Under the pressure of mobilisation, as John Godfrey has noted, the number of workers employed in war industries rose from 50,000 in August 1914 to 1.7 million in September 1917,[20] and the men employed in powder factories from 7000 to about 120,000 in September 1917.[21] Vast quantities of shells of higher calibres became greatly in demand, as trench warfare and the struggle at Verdun rose to epic proportions.

To keep pace with artillery and shell production, the production of chemical munitions proved to be a huge challenge. Before the war, France imported most of its chemicals from Germany. In September, within weeks of the onset of fighting, the Service des Poudres recognized that France would soon be short of explosives and intermediates, and signed contracts with British and American companies

Table 1. French Heavy Industry, 1913–18 (in millions of tons)

| | Coal | | Pig Iron | | Steel | |
|---|---|---|---|---|---|---|
| | Production | Imports | Production | Imports | Production | Imports |
| 1913 | 40.8 | 23.9 | 5.2 | – | 4.7 | −0.4 |
| 1914 | 27.5 | 17.2 | 2.7 | – | 2.7 | −0.2 |
| 1915 | 19.5 | 19.6 | 0.6 | 0.2 | 1.1 | 1 |
| 1916 | 21.3 | 20.3 | 1.5 | 0.6 | 2 | 2.6 |
| 1917 | 28.9 | 17.3 | 1.7 | 0.7 | 2.2 | 2.6 |
| 1918 | 26.3 | 15 | 1.3 | 0.4 | 1.8 | 1.8 |

Negative figures denote a surplus of exports. Dashes indicate amounts less than 50,000 tons.
Source: G. Hardach: 'Industrial Mobilization in 1914–1918: Production, Planning and Ideology', in Patrick Fridenson (ed.), *The French Home Front, 1914–1918* (Providence: Berg, 1992), 63.

Table 2. French Imports of Sulfuric Acid (metric tons)

| 1913 | 1914 | 1915 | 1916 | 1917 | 1918 | 1919 |
|------|------|------|------|------|------|------|
| 9901 | 5562 | 23,167 | 63,337 | 50,165 | 918 | 1552 |

Source: A. Fontaine, *L'Industrie Française pendant la Guerre* (Paris/New Haven: Publications de la Donation Carnegie, P.U.F/ Yale University Press, 1925), 233.

for the delivery of large supplies of benzene and sulfuric acid.[22] France also imported from the USA (especially Du Pont de Nemours) black powder and nitrated explosives. By contrast, in mid-1916 poison gas was largely produced by French firms, sometimes in cooperation with British companies.[23]

How can we explain the differences between chemical and shell production? First, the iron and steel-making industry was well integrated,[24] whereas the chemical industry was profoundly divided. In managerial terms, Taylorism was diffused throughout the steel industry.[25] Moreover, while the steel industry could design and produce new varieties of shells as required, the chemical industry had to begin work in, what were for them, new fields, and in huge volumes. For example, the only French company to produce phenol – the Société Chimique des Usines du Rhône – had to scale up production massively. This took time. The SCUR produced only sixty-four tons in December 1915. However, by June 1917, it was producing 200 tons per month.[26] In this context, Kuhlmann's serious contribution to the war effort began in October 1915, but really picked up steam in 1916, when the company took over the production of two chemicals, needed for the manufacture of explosives, which it had technically (if not economically) mastered before the war – namely, oleum (concentrated sulphuric acid) and carbon tetrachloride.

OLEUM

At the end of 1916 – after two years of war, and with no end in sight – the largest French producer of oleum was the Compagnie de Saint-Gobain – whose factories accounted for 13,400 of the 18,480 tonnes produced in France that year. This was equivalent to 72.5 per cent of the country's total output. By contrast, Kuhlmann produced only 450 tonnes, just 2.4 per cent of the total.[27] French war plans in 1913/14 entailed a 'mobilisation generale' of every man, which meant that many trained engineers and other workers vital to industrial manufacturing were called to the colours. Following vigorous debate, in June 1915, the so-called Dalbiez Law was passed to enable selected firms to recall specialist workers. This, together with ever-growing demands for explosives, helps explain why the State, in October 1915, contracted with Kuhlmann to build a factory at Port-de-Bouc, close to Marseilles. The government paid construction costs, and the company guaranteed output at a fixed price.[28] After the war, according to the terms of the contract, the factory was to remain the property of the State, while the company was given a rental option.

According to John Godfrey, Saint-Gobain tried everything in its power to prevent its rival from acquiring this property, and only government intervention stopped Saint-Gobain from acquiring the land in question.[29] But Kuhlmann won, and between 1915 and 1918, expanded rapidly, acquiring a number of chemical companies in different parts of France. The first of these, – acquired in early 1916 – was the Société de Pennaroya's factory in Estaque, near Marseille. This plant specialized in the production of sulphuric acid and chlorine. Among other acquisitions in late 1916 were Les Fils de Salles, with acid and superphosphorate plants in Bordeaux and Ivry-sur-Seine; and the two Société des Colles Tancrède et Collette factories in Aubervilliers and Nevers. The latter produced sulphuric acid and superphosphates. As Kuhlmann's commemorative centenary brochure in 1925 puts it: 'the…company thus had plants in the four large ports (Marseilles, Bordeaux, Nantes, and Rouen) that served the main agricultural areas of France'.[30] The war gave Kuhlmann a chance to break out of the confines of northern France, and to become a truly national entreprise.

Kuhlmann's strategy included an attempt to gain guaranteed access to raw materials. In late 1916, an agreement was signed with the Compagnie Royale Asturienne des Mines to build in Tonnat-Charente a factory to roast zinc sulphide, one of the intermediates in the production of sulphuric acid. An increase in capital through a share issue was sanctioned to pay for this operation, and two Compagnie Royale Asturienne directors were co-opted to unit Kuhlmann's Board. These were Louis Hauzer, managing director of Compagnie Royale Asturienne, and François Roussel (who, following his death, was replaced by Louis Boudenoot, a senator and chairman of Mines de Carvin).[31] Then, in late 1917, the Société des Mines et Fonderies de Zinc de la Vieille-Montagne, which operated pyrite mines outside France, joined Kuhlmann's empire.[32] Edgar de Sinçay, director of Vieille-Montagne, was appointed to Kuhlmann's board, a move that broke France's dependence upon the country's only pyrites mine, which was owned by Saint-Gobain and located at Sain-Bel, 20 km east of Lyon. While some of these acquisitions were funded by shares bought by Compagnie Royale Asturienne des Mines and the Société des Mines et Fonderies de Zinc de la Vieille-Montagne, we know nothing about how Kuhlmann's acquisitions were paid for. We know more about the second factor behind Kuhlmann's renewal – its mastery of rare compounds needed for the development of chemical weapons.

KUHLMANN AND THE MANUFACTURE OF POISON GASES

The first gas attack on the Western Front, launched by the Germans on 22 April 1915, took the Allies by surprise. Once over the shock, however, the French government immediately appointed a commission of inquiry, comprising military officers and scientists,[33] which in June 1915, became the Commission des Etudes Chimiques de Guerre under the chairmanship of Georges Weiss.[34] This was reorganized on 18 October 1915, when, under the Direction Chimique du Matériel de Guerre (DMCG), it became a fully-fledged Direction, the 13th, headed by General

Ozil. The new DMCG was made up of three sections.[35] The first, was the Inspection des Etudes et des expériences chimiques (IEEC) under General Perret, which ran the Technical and Industrial Section, which supervised the industrial production of gases. This section was run by Ernest Cuvelette, who was replaced in 1917 by Emile Bollaert. The two men had similar professional backgrounds, both coming from Mines de Lens. Cuvelette was a mining engineer and a graduate of the Ecole Polytechnique and had close connections to the Kuhlmann family.[36] The third section of the DMCG was the Central Office, headed by Papon, which was responsible for purchasing, quality control, and supply.

Early analyses of unexploded German gas shells quickly revealed the presence of chlorine. In preparing to retaliate, however it was found that only small quantities of liquid chlorine were produced in France – in two rundown factories, one of which was close to the front line.[37] However, a commission of experts quickly came up with a formula for a toxic gas that contained chlorine and carbon tetrachloride, a solvent used to ensure that the toxic mixture did not freeze in low temperatures.[38] It so happens that the only French company producing carbon tetrachloride before the war was Kuhlmann. In January 1916, Kuhlmann also acquired the only company operational in the production of liquid chlorine, the Compagnie de Pennaroya, at Marseilles l'Estaque. Several of their directors joined Kuhlmann's board. Around the same time, two processes to produce phosgene were invented. The first involved reacting oleum with carbon tetrachloride, while the second involved combining chlorine with carbon oxide in the presence of a catalysing agent to produce sulphuric protochloride.[39] To its great advantage, Kuhlmann was to remain the sole French supplier of both carbon tetrachloride and sulphuric protochloride until June 1918.[40]

The production of gas weapons per se did not fall within Kuhlmann's remit. Rather, the laboratories of the IEEC and of the Société des Usines Chimiques du Rhône (SCUR) were responsible for developing yperite gas (dicholorodiethyl sulphide).[41] Two army officers – Jacques de Kap-Herr and Joseph Frossard[42] – were seconded by the Ministre de l'Armement to the Société des Usines Chimiques du Rhône to hasten this development. Meanwhile, given its widening experience, Kuhlmann began to devote its energies to establishing an industrial dye business.

ÉTABLISSEMENTS KUHLMANN AND THE COMPAGNIE NATIONALE DES MATIÈRES COLORANTES

Before 1914, the French industrial dyes sector was almost totally dependent upon German know-how, both in terms of imports and assembly plants on French soil. Lutz Haber estimates that France – in common with most industrialized countries of Europe – depended upon Germany for 85 per cent of its needs.[43] Only one French company, the Société Anonyme des Matières Colorantes de Saint-Denis (Société de Saint-Denis), was competitive. The war spelt an end to German imports and to shortages of dye products, especially aniline and alizarin. The French government reacted to this situation in two ways.

First, the Office des Produits Chimiques et Pharmaceutiques established in the beginning of the 1915, together with a number of private firms, decided to produce industrial dyes. However, despite the confiscation of German-owned factories, there was a shortage of material for the manufacture of synthetic indigo, since compounds such as phenylglycine and monochloracetic acid traditionally came from Germany. In fact, when the phenyglycine ran out, the French production of indigo stopped. In order to resume production, the producers needed aniline, which was produced in France, and monochloracetic acid, which was not.

This problem was overcome in January 1915 when an agreement between the Office des Produits Chimiques et Pharmaceutiques and the Société Camus et Duchemin – and between the research teams of Professor Auguste Behal and R.-P. Duchemin – led to the discovery of a synthetic indigo substitute.[44] Camus and Duchemin delivered its entire production to the Compagnie Parisienne des Couleurs d'Aniline, located at Creil.[45]

Duchemin was one of only two French engineers trained before the war at the Badische Anilin (BASF) factory in Ludwigshafen. He was highly recommended by Professor Albin Haller (Director of the Ecole de Physique et de Chimie de Paris, and a professor at the Sorbonne) as 'particularly…able to furnish the desired plans' for aniline production,[46] and went on to become a key figure in the dyestuffs industry in France. The other was René Masse, a Polytechnician, whose company, the Société d'Eclairage, Chauffage et Force Motrice, based in Genevilliers, in the suburbs of Paris, developed red alizarin.[47]

The second offensive on the industrial dyes front came with the creation of a large company along the lines of British Dyes, Ltd, established by the British government in July 1915.[48] Etienne Clémentel, Ministre du Commerce, wanted a syndicate to bring together producers and consumers, so in February–March 1916, held a series of consultative meetings in Paris and in the provinces, involving several captains of industry. These consultations – chaired by Denys Cochin, Ministre d'Etat (sans portefeuille), sous-secrétaire d'Etat au 'blocus' after 12 December 1916 – led to the establishment of a Commission des Matières Colorantes. This commission held its first meeting on 30 April 1916. Clémentel was chairman and Cochin was deputy, and the commission included Auguste Béhal and Henri Blazeix, technical section heads at the Ministère du Commerce[49]; Henri Mauclère, managing director of the Service des Poudres; Ernest Cuvelette, head of the technical and industrial section of DMCG, and director of Mines de Lens; Edmond Ledoux, an engineer; and Edouard Donat-Agache, chairman of the board of Kuhlmann. Later, Georges Patart, inspecteur général of the Service des Poudres, joined the Commission.[50]

There was, however, not a single industrialist specialising in dyes on this Commission, neither from the Société de Saint-Denis nor any other, smaller company, such as Steiner (based in Vernon) nor from Maboux et Camell (based in Lyons). Instead, and remarkable for their presence, were Kuhlmann, which was not active in dyes, and Cuvelette, who before the war had planned to build dye factories. In 1911, Mines de Lens had linked up with Compagnie de Commentry-Fourchambault to set up the Société Métallurgique de Pont-à-Vendin. According to

M. Gillet, apart from the intrinsic interest of holding a stake in this joint-venture, Mines de Lens 'intended to use the newly built factories to ensure the rapid development of coal refining, with an eye to creating a huge dye factory'.[51]

Following its initial meeting in April 1916, it was decided that the Commission des Matières Colorantes should become the forerunner of a new Syndicat des Matières Colorantes.

## THE SYNDICAT NATIONAL DES MATIÈRES COLORANTES (SNMC)

The financial structure of the Syndicat National des Matières Colorantes (SNMC) was mapped out in November 1916. The Syndicate, which was to issue 20,000 shares, was to be the pivot of a new Compagnie Nationale des Matières Colorantes (CNMC) – created in January 1917 – which was to issue 80,000 shares. The constitution of the Syndicat was handed by the Banque de Paris et des Pays-Bas and Établissements Kuhlmann. They formed the first nucleus of major shareholders (Table 3), subscribing the first 20000 shares at 500 francs each, for a total of F10 million. The other 60,000 shares were sold to the public,[52] including 12,000 shareholders who were to get only a few shares each.[53]

Table 3. Compagnie National des Matières Colorantes, 1917 Main share-holders and members of the board

| Nom de la compagnie | Représenté | Type d'industrie | Nombre d'actions |
|---|---|---|---|
| Mines de Vicoigne et Noeux | Eugène Barthelemy | Charbon | 1000 |
| Mines de Lens | Felix Bollaert | Charbon | 500 |
| Forges de Firminy | Jean Neyret | Acier | 400 |
| Cie des Forges et Aciéries de la Marine et Homécourt | Théodore Laurent | Acier | 1000 |
| Schneider et fils | Schneider et Ernest. Métivier | Acier | 1000 |
| Société d'Eclairage, Chauffage et Force Motrice | Réné Masse | Gaz | 1000 |
| Société Norvégienne de l'Azote et des Forces Motrices | Comte de Germiny | Industrie chimique | 1100 |
| Etablissements Kuhlmann | Donat-Agache and Emile Lambert | Industrie chimique | 900 |
| Chiris et Jeancard | Paul Jeancard | Industrie chimique | 600 |
| Banque de Paris et des Pays-Bas | Gaston Griolet, Gaston Guiot, Raphaël Georges-Levy, Louis Lion | Banque | ? |
| Blanchisserie et Teinturerie de Thaon | Paul Lederlin | Textile | 600 |

Source: Etablissements Kuhlmann, *Cent ans d'industrie chimique. Les Etablissements Kuhlmann.1825–1925* (Paris: Compagnie Nationale des matières Colorantes et Manufactures de Produits Chimiques du Nord reunites, 1926), 138 p.

The future CNMC had several obligations. The contract signed on 11 September 1916 between the Government and the SNMC specified that, during peacetime, the company had to produce colouring materials and chemical products 'that could be requested in large quantities by industrialists and French merchants, and sell them at an affordable price'. Moreover, the Ministre de la Guerre and the Ministre du Commerce, 'if judging necessary will indicate to the company the production of all necessary explosives, chemical products and dyestuffs'.[54] Finally, scientific goals were set, in that the company should 'run a laboratory devoted to chemical researches concerning its own industry, suitably furnished, and should employ twenty researchers enrolled after a state directed examination'.[55] The company's vice-president would be appointed by the State, which would share in the profits.[56] In exchange, after the war, the State would give the company Oissel power-plant – worth FF 15 million – and 'any other plant built during the war by the Service des Poudres, [which] the Service thought it was necessary to keep for future use but not to keep running after the war'.[57] The new company was therefore, to a certain extent, a public company dependant on public funds. In John Godfrey's view, the company was purely an emanation of the government and of Clémentel,[58] in which the latter played one industrial group off against another, as the composition of the company's board of directors suggests.[59]

During the first hearings of a joint Assemblée Nationale's Commission du Budget et des Contrats d'Etats, Budget and State Contracts Commission, Albert Thomas was asked by a certain M. Simyan 'whether or not he thought that the agreements signed were a hidden subsidy to dye manufacturers?' The agreements stipulated that the CNMC would have use after the war of a plant — the converted explosives factory in Oissel, near Rouen, worth FF 15 million — for a modest annual fee. Thomas' response reads as follows:

> In the strict sense of the term, my answer is no, although there is quite obviously an element of State aid. That's what the current court case is all about.... But you can find legal experts to contradict other legal experts in every administration. However, that was not the problem. It was a question of subsidising, of providing broad aid to an industry that had already been helped in England, Japan and every other country, and we were asking ourselves what was the best way to go about helping it here in France.[60]

These remarks go to the heart of the matter — the extent to which the French dye industry was mapped out by political decisions. The Société de Saint-Denis, excluded from the SNMC by the Ministre du Commerce, was well aware of the extent of state favouritism towards Kuhlmann. So, the company decided to launch a double counter-attack. Alcide Poirrier and his son-in-law, André Cotelle, decided to denounce their exclusion. Cotelle first denounced the government's methods, then denounced public intervention in private business, and presented himself as a defender of liberalism.[61] The early increase in the capitalization of the Société de Saint-Denis – from FF 4,375,000 to FF 7 million in 1915 – spoke not of the emergence of a powerful industrial and chemical company, but rather of downstream vertical integration among textile companies.[62] Moreover, Saint-Denis historically suffered a number of failures in making synthetic indigo, which led to a diminution of its capital.[63] From an industrial point of view, Saint-Denis did not manage to reach an agreement with a benzene producer like Mines de Lens, a company which held a monopoly in France.

However, it seems there were sound business reasons for favouring the SNMC, over the Société de Saint-Denis. In fact, during the inquiry that took place in the first months of 1917, following the expansion of the SNMC into the CNMC, the CNMC supporters, including Kuhlmann, did not deny the support they had received from the Ministries of Trade and Armaments, but rather argued that this support flowed from the government's industrial policy.

Private archives – especially those of BNP-Paribas – afford valuable insights into this affair. First, it appears that none of the SNMC's board of directors was actually appointed by the government.[64] Second, the SNMC'S board of directors was an assembly of industrial groups. The first group was made up of Donat-Agache Kuhlmann, which was represented by three board members, including Félix Bollaert, director of Mines de Lens, and Emile Lambert.[65] This group was certainly a family business, but it also enjoyed extensive relations with Mines de Lens, the Compagnie d'Anzin, and the Compagnie de Noeux et Vicoigne.[66]

The second group was made up of coal and iron and steel producers which, according to a law of 29 November 1915, were required to recover benzol given off by gas lighting and their own activities. The most enthusiastic response to these provisions came from Schneider[67] and Mines de Lens, which was the main producer of benzol before the war.[68] Other chemical companies joined the CNMC board, including the Société Norvégienne de l'Azote et des Forces Motrices, founded by the Banque de Paris et des Pays-Bas in 1905,[69] and the Société d'Eclairage de Chauffage et de Force Motrice, which was involved in the alizarine factory run by Réné Masse. This strategy of vertical integration embraced the dyeing industry both upstream and downstream, as the SNMC board included representatives of textile firms, such as Paul Lederlin, director of Blanchisserie et Teinturerie de Thaon. The Agache-Kuhlmann company also owned many textile factories in the north of France.

From 1917 the combination of Kuhlmann-CNMC thus became an example of vertical integration from benzine to chlorine, acid to textiles.[70] Kuhlmann owned 8,100 of 60,000 shares,[71] which seems an extremely low level; but nine of sixteen members were faithful to the Kuhlmann-Mines de Lens-Banque de Paris et des Pays-Bas group. Georges Patart was appointed Director of CNMC, and so combined the interests of the government an the Service des Poudres.[72] Thus for Kuhlmann, the war was a lever: the company acquired new technology and a new and powerful network.

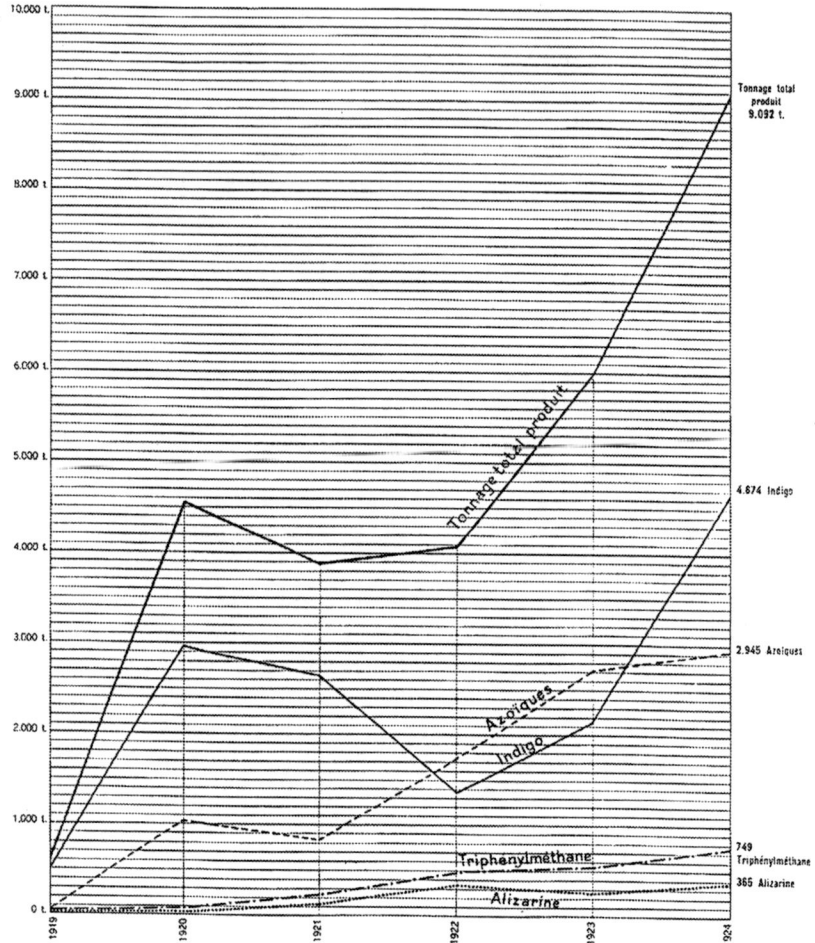

Figure 1. Établissments Kuhlmann: Annual Production, 1873 to 1924

*Source*: Etablissements Kuhlmann, *Cent ans d'industrie chimique. Les Etablissements Kuhlmann. 1825–1925* (Paris: Compagnie Nationale des matières Colorantes et Manufactures de Produits Chimiques du Nord reunites, 1926), 138 p.

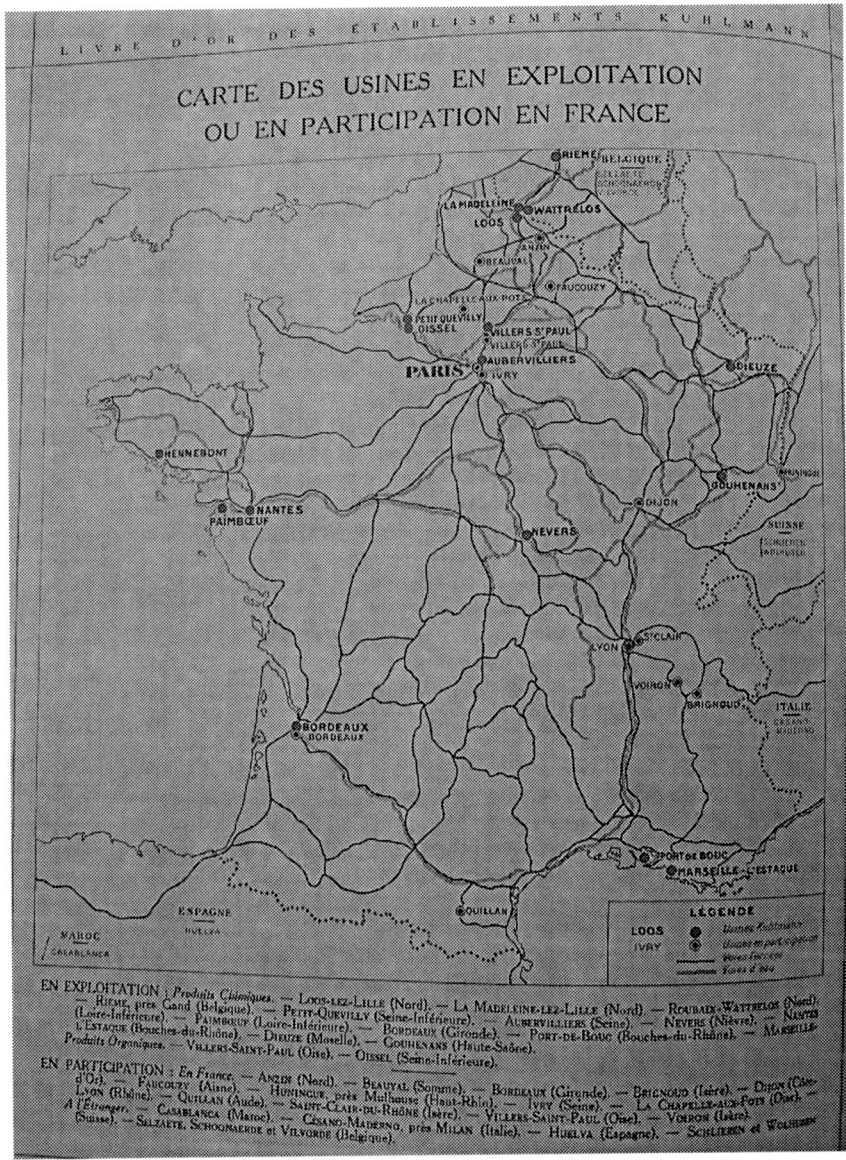

Figure 2. Établissments Kuhlmann:. Dyestuffs Production, 1919 to 1924

*Source*: Etablissements Kuhlmann, *Cent ans d'industrie chimique. Les Etablissements Kuhlmann. 1825–1925* (Paris: Compagnie Nationale des matières Colorantes et Manufactures de Produits Chimiques du Nord réunies, 1926), 138 p.

Table 4. Établissements Kuhlmann-CNMC, 1925

Conseil D'Administration en 1925
*Président*
M. AGACHE-KUHLMANN (DONAT). Administrateur de la Compagnie des Chemins de fer du Nord et des Mines d'Anzin.

*Vice-Présidents*
MM. CUVELETTE (ERNEST), C., Administrateur-Directeur Général de la Cie des Mines de Lens.
GRIOLET (GASTON), G. O. Président du Conseil d'Administration de la Banque de Paris et des Pays-Bas.
LAMY-KUHLMANN (ÉDOUARD). Président d'Honneur de la Société Industrielle d'Amiens.
MASSE (RENÉ). C. Vice-Président-Directeur de la Société de Chauffage, d'Éclairage et de Force Motrice.'

*Secrétaire*
M. GRANDEL (PAULIN). ancien Directeur Général des Établissements Kuhlmann.

*Administrateurs*
MM. BOLLAERT (FÉLIX), 0. Président du Conseil d'Administration des Mines de Lens.
CHABAUD LA TOUR (RAYMOND DE), Administrateur des Mines de Nceux.
CHASTEL (ANDRÉ)., Directeur Général de la Société Minière et Métallurgique de Penarroya.
CRESPEL (ALBERT).:, Administrateur des Mines de Lens.
DELEMER (JEAN), Administrateur-Délégué de la Société Anonyme de Pérenchies.
DELSOL (ALBERT), Administrateur de la Société des Mines de Carvin.
DUPONT-DESCAT (ALFRED), Administrateur des Mines de Courrières.
DURAND-GASSELIN (HIPPOLYTE), Industriel.
HAUZEUR (Louis), 0. Président du Conseil d'Administration, Directeur Général de la Compagnie Royale Asturienne des Mines.
HEURTEAU (CH-ÉMILE), Président du Conseil d'Administration, de la Société Minière et Métallurgique de Penarroya.
KIENER-KUHLMANN (ÉMILE), 0. Chef d'Escadrons de Cavalerie en retraite.
LAMBERT (ÉMILE), 0. Administrateur-Délégué de la Société Lambert-Rivière.
LAURENT (THÉODORE), C. Administrateur-Directeur Général de là Compagnie des forges et Aciéries de la Marine et d'Homécourt.
LAVAURS (ANDRÉ). Attaché à la Direction de la Société Miniére et Métallurgique de Penarroya.
LECOEUVRE (FRANCIS), Administrateur-Directeur Commercial de la C[10] Royale Asturienne des Mines.
LEDERLIN (PAUL), 0. Administrateur-Délégué des Blanchisseries de Thaon.
LEDOUX (EDMOND), Administrateur-Délégué de la Société pour l'Utilisation des Combustibles.
LEDOUX (FRÉDÉRIC), Administrateur-Délégué de la Société Minière et Métallurgique de Penarroya.
DE SINÇAY (EDGAR), Administrateur de la Société des Mines et Fonderies de Zinc de la Vieille-Montagne.

*Comité De Direction*

*Président*:
M. AGACHE-KUHLMANN (DONAT).
*Vice-Présidents*:
MM. CUVELETTE (ERNEST).
(FRÉDÉRIC).
MASSE (RENÉ).
LAMY-KUHLMANN (ÉDOUARD).

*Administrateurs*
MM. LAMBERT (ÉMILE).
HAUZEUR (Louis).
LEDOUX
DUCHEMIN, (RENÉ).
*Directeurs génér*
BERR (Raymond)
FROSARD (René)
*Secrétaire*
M. GRANDEL (Paulen)

Source: Etablissements Kuhlmann, *Cent ans d'industrie chimique. Les Etablissements Kuhlmann.1825–1925* (Paris: Compagnie Nationale des matières Colorantes et Manufactures de Produits Chimiques du Nord reunites, 1926), 138 p.

## THE CNMC-KUHLMANN MERGER

In 1924, six years after the Armistice, Établissements Kuhlmann, following the logic of its expansionist strategy, succeeded in absorbing the CNMC. At the time, the CNMC was producing 70 per cent of French colouring material. Although it no longer had links to the State, it continued to represent itself as at the service of the nation:

> One notes with pleasure, but without surprise, that the feared over-production of chlorine has been avoided, so that should we be forced to re-live the Dark Hours of the recent past, we can rest assured that France has at its disposal a number of factories that are fully equipped to produce gas and explosives.[73]

Looking at Kuhlmann's management in 1925, (see Table 4) we find that of twenty-four directors, twelve or thirteen were representatives of the mining and raw materials sectors, including the two deputy chairmen, Ernest Cuvelette and Réné Masse. Three or four other directors were members or representatives of the Kuhlmann family. One notes also the presence of G. Griolet, chairman of the board of the Banque de Paris et des Pays-Bas. Kuhlmann also recruited a Polytechnique graduate, Raymond Berr. Joseph Frossard and Jacques de Kap-Herr, heroes of French chemical weapons development, headed the organic products department. Also on Kuhlmann's board was René-Paul Duchemin, formerly head of Camus et Duchemin and R. Masse. Thus, Kuhlmann managed to bring together the experts in gas weapons, as well as those who had managed to kick-start the production of dyes. Thus, Kuhlmann became a national firm, dominating the dyestuffs market, but also having the capacity, through its staff and materials, to lead French industry in both war and peace.

## CONCLUSION

The 'renaissance' of Établissements Kuhlmann after the early wartime losses of 1914 was due to its rapid development of basic chemical products, combined with its monopoly of the chemicals needed for gas warfare. But its conquest also owed much to its effective strategy of mobilization. This took place along four lines: *industrial*, through a strategy of vertical integration; in *personel*, through the mobilization of experts such as R.-P. Duchemin and Réné Masse; *financial*, through the establishment of a new, state-subsidized company, the CNMC; and *political*, by gaining the support of leading figures in government, including Ernest Cuvellette, Albert Thomas, and Etienne Clémentel.

For the historian of chemistry many questions remain. The history of Kuhlmann suffers from a lack of surviving primary sources.[74] The picture given by secondary sources remains incomplete. We do know, however, that Kuhlmann became as much a part of wartime industry as of the peacetime economy, both in terms of its productive capacity and its links with the State. It was also the only French company that (unlike the Société de Saint-Denis) managed to obtain continuing supplies of raw materials. This war was a compelling moment at which close cooperation between the State and industry blurred differences between public and private. The history of the CNMC shows that a company could represent State interests, and that politicians could act on behalf of a company.

If the war was a golden opportunity for Kuhlmann, it was equally important for the French chemical industry – which tried, but without the same degree of success, to reshape itself. Some have argued that, both during and after the war, France pursued a 'plan sidérurgique', aimed at the Saar coal mines and the occupation of the Ruhr.[75] But there was also a 'plan chimique', uniting industrialists, top civil servants and politicians, who were impatient to revitalize a declining national industry. The birth of the CNMC was a successful part of this plan. The return of Alsace and Lorraine gave France an advantage in the field of potash. The mines near Mulhouse were barely exploited by the Germans, who instead favoured the Ruhr and the German cartel. After the war, their production increased a great deal: from 325,000 tons of potash in 1914 to 1.7 million tons in 1924.

After the war, the development of an atmospheric nitrogen complex, based on Fritz Haber's discovery, caused a great stir among French chemical firms. The government favoured a national trust to exploit the Haber patent, but industry refused to cooperate. During the spring and summer of 1919, Saint-Gobain, in agreement with l'Air Liquide, decided to create a company, la Société Chimique de la Grande Paroisse, based on the alternative Casale process.[76] At the same time, the relations between Kuhlmann and the State change. Kuhlmann went 'back to business' and relations with the State deteriorated. Kuhlmann, allied with Mines de Lens, decided not to accept the government's proposal to develop an atmospheric nitrogen complex in Toulouse, it was considered too far from their northern basis. In 1924, pressed by the Service des Poudres, the government decided to create a state-directed office, the Office Nationale Industriel de l'Azote, which produced fertilizers. Its first president was Georges Patart, who had become head of the Service des Poudres in the spring of 1919, when he left the CNMC. Patart replaced Général Lheure, who became head of the Société Chimique de la Grande Paroisse.[77] Once again, we see the importance of the men of the Service des Poudres, and of graduates of the Ecole Polytechnique, in the chemical industry.

The 'plan chimique' of the post-war period did have some long-term benefits. One by-product of Germany's defeat was the passage to France of German industrial interests in Iraq, along with 25 per cent of the Turkish Petroleum Company.[78] In 1924, the government led by Raymond Poincaré began discussion that in 1928 led to the foundation of the Compagnie Française des Pétroles (CFP), headed by industrialist Ernest Mercier (leader of the right-wing movement, 'Le redressement français'). By 1929, the State owned 25 per cent of its shares.[79] The CFP's sister company, the Compagnie Française de Raffinage, was protected by a 1928 law that gave the State the right to regulate the importation and distribution of oil in France. This new industry was also protected by the State. So, if not completely successful in modernizing industry, the 'plan chimique' at least gave France the basis of its

Table 5. Établissements Kuhlmann Finances, 1900 to 1925

| | Résultats Financiers Des Établissements Kuhlmann DE 1900 A 1925 | | |
|---|---|---|---|
| Exercises | Capital Social | Benefits | Ratio |
| 1899–1900 | 6,000,000 | 735,621,34 | 10 |
| 1900–1901 | 6,000,000 | 486,303,79 | 10 |
| 1901–1902 | 6,000,000 | 1,013,578,29 | 10 |
| 1902–1903 | 6,000,000 | 1,146,574,54 | 14 % |
| 1903–1904 | 6,000,000 | 737,005,08 | 14 |
| 1904–1905 | 6,000,000 | 1,040,304,99 | 10 |
| 1905–1906 | 6,000,000 | 1,298,460,32 | 10 |
| 1906–1907 | 6,000,000 | 1,874,770,67 | 12.50 |
| 1907–1908 | 6,000,000 | 1,535,244,49 | 12.50 |
| 1908–1909 | 6,000,000 | 1,646,393,82 | 12.50 |
| 1909–1910 | 6,000,000 | 1,187,351,27 | 12.50 |
| 1910–1911 | 6,000,000 | 1,409,726,22 | 15 % (1) |
| 1911–1912 | 6,000,000 | 2,267,689,20 | 16.66 |
| 1912–1913 | 6,000,000 | 1,639,798,87 | 16.66 |
| 1913–1914 | 6,600,000 | 1,450,676,05 | 6 %(2) |
| Période du 1 er Mai 1914 an 31 Décembre 1918. | 60,000,000 | 8,442,955,12 | 12 % |
| 1919 | 60,000,000 | 9,427,944,50 | 12 |
| 1920 | 81,450,000 | 11,760,443,25 | 12 % |
| 1921 | 90,000,000 | 131,935,93 | 0 |
| 1922 | 100,000,000 | 6,838,062,00 | 8:0 |
| 1923 | 150,000,000 | 20,315,124,20 | 12 |
| 1924 | 180,000,000 | 21,863,252,03 | 12 |
| 1925 | 180,000,000 | 24,239,812,70 | 12 |

(1) Dès 1910, le capital était complètement amorti; (2) Un acompte seul a pu étre versé, en raison de la guerre.

Source: Etablissements Kuhlmann, *Cent ans d'industrie chimique. Les Etablissements Kuhlmann.1825–1925* (Paris: Compagnie Nationale des matières Colorantes et Manufactures de Produits Chimiques du Nord reunites, 1926), 138 p.

future 'energy independence'. In many respects, its new managers belonged to that post-war generation described by M. Levy-Leboyer who grew up in the idea:

> A good number of them [company bosses] had acquired experience in the upper echelons of the State's administration, whether in the office of Loucheur, who took over the Armaments Ministry in 1916, or in one of the numerous departments responsible for military supplies. In this respect, the war actually knit closer together a rather technocratic generation of bosses. Their experience during the war also meant their influence grew afterwards. This is especially true of those who engaged in the 'Redressement français' movement or took over key posts in large groups such as Lyonnaise des Eaux, Union d'Eléctricité, Alsthom, Péchiney and Kuhlmann.[80]

Whatever its services to the State, this was a generation that certainly had a very good understanding of its own interests as well.

## ACKNOWLEDGEMENTS

The author wishes to express his appreciation to Roy MacLeod for his assistance in preparing a final version of this paper.

## SOURCES

Owing to the destruction of Kuhlmann's archives in the 1940 invasion, and the closing of the company in 1982, it has been difficult to locate company records. This paper is therefore based mainly on subsidiary sources, including *technical archives* (patents from the Institut National de la Propriété Industrielle, Paris and the Service Historique de l'Armée de Terre, Vincennes); *banking archives* (Crédit Lyonnais, Paris; BNP-Paribas); *government archives* (Service des Archives Economiques et Financières of Ministère de l'Economie et des Finances, at Savigny-le-Temple; and the Ministère de l'Armement and the Ministère du Commerce, Archives Nationales, Paris); *private archives*: (the Saint-Gobain archives in Blois); and the archives of certain *grandes écoles* (the Ecole Polytechnique, the Ecole Centrale, and the Ecole de Physique et de Chimie de la Ville de Paris).

## NOTES

[1] See G. Rousseau, *Etienne Clémentel (1864–1936): Entre Idéalisme et Réalisme, Une Vie Politique (Essai Biographique)* (Clermont Ferrand: Archives Départementales du Puy-de-Dôme, 2000). Clémentel was Ministre du Commerce (Minister of Trade) between November 1915 and November 1919.

[2] Albert Thomas, a socialist, was Sous-Secrétaire d'Etat pour l'Artillerie et l'Equipement militaire (Under-Secretary of State for Artillery and Military Equipment) from 18 May 1915; Sous-secrétaire d'Etat pour l'Artillerie et les Munitions (Under-Secretary of State for Artillery and Munitions) from 29 October 1915; and Ministre de l'Armement (Minister of Armaments) from 12 December 1916 to September 1917. Jean-Jacques Becker and Serge Bernstein, *Victoire et Frustrations, 1914–1929* (Paris: Le Seuil: Collection Nouvelle Histoire de la France Contemporaine, 1990), 72.

[3] Including the use, after the war, of an explosives factory valued at FF 15 million, located at Oissel, near Rouen.

[4] Kuhlmann's capital rose from FF 6 million in 1914 to FF 60 million in 1919. See *Cent ans d'industrie chimique: Les Établissements Kuhlmann (1825–1925)*, (Centenary brochure, 1925), 25. From 1916 to 1924, Kuhlmann increased its capital 10 times. The first issue of capital in 1916 was privately subscribed by Pennaroya. Pennaroya took 5, 4 M francs of the new 12 M francs company (i.e, 45% of the company). See Archives Départementales de Paris. Dossier D31U31565. For Kuhlmann tried to combine all the resources of their large family and investments from corporate companies (mining companies such as Pennaroya and Vieille-Montagne and even banks such as Crédit Commercial de France). This information comes from a compulsory procedure, les 'modifications de statuts' which are kept in a special public archive. Each time a company increased its capital, it must modify its statutes and list its new shareholders. However, this archive presents two flaws: on the one hand, it's a picture of a situation at just one moment: if a shareholder decides to sell his shares in the future, it will not appear. The second limitation of the archive derives from the increasing number of people buying shares during the twenties. In 1918, 800 people or company bought the new shares in Kuhlmann, this number rose to 20,000 in 1928.

[5] *Cent ans d'industrie chimique, op. cit.* note 4, 3.

[6] J.-E. Leger, *Une Grande Entreprise dans la Chimie Française: Kuhlmann, 1825–1982* (Paris: Nouvelles Éditions Debresse, 1982), 27.

7 A. Thépot, 'Frédéric Kuhlmann, Industriel et Notable du Nord, 1803–1881', *Revue du Nord*, 68 (265), (1985), 533. At the close of his career, Kuhlmann was also a professor at the École Centrale de Paris.

8 F. Barbier (ed.), *Le Patronat du Nord sous le Second Empire: une Approche Prosopographique* (Genève: Librairie Droz, 1989), especially, 'Charles-Frédéric Kuhlmann', 257–261.

9 Myriame Provence, Frédéric Kuhlmann, 'Chimiste et Grand Capitaine d'Industrie (1803–1881)', *Généalogie-Magazine*, 126 (1994), 29–34.

10 Myriam Provence 'La Descendance de Frédéric Kuhlmann', *Généalogie-Magazine*, 126 (1994), 35–36.

11 Barbier, *op. cit.* note 8, 'Edouard Donat-Agache', 78–80.

12 Barbier, *op. cit.* note 8, 'Théodore Barrois', 83–85.

13 *Cent ans d'Industrie Chimique, op. cit.* note 4, 22.

14 M. Gillet, *Les Charbonnages du Nord de la France au XIX ème Siècle* (Paris-La haye: Mouton, 1973), 106.

15 Kuhlmann's capital rose from FF6 million in 1914 to FF60 million in 1919. See *Cent Ans d'Industrie Chimique, op. cit.* note 4, 25.

16 Léger, *op.cit.* note 6, 44–45.

17 See *Annuaires des Anciens Elèves de l'Ecole Polytechnique et de l'Ecole Centrale*, 1913–1925.

18 *Cent ans d'industrie chimique, op. cit.* note 4, 22.

19 G. Hardach, 'Industrial Mobilization in 1914–1918: Production, Planning and Ideology', in Patrick Fridenson (ed.), *The French Home Front, 1914–1918* (Providence: Berg, 1992), 62.

20 John F. Godfrey, *Capitalism at War: Industrial Policy and Bureaucracy in France,1914–1918* (Leamington Spa: Berg, 1987), 186–187.

21 Art. L. Tissier, 'Les Poudres et les Explosifs', in J. Gerard (ed.), *Chimie et Industrie: Dix Ans d'Efforts Industriels* (Paris: Chimie et Industrie, 1926), 1344.

22 In 1916, the Service des Poudres signed contracts with American firms for a daily production of sixteen tons of chlorobenzine. See L. Tissier 'Les Poudres et Explosifs', in Gerard (ed.), *op. cit.* note 21, 1332 and 'Le Service des Poudres pendant la Guerre de 1914–1918', *Revue historique de l'armée*, 3 (1958), 140.

23 E. Vinet: 'La Guerre des Gaz et les Travaux des Services Chimiques Français', *Chimie et Industrie*, 2 (11–12), (1919), 1387.

24 See M. J. Rust, 'Business and Politics in the Third Republic: The Comité des Forges and the French Steel Industry, 1896–1914' (Unpublished PhD dissertation, Princeton University, 1973).

25 Patrick Fridenson, 'Un tournant taylorien dans la société française', *Annales ESC*, 42, (1987), 1931–1960; A. Moutet, *Les Logiques de l'Entreprise. La Rationalisation dans l'Industrie Française dans l'Entre-deux-guerres* (Paris : Editions de l'EHESS, 1997).

26 'Le Service des Poudres pendant la Guerre de 1914–1918', *op. cit.* note 22, 137.

27 Tissier, *op. cit.* note 21, 1341.

28 Service Historique de l'Armée de Terre (SHAT), Vincennes, 10 N 185, (Commission of Inquiry into State Contracts, Chemical Products, File on sulphuric acid, 65–66°. Contract between the State and Kuhlmann on 24 October 1915). Établissements Kuhlmann declared that it had taken out an option to acquire 16.5 hectares of land situated on the seafront at Port-de-Bouc. The cost of setting up operations was estimated at FF6 million.

29 Clearly, Saint-Gobain was intent on crushing its neighbour by charging lower prices and blocking off any potential growth for that factory', Godfrey, *op. cit.* note 20, 161.

30 'Cent Ans d'Industrie Chimique: les Etablissements Kuhlmann, 1825–1825', *op.cit.* note 4, 32.

31 *Ibid.*, 30–31.

32 *Ibid.*, 32–33.

33 Quoted in Vinet, *op. cit.* note 23, 1377.

34 Georges Weiss was then Directeur des Mines, Director of Mines at the Ministry of Public Works.

35 See Olivier Lepick, *La Grande Guerre Chimique, 1914–1918* (Paris: Presses Universitaires Françaises, 1998). For the French riposte to the initial German gas attacks, see 105–110.

36 'Directors of the company [the Compagnie de Lens] who were originally from Lille like the managing director, Elie Reumaux, and the deputy managing director, Ernest Cuvellette (who became

part of the Kuhlmann family in 1919), were all committed to policies aimed at adding value to coal; had not Cuvellette himself said that if he had his way, not one tonne of coal would leave Lens unrefined?' Gillet, *op. cit.* note 14, 106.

[37] SHAT, 16 N 904. Cuvelette, in a report to Colonel Lenfant, director of Materiel Chimique de Guerre, (Wartime Chemical Material), concerning measures necessary to implement a programme to manufacture chlorine, dated 10 July 1915, notes that the Motte-Breuil factory in the Oise département and the Volta factory in Moutiers were the only two in a position to produce chlorine. But, the first was 'under sequestre' by the Service des Poudres, and could not be bought, so the Volta factory being enlarged.

[38] Tissier, *op. cit.* note 21, 1335: 'Carbon tetrachloride had many uses, from the manufacture of chloroform to the preparation of mustard gas, from use in fire extinguishers to the cleansing of shell cases.... Before the war, it had started to be used as a solvent for extracting fatty substances or flower essence...and from its factory in Estaque, the Société Kuhlmann was able to deliver an average of 1–1.5 tonnes of carbon tetrachloride per day through the processing of chlorine by the Weldon method. Deliveries increased after Kuhlmann extended its premises. In all, the company supplied 1030 tonnes of carbon tetrachloride.'

[39] Quoted in E.Vinet, *op. cit.* note 23, 1442.

[40] SHAT, 10 N 185. Commission of Inquiry into State Contracts, Chemical Products, 28 October 1919. Supporting evidence, no. 2312, contract awarded to the Établissements Kuhlmann involving the supply of sulphur deprotochloride. The reply to the Commission's question regarding the invitation to competing firms to tender for this contract reads as follows: 'Sulphur protochloride is to be used in the production of mustard gas; there is no case for inviting competing offers, since manufacture of this substance is already undertaken by the Établissements Kuhlmann. Other factories are due to commence producing this substance for use in the manufacture of mustard gas.' See also note contract number 2313, agreement signed with Établissements Kuhlmann regarding the supply of carbon tetrachloride. There was no invitation to tender. This means that Kuhlmann had the monopoly of production of sulphur protochloride. (Whether mustard gas could have been made without sulphur protochloride is another question).

[41] Vinet *et al., op. cit.* note 23, 1442. Theoretical research continued in the IEEC and the laboratories of the Société des Usines Chimiques du Rhône.

[42] Vinet, *op. cit.* note 23, 1413, writes: 'Thanks to the impetus provided personally by M. Loucheur, Armaments and War Materials Minister, as well as by the two officers responsible for organizing production — Lieutenant J. Frossard from the Armaments Ministry and Lieutenant de Kap-Herr from the industrial techniques section...the French army was able to use mustard gas as a vital weapon in its efforts to put a final stop to the German offensive of spring 1918.' Louis Loucheur, was a Polytechnician, businessman, and head of the Société Générale d'Entreprise, became an important shell and gas manufacturer in 1915, and replaced Albert Thomas as Ministre de l'Armement Armaments in December 1916. See S. D. Carls, *Louis Loucheur (1872–1931): Ingénieur, Homme d'État, Modernisateur de la France* (Paris: Presses Universitaires Françaises, 2000).

[43] L.F. Haber, *The Chemical Industry, International Growth and Technological Change, 1900–1930* (Oxford: Clarendon Press, 1971), 157.

[44] AN F 12 7710. Historical Record of the Production of Monochloracetic Acid by Camus et Duchemin et Compagnie at Ivry-sur-Seine, 11 October 1915; AN F 12 7710. Contract signed on 14 August 1915.

[45] Synthetic indigo was produced by the Compagnie de Couleurs d'Aniline in Creil, a subsidiary of Hoechst in France before the war, see *Hoechst in Frankreich, 1881–1914. Zur Geschichte des Werkes Creil. La Compagine Parisienne des Couleurs d'Aniline* (Frankfurt am Main: Farbwerke Hoechst-AG, 1990), 97.

[46] Ecole Supérieure de Physique et de Chimie Industrielle, Fonds Albin Haller, carton 32, H 32–78, Eugène Kuhlmann to Albin Haller, received on 7 April 1915.

[47] Charles Moureu, *La Chimie et la Guerre: Science et Avenir* (Paris: Masson, 1920), 173.

[48] M.R. Fox, *Dye-Makers of Great Britain, 1856–1976: A History of Chemists, Compagnies, Products and Changes* (Manchester: Imperial Chemical Industries, 1987), 86: 'The new company had an authorized share capital of £2 million and a debenture loan from the Government of £1.5 million at 4% interest.'

49 Before and during the war, Cochin was also on the board of Compagnie of Saint-Gobain. See Saint-Gobain Archives, Blois, Minutes of Board Meetings, 1915 and following years.

50 AN F 12 7708 (Ministère du Commerce). Notes on the Commission des Matières Colorantes, first meeting, 30 April 1916. Two versions of these notes exist — one handwritten, the other, typed. The latter version systematically avoids mention of the companies whence the Commission's members came.

51 Gillet, *op. cit.* note 14, 106.

52 Two public offers were made. The first was cancelled because of a suspicion that there was speculation by Germans acting through neutral shareholders. The second was issued in December1916, with the stipulation that only French citizens could be shareholders.

53 AN F 12 7708, Bill for the Ratification of the Contract concluded 11 September 1916, between the Ministre de la Guerre and the Syndicat National des Matières Colorantes, 18: 'The 40 million subscribed are divided between 12,500 holders: among them, 12,000 hold 62.5 per cent of the capital shares. They individualy owned less 20 shares, i.e 10,000 Francs)'.

54 *Ibid.*, Article 7.

55 *Ibid.*, Art. 8.

56 *Ibid.*, Art. 11.

57 *Ibid.*

58 Godfrey, *op. cit.* note 20, 169.

59 *Ibid.* In October 1916, the board of directors of the new Société Nationale des Matières Colorantes were named. The President was to be René Masse, president of the professional association of the French gas industry. Other directors included Donat-Agache of Kuhlmann and seven other representatives of smaller chemical companies that consumed their products, such as dyeworks. The others membrers were Alfred Bozon (Banque de l'Union Parisienne), Gaston Griolet (President of the board of directors of the Banque de Paris et des Pays-Bas), three metallurgists, Théodore Laurent (Forges et Aciéries de la Marine et d'Homécourt), Ernest Métivier (Schneider), Jean Neyret (President of the Société des Aciéries et Forges de Firminy) Gaston Guiot (minister plenipotentiary), and Raphaël-Georges Lévy (a member of the Institut) were also named.

60 AN C 7559 B. Cahier 2252. Hearings of Ribot, Clémentel and Albert Thomas, 16th February 1917 at the Commission du budget et des marchés réunies.

61 AN F 12 7708, Cotelle letter to Minister of Trade, 11 April 1917. Cotelle letter denounced 'la constitution d'un corps de vingt chimistes choisis par concours autant par l'État que par la société pouvant dans ces conditions être des représentants exclusifs des intérêts sociaux...l'obligation pour la société de produire pour l'injonction du Ministère du Commerce des marchandises que la société, laissée à son initiative, pourrait être amenée à ne pas fabriquer..., le contrôle de l'État, auquel une société comme la sienne ne pouvait être amenée à se soumettre'.

62 AN F 12 7708. Cotelle to Ministre du Commerce, 11 April 1917. Sénator Boucher, manufacturer of Gerardmer (Vosges), Turpin (a merchant in Rouen), and Léo Vignon (chemistry professor in Lyon) joined the board in 1915. Saint-Gobain held no Saint-Denis shares until 1924.

63 From a F 10 million capitalisation in 1881, the Société de Saint-Denis fell to F 9 million in 1889, F7 million in 1901, and F 4.375 million in 1903. See Saint-Gobain Archives, Blois, juridic files, Files 281 and 489.

64 BNP-Paribas Archives, Paris, SNMC, box 474/8, Minutes of the meeting held on 17 October 1916 at the Banque de Paris et des Pays-Bas, during the course of which Griolet, Lévy, Guiot and Alfred Bozon were nominated.

65 Emile Lambert's Société Lambert, Rivière et Cie was responsible for marketing and selling Kuhlmann group chemical products. Edouard Kuhlmann was a member of the board of Établissements Kuhlmann in 1873, and president in 1897, and a member of the board of Compagnie d'Anzin. See Gillet, *op. cit.* note 14, 99.

66 The three compagnies had common administrateurs.

67 'In an agreement signed by the War Ministry, represented by Mauclère (deputy managing director of the Service des Poudres), starting on 1 November 1915, Schneider et Cie was to supply the Ministry with 90 tonnes of benzene and 30 tonnes of toluene recovered from its benzol recovery plant in Creusot. Once this initial delivery was complete, further consignments of at least 28 tonnes a month were to be

made until the Armistice was signed. This market became so important that, by July 1916 at the latest, a technical gunpowder and explosives section had been set up in Schneider et Cie's head office.' Quoted from Agnès D'Angio, *Schneider et Compagnie et la Naissance de l'Ingénierie: Des Pratiques Internes à l'Aventure Internationale, 1836–1949* (Paris: CNRS Éditions, 2000), 144–145.

[68] SHAT, 10 N 183. Report on contracts for nitrochloric benzene concluded by the Service des Poudres, report for the Commission of Inquiry set up by the Chamber of Deputies.

[69] E. Bussière, 'Paribas, l'Europe et le Monde' (unpublished manuscript, 43). Gaston Griolet was the head of the Banque de Paris et des Pays-Bas and member of the board of the Norvégienne de l'Azote (Norsk Hydro). The Banque de Paris et des Paris-Bas assisted French State negotiations for avoiding BASF to take control of the norwegian company.

[70] Despite the acquisition of equipment in the USA, the CNMC did not produce dyes during the war. *Cent ans d'industrie chimique, op. cit.* note 4, p. 48

[71] Or a total of 13.5 per cent. But one has to notice two things: the extreme dispersal of the capital and the fact that the board of Saint-Gobain, even with the alliance of their cousins, held only 3 to 4 per cent of the capital. See J.-P. Daviet, *Un Destin International: la Compagnie de Saint-Gobain (1830–1939)*, (Paris: Editions des Archives Contemporaines, 1988), 648.

[72] *Cent ans d'industrie chimique, op. cit.* note 4, 45.

[73] Banque de Paris et des Pays-Bas Archives, box 474/8, SNMC, Minutes of the General Meeting of Établissements Kuhlmann, 16 June (probably 1924, as there is mention of the coming merger with CNMC).

[74] According to the Saint-Gobain Archives, 'regarding the 1916 merger, Pennaroya received 33 per cent of the capital of the new Kuhlmann company, which expanded from 6 million francs (1200 shares of 500 francs) to 25 million (10000 shares of 250 francs), with four members of Pennaroya being appointed to the board'. Daviet, *op. cit.* note 71, 568. We also know that in 1923, whereas the capital of Kuhlmann was F 100 million (400,000 shares of F 250), the Société de Pennaroya held 48,118 shares, or 12 per cent, and the Compagnie Royale Asturienne had 25,998 shares, or 6.5 per cent. See Archives of the Banque de Paris et des Pays-Bas, SNMC, carton 474/8, Minutes de l'Assemblée Générale extraordinaire, 27 December 1923.

[75] J. Bariéty, *Les Relations Franco-allemandes après la Première Guerre Mondiale, 10 Novembre 1918–10 Janvier 1925, de l'Exécution à la Négociation* (Paris: Publications de la Sorbonne, 1977).

[76] Daviet, *op. cit.* note 71, 485 ff.

[77] *Ibid.*, 495.

[78] Becker and Bernstein, *op. cit.* note 2, 222.

[79] D. Woronoff, *Histoire de l'Industrie en France, du XVI ème siècle à nos jours* (Paris: Editions du Seuil, 1998), 372.

[80] Interview with M. Levy-Leboyer, 'Des Héritiers aux Patrons', in l'Histoire (ed.), *Puissances et Faiblesses de la France Industrielle, XIX–XXème Siècles* (Paris: Le Seuil, 1997), 365.

# ORGANIZING FOR TOTAL WAR: DUPONT AND SMOKELESS POWDER IN WORLD WAR I

The war taught us that America can organize, train, and transport troops of a superior sort at a rate which leaves far behind any programme for the manufacture of munitions. It upset the previous opinion that adequate military preparedness is largely a question of trained man power...[We had] no plan for the equally important and equally necessary mobilization of industry and production of munitions, which proved to be the most difficult phase of the actual preparation for war. The experience of 1917 and 1918 was a lesson in the time it takes to determine types, create designs, provide facilities, and establish manufacture. These years will forever stand as the monument to the American genius of workshop and factory, which in this period insured the victory by insuring the timely arrival of the overwhelming force of America's resources in the form of American munitions.
— Benedict Crowell, Assistant Secretary of War, Director of Munitions, 1919.[1]

We have all heard much of the rapid strides made by various American industries during the War, notably shipbuilding and steel manufacture...but there has been little said or written about the explosives industry. The average citizen does not realize that the duPont Company was no more prepared to meet the demands that they (sic) later met than the shipbuilders...and not as well prepared perhaps as the steel manufacturers.... The way in which the organization was enlarged, new plants built, methods of manufacture improved and production increased, constitutes one of the greatest achievements in American industrial history. Incidentally, it might be added that the Company also found time to put their (sic) peace time activities on a more substantial basis and to develop the manufacture of dye stuffs to such an extent to make this Country largely independent of Germany.
— DuPont Company Report, 1919.[2]

## INTRODUCTION

It seems quite remarkable that given the industrial muscle of the belligerent countries in the First World War that the course that the war took was completely unanticipated in spite of years of military planning. No one had incorporated the fact that industrialization would alter the scale and intensity of modern war. Modern industrial technology, methods of organization, and newly acquired innovative capabilities allowed prodigious quantities of explosive energy to be expended against unprecedented numbers. War itself came to resemble a far flung industrial operation involving the movement and deployment of vast resources, human and physical. Fighting the war required the nations of Europe to mobilize these resources and to coordinate their efforts with allies. As the scale of warfare escalated over the first three years, United States corporations sold the Allies massive quantities of munitions. However, when the United States declared war on Germany in April 1917, very little planning had been done with regard to sending a large American army to France. In fact, it was not until the autumn of 1917 that the Wilson Administration realized that this would be necessary. At that point plans were made to have a two million man American army in France for an offensive in the spring of 1919. To support that effort, US production of munitions had to be at least doubled — a feat of which American industry proved itself capable.

## THE AMERICAN MUNITIONS MACHINE

To keep themselves supplied with war *matériel*, the Allies — and later the United States — depended heavily on a few large American firms. Three firms accounted for one-third of the $3 billion of purchases that J.P. Morgan of New York orchestrated for the Allies. This in part reflected Morgan's preference for dealing with large reliable firms, but also recognized the ability of large firms to handle large orders.

The leading suppliers of the Allies were DuPont ($425 million), Bethlehem Steel ($298 million), and American Smelting and Refining ($242 million).[3] The US Army Ordnance Department estimated that Dupont had 'supplied over 40 per cent of the standard explosives used on the Allied side, plus millions of pounds of specialized explosives, caps and fuses'.[4] Bethlehem produced 3570 finished guns, over 65 million pounds of gun forgings, 70 million pounds of armour plate, and 1.1 billion pounds of shell steel and made or assembled 20 million rounds of artillery shells. In terms of total American war production, this represented 60 per cent of finished guns, 65 per cent of gun forgings, and 40 per cent of finished ammunition.[5] American Smelting and Refining was the largest producer of copper during the war. Without the output of these three firms, the Allies could probably not have sustained their war effort beyond 1915.[6]

In the United States, the capacity to organize and carry out large scale enterprises was developed in the private sector, initially to build a continental network of railroads after the Civil War. These capabilities, which developed in response to the unprecedented complexities of railroad finance and management, soon spread into other corporations that emerged after the railroads had created a national market.[7] Large companies — such as Carnegie Steel and Standard Oil — built very large plants to lower production costs, owned raw materials, and invested in transportation networks. To oversee and coordinate these diverse functions, corporations built many-tiered managerial hierarchies. According to business historian Alfred

D. Chandler, Jr., it was these organizational capabilities that gave large firms their competitive advantage. These firms became proficient at planning and carrying out expensive and complex projects, in addition to managing normal operations.

Again beginning with the railroads, corporations began to hire increasing numbers of professionally-trained scientists and engineers who at first sought to rationalize processes and products. Soon after the turn of the twentieth century, corporations gave some of these technical laboratories a separate box on the organization chart, thus creating the modern industrial research laboratory. By 1914 many companies had established laboratories, notably General Electric, DuPont, American Telephone and Telegraph, and Kodak. In addition to routine monitoring of process efficiency and product consistency, these laboratories also took on the job of assimilating new technologies developed elsewhere, and made modest attempts at innovation.[8] Generally, large corporations with their manufacturing, engineering, and research capabilities formed the basic units of the American munitions machine.

Once the United States entered the war, organizing and directing production by the government and the military presented many new challenges. However, America was not completely unprepared for modern warfare. During the 1880s, the United States had decided to build a modern steel navy in response to the emerging naval arms race in Europe. Bethlehem and Midvale steel companies became major contractors for armour plate, guns, and shells.[9] For the new propellant — smokeless powder — and high explosives, the navy worked closely with DuPont, a gunpowder and dynamite manufacturer.[10] These American firms became proficient enough to join international cartels organized by their counterparts in France, Germany, and Britain. In spite of the increasing naval competition between these countries, their respective munitions companies formed international cartels that divided up world markets and shared profits. An important component of this cooperation was the exchange of technology between firms, thus eliminating any potential strategic advantages of one country over another. These contacts helped American firms catch up to European practice and kept them abreast of new innovations. According to W.J. Reader,

> When Great Britain went to war with Germany,....the explosive makers — British, American, and German alike — were taken completely by surprise. As late as 7 July they were discussing amongst themselves a proposal for a joint undertaking in America, and there is no hint in the documents of any discord or hanging back. When the war did break out, the instinct of the British and Americans (we have no evidence from the German side) was to deplore it and play down its importance.[11]

None of the firms at that point realized what a bonanza the war would be for them.

After the opposing armies dug into their well-fortified trenches, conventional offensive tactics produced little more than unprecedented numbers of casualties. Machine guns provided enough firepower to make 'no man's land' an appropriate name for the terrain between the armies. Military leaders hoped that massive artillery barrages could weaken enemy defensives enough to make infantry advances possible. At the Battle of the Somme in 1916, the British fired 4 million shells at the German lines in seven days. By the end of the war, Allied artillery was lobbing an average of forty thousand shells per day. This was accompanied by twenty five million small arms rounds fired daily from rifles and machine guns.[12]

The wide variety of munitions used in the war reflected the application of systematic investigation of explosive energy. The two major variables in characterizing explosives were ease of detonation and rate of combustion. High explosive shells, which caused more casualties than any other weapon in the war, contained six types of explosives. First, there is the primer, which is very sensitive and explodes when struck by the firing pin. The primer charge then sets off the igniter, which is usually old-fashioned black powder that ignites the large smokeless powder propellant charge, sending the shell out the gun barrel. Upon impact, the fuse ignites a highly sensitive detonator charge which sets off the booster charge, often an explosive called tetryl, which in turn detonates the bursting charge, the main component of which was TNT.[13] By 1914, American companies either made these explosives — DuPont began producing TNT as a commercial explosive several years before the war — or else had the technical capabilities to make them.[14]

Massive production of explosives depended on utilizing America's resources with a degree of efficiency uncharacteristic of the resource-abundant nation. America's huge steel industry used enormous quantities of coke made by boiling out the volatile compounds of coal. Until the war not much of the ammonia and aromatic organic compounds was recovered. When production of explosives was hindered by limited supplies of raw materials, especially toluene for TNT, American firms began to capture these by-products. US production of toluene increased by an order of magnitude from the beginning of the war to April 1917. American war plans called for a tripling of that capacity by 1919. Because US steel capacity doubled during the war, there would be plenty of toluene to meet that goal.[15]

When the industrial production of TNT fell below battlefield demand, innovative explosives were substituted. More plentiful ammonium nitrate was mixed with TNT to produce a high explosive called 'amatol'. Entirely new explosives were developed from nitrating other organic materials such as xylenes and starch.[16] In this case the military relied on the expertise of industrial chemists to find substitutes for scarce resources.

The largest munitions project of the war was the increase in smokeless powder capacity by a factor of 100 between 1914 and 1919.[17] At the beginning of the war, only DuPont and the US military had the capability to produce smokeless powder. During the war, it proved difficult to transfer that knowledge to other organizations. But more important than technical capability was the ability to organize very large-scale production.[18] The overwhelming responsibility fell on DuPont, which supplied five-sixths of America's powder, and one-quarter of its high explosives.[19] Overall, DuPont produced 1.4 billion pounds (640,000 metric tons) of smokeless powder which sold for about $850 million, accounting for 80 per cent of the company's total sales.[20]

Although the manufacture of smokeless powder involved chemical processes, they were far from the large scale continuous processes used to make the nitric and sulphuric acid reagents. Its principal raw material — short cotton fibres — remained in solid or plastic form throughout the manufacturing process. Thus, the efficiencies associated with the ease of moving and processing gases and liquids were unavailable. Batch sizes were quite small — only two pounds of cotton was nitrated in each pot — and consolidated 40-pound blocks were manually moved from one operation to another.[21]

In spite of the massive scale-up of total production, costs did not reflect the significant economies of scale found in chemical manufacture. Indeed, production costs appear to have declined only by about 20 percent during the war.[22] Early in the war, as DuPont was anticipating the need to expand its capacity, the head of its smokeless powder division, H. Fletcher Brown, observed that, 'The manufacture of smokeless powder is a difficult manufacture. It consists of a large number of operations which have to be performed with great exactness....'[23] The process included about two dozen operations, many of which involved presses the size of ordinary machine tools. Although the production process involved relatively simple operations, control had to be quite sophisticated to produce a safe and standardized powder.

Several decades of development were necessary to make smokeless powder a satisfactory military explosive. It had the potential to be a more powerful and clean burning propellant than black powder. Experiments indicated that over half of the gunpowder used in a rifle or cannon did not combust. It remained in solid form, fouling and eroding gun barrels while also creating large clouds of smoke. Using black powder in big guns produced hundreds of pounds of particle smoke on each firing. Large scale ordnance, made possible by the development of high quality open hearth steel, thus required a new kind of propellant. The major problems with smokeless powder were in developing a form that would burn at a rate to produce moderate and sustained gas pressure, and in stabilizing it so that it would not deteriorate and explode.[24]

The smokeless powder manufacturing process was shaped by these criteria. In the first step, several pound batches of cleaned and purified cotton linters were dipped into pots of mixed nitric and sulphuric acids. After about thirty minutes, each cellulose molecule would have added somewhat less than the maximum three nitrate groups, giving it the specified 12.6 to 12.8 per cent by weight of nitrogen. To accomplish this, however, the cotton absorbs about ten times its weight in acid. All of this unreacted acid has to be removed from the cotton. To accomplish this task requires centrifuging the cotton fibres, and then boiling them for forty hours with at least four changes of water. Additional pulping, washing, neutralizing, centrifuging and pressing produces purified guncotton (nitrocellulose, containing a small amount of water). This material retains the fibrous structure of cotton and burns so explosively that it is unsuitable as a propellant. To convert fibrous guncotton into plastic smokeless powder, first the water is displaced by ethyl alcohol in a high pressure press, and then the guncotton is mixed with an equal weight of ether for about forty-five minutes. [25]

At this point a stabilizing chemical, usually diphenylamine, is added. This chemical performs the critical task of scavenging any acid-forming gases released from the nitrocellulose.[26] DuPont had begun to buy this stabilizer from Germany in 1908. With the onset of the war, DuPont chemists developed an improved process to make this chemical, which was similar in structure to dyestuffs' intermediates. The company's success with diphenylamine encouraged DuPont to consider going into the dyestuffs field.[27]

In the smokeless powder process, the ethanol-ether mixture turns the nitrocellulose into a plastic material that is extruded through a fine-meshed screen called a 'macaroni press' to remove any insoluble material. The extruded bead is then compacted into forty-pound blocks. The smokeless powder is then once again extruded, this time into the final ammunition shape, usually in the form of a perforated cylinder. The final major and tedious step is the removal of the solvents from the powder. This has to be done slowly, or else the smokeless powder charge may develop cracks. Much of the solvent is recovered in a system in which the powder is placed on mesh screens, and solvent vapors are condensed by refrigeration. Maximum solvent recovery took eight to ten days. At this point the powder was still ten to fifteen per cent solvent, which was removed by air-drying. In 1914, this process took anywhere from two weeks to nine months, depending on the size of the piece. Drying was considered to be the major bottleneck in the production of smokeless powder and alternative processes were developed. One involved putting the powder on trays loaded on trucks that were driven through wind tunnels. A more effective approach turned out to be immersing the powder in warm water, which displaced the solvents. The water was removed by air-drying. With the exception of this important drying step, very little innovation appears to have occurred in the process. In fact, it remained essentially the same during the Second World War. The only significant change was an increase in the size of the cotton dipping charges from two to thirty-two pounds.

Before 1914, no one at DuPont imagined that this difficult-to-manufacture product would be the agent that transformed the company into a large diversified chemical company. In fact, the company management had stopped investing in smokeless powder plants, and had begun to find other outlets for its existing nitrocellulose capacity.[28] The major reason for this was continuing discontent in Congress about having the US military dependent upon the 'Powder Trust'. Governmental relations had already been strained by an antitrust suit that in 1913 led to the spin-off of two new dynamite and black powder companies. Ironically, the military interceded with the court on DuPont's behalf, asking that the company be permitted to keep all three of its smokeless powder plants instead of giving one to each company. The US military, through decades of working closely with DuPont, had come to respect the company's abilities in production and research. When the war began, it soon became clear that DuPont had developed a capability that turned out to be indispensable, initially for the Allies, and eventually for the United States as well.

On 26 October 1914, H. Fletcher Brown informed the DuPont Executive Committee that:

> Orders for guncotton and for smokeless powder have been coming in so rapidly and the amount of additional business in sight is so large that rapid changes in plans for new construction have been necessary during the past two weeks.[29]

The first big order — eight million pounds of cannon powder for France — was nearly equal to the company's annual capacity. Pierre du Pont, however, saw war orders as more a risk than an opportunity. Everyone expected a short war, similar to the Franco-Prussian conflict of 1870. The cessation of hostilities would inevitably lead to abrupt cancellation of munitions contracts, as had happened to DuPont in the Spanish-American War. Pierre decided that to reduce DuPont's financial risk, the capital to build new plants would have to be paid for by the Allies, and that the plants would have to be amortized immediately. Thus, when the war ended, DuPont would have no worthless investment on its books.

Before the war, the company had been selling smokeless powder to the government for 53 cents per pound.[30] DuPont's actual costs are not a matter of record, and are not stated even in Pierre's voluminous financial calculations. Apparently, this was a top-secret number, never to be divulged. It would become a matter of considerable political controversy when the United States entered the war. Pierre decided that the cost to the Allies should be $1 per pound, nearly double the pre-war price. And DuPont wanted half the money 'up front'. Whatever the French, British, Russians, and Italians thought of Pierre's pricing, they apparently had little choice but to accept DuPont's terms. By 1 May 1915, the company had contracts for more than 100 million pounds of powder, as well as orders for a variety of other explosives, the most important of which was TNT.[31] To meet the demand for powder, DuPont made major additions to its three existing plants and built a massive new guncotton plant in Hopewell, Virginia.

It was at Hopewell that the DuPont engineering organization, under Major William G. Ramsay, demonstrated its remarkable abilities in planning and construction. The plant was to produce one million pounds of guncotton per day; the final processing into smokeless power would be done at the existing facilities. Ground was broken on 30 October 1914, and the first unit began operation three months later. Because of limits on the size of production equipment, the plant was built as a series of independent units, eventually fifteen in all. The entire plant was finished sixteen months later on 1 June 1916. At the peak of construction, 25,000 worked with the contents of 425 railroad cars of *matériel* delivered every day. In addition to the plant itself, DuPont turned the formerly forested area into a small city, building houses and apartments for 1850 families and quarters for unmarried men. Unfortunately, much of the town had to be rebuilt after a disastrous fire in June 1915. The plant and town used enormous quantities of water — seventy million gallons of filtered water per day — enough to supply the city of Boston. In the power plant, ninety-three steam engines burned ninety railroad car loads of coal per day. These statistics were compiled to publicize what DuPont considered to be an achievement of near pharaonic proportions.[32]

As the money began to pour into DuPont, even the normally meticulous Pierre could not accurately determine the profit figures. In 1914, DuPont sales and profits had been $25 million and $5.6 million dollars, respectively. The next year brought sales of $131 million, and after amortization, profits of $57.1 million.[33] After the British in November 1915 ordered an unprecedented sixty-five million pounds of cannon powder — much of it destined for the Somme offensive the following summer, in which four million shells were fired in seven days — the next seven months brought only twenty million pounds of additional contracts. Of course, during these seven months occurred the massive confrontations at Verdun and the Somme.[34] Pierre began to think that one of these offensives might succeed and end the war.[35] Then, in the wake of horrendous losses with little to show for them, he thought a peace settlement might follow. But these hopes were dashed when the French ordered fifty million pounds of cannon powder in late July and the British contracted for an astounding 181 million pounds in September. Obviously, neither country had lost its commitment to continuing the conflict.

Not including these giant requests, DuPont also garnered about $120 million in extra payments that paid for the Hopewell plant and capacity increases in the other plants.[36] DuPont did not treat this money as profit, but used it to write off the value of the plants. Without this immediate amortization charge, the company's astonishing performance becomes even more incredible: for 1915 and 1916 combined, on sales of $450 million, the gross profit was approximately $250 million.

By mid-1916, having amortized his plants, the ever-rational Pierre now was willing to lower his prices beginning with the large French and British cannon powder orders. He cut the price of cannon powder in half to 50c per pound, while rifle powder declined to 65c per pound.[37] This resulted in savings of several hundred million dollars for the Allies and the United States. Politically, it was probably not possible for DuPont to continue to charge such high prices while publishing its sales and profit figures in the annual report.

As relations between the United States and Germany deteriorated, DuPont began to consider how to increase its smokeless powder capacity if America declared war. In February 1917, company employees began scouting for sites that were at least 200 miles from the coast — a condition stipulated by the War Department — with a good and abundant water supply, adequate waste disposal, and a good labour market.[38] In March, in an unsolicited report, DuPont informed the War Department and the Navy that the company was fully contracted to supply the Allies, and with its current plants could not supply large amounts of powder to the United States. Soon after the declaration of war on 6 April, the Chief of Army Ordnance, General William Crozier, informed DuPont that he expected the Army's powder needs over the next year to be 78 million pounds, which was only about one quarter as much as was being made for the Allies. DuPont believed that this modest amount could be supplied without major new construction.[39]

For nearly six months after declaring war, the role of the American government was limited primarily to cheerleading for the Allies. It was not until the autumn of 1917 that the urgent need for American *matériel* and men became apparent to the Wilson administration. The Allies wanted one million American men by July 1918, and another million by the following spring. The Allies also expressed an immediate demand for large quantities of propellants, high explosives, and shells.[40]

In response, the Ordnance Department contacted DuPont in early October, and solicited a bid for a new smokeless powder plant capable of producing either a half million or a million pounds per day. Within three weeks, DuPont submitted a proposal for a million pound per day plant, with the first unit scheduled to come on stream in July, and with project completion by April 1919. The Ordnance Department accepted the proposal, and a contract was signed by General Crozier, who passed it to Secretary of War Newton Baker for ratification. The fact that DuPont had already profited so much from the Allies, and stood to so much more from building and operating the new plant, troubled Baker, who referred the contract to the Explosives Committee of the War Industries Board, one of many ad hoc organizations formed to oversee American mobilization.[41]

Several members of this committee were critical of DuPont's profiting from America's war effort. What Pierre had asked for — and had received — in the original contract, was $13.5 million in construction commission for building a $90 million government-owned plant. In addition, DuPont would receive 5c per pound for every pound of powder delivered. And if the cost of production fell below 44.5c per pound, DuPont would keep half of the savings. One member of the committee, St. Louis businessman Robert Brookings, believed that DuPont's actual production cost was about 35c per pound. If this were accurate, the company would earn an additional $30 million on the first government production order for 450 million pounds of powder. On the advice of the War Industries Board, Newton Baker cancelled the contract.[42]

This matter was not resolved for three months. Whether it had an effect on the war effort is unclear. Apparently, Baker was willing to wait to get better terms. He offered DuPont $1 million for building the plant, and deferred consideration of production-based payments. DuPont countered with an offer of $9 million for construction and 15 per cent of production value. (In the original contract the latter figure would have been equivalent to a production cost of 40c per pound.) When negotiations reached an acrimonious impasse, Baker tried to find someone else to build a plant. He invited the Thompson-Starrett construction company, one of the nation's largest, which reluctantly agreed to submit a bid. In mid-December, Baker asked Daniel C. Jackling, head of American Smelting and Refining, one of America's largest corporations, and the major supplier of copper to the Allies, to take on the task.

Jackling began conversations with Thompson-Starrett, but also sought help from DuPont because his expertise was in copper, not smokeless powder. Jackling got Thompson-Starrett to agree to build a half million pound per day plant at a site to be called Nitro, near Charleston, West Virginia. He also suggested to Baker that DuPont be asked to provide the other half million pounds from a proposed plant called 'Old Hickory', in honour of Andrew Jackson's nearby home outside Nashville, Tennessee. Perhaps now feeling pressure, DuPont finally acceded to an agreement, whereby Pierre surrendered all the construction commission; however, a production payment of 3.5c per pound was included, as was the cost reduction incentive. Soon, the size of the plant was increased to 900,000 pounds per day.[43]

When the contract was finally signed, DuPont's representative in Washington cabled to Pierre, 'Contract signed by all concerned.... Turn loose the wolves'.[44] With a sense of wounded pride, the DuPont Company, especially its engineering division, set out to show what it could do. In many ways, the 'Old Hickory' plant represented a repeat of the Hopewell project, although the new plant would produce finished powder, and not just guncotton. Powder was shipped from the first unit on 2 July 1918, the month stipulated in the original October contract. To celebrate this feat, employees were given a holiday on 4 July. The plant was over 90 per cent complete when the war ended on 11 November. DuPont far surpassed Thompson-Starrett in West Virginia.[45]

To document the company's heroic achievement, DuPont published a pamphlet, '"Old Hickory": The World's Greatest Powder Plant', The Building of Which Broke All-Speed Records for Wartime Construction'.[46] It begins with the modest claim that, 'The story of Old Hickory is a story to be written in superlatives'. In this and other reports, DuPont piled superlative upon superlative. Old Hickory was the largest construction project of the First World War. The plant included the largest nitric acid, sulphuric acid, and refrigeration plants in the world.[47] The powerhouse would provide 60 million pounds of steam per day by burning 5000 tons of coal, or two full train-loads. The maximum work force was 40,000, but nearly 250,000 worked on the site during its construction. Up to 20,000 workers daily commuted eighteen miles by train from Nashville. The blue prints for the plant covered 780,000 square feet, or 18 acres. The plant was built from the contents of 30,000 boxcars. A village to house 30,000 people was constructed, which included a worker-administered outdoor amphitheater to provide nightly entertainment for 6000 people. It is ironic that the purpose of this grand undertaking was to raise to a new level the mayhem of trench warfare. Fortunately, the war ended before the new plants doubled America's capacity for destruction.

CONCLUSION

After only nineteen months offically at war, America was still mobilizing when the Armistice came. Nevertheless, the United States managed to send enough men and *matériel* to tip the balance. Several lessons would be incorporated into the country's response to the threat of war two decades later. As 'Old Hickory' project demonstrated, government and industry had found a way to cooperate with a degree of mutual understanding. Government had come to recognize that corporations had organizational and technical skills that could not be easily duplicated. Industry had been willing to set aside the immediate interests of stockholders, in the service of the nation.

DuPont emerged from the war with a full treasury and the self-confidence to use that money to become the nation's, if not the world's, leading chemical company. The war forced DuPont to begin a process of diversification, if only to maintain a much larger scale of operations. At the time, the company's massive effort to become a dyestuffs producer — before the war almost all of the dyes used in America were made in Germany — seemed the most important new initiative. However, the company's future direction and success were founded upon smokeless powder. A nylon plant has much more in common with a smokeless powder facility than a dyestuffs plant. To capitalize on its smokeless powder know-how, DuPont began to invest in closely related cellulose technologies — including celluloid plastics, lacquers, photographic film, rayon fibres, and cellophane film. In the 1920s, a research programme to explore the nature of cellulose and other polymers established the company as a pioneer in synthetic polymers, and led DuPont into markets that it continues to enjoy today.[48]

## NOTES

[1] Benedict Crowell, *America's Munitions, 1917–1918* (Washington, DC: Government Printing Office, 1919), 17–18.

[2] Hagley Museum and Library (Wilmington, Delaware), Accession 1793, 'Production of Ordnance Material by the DuPont Company during the European War', 14.

[3] Mark Reutter, *Sparrows Point: Making Steel* (New York: Summit Books, 1988), 124–125.

[4] William Bradford Williams, *Munitions Manufacture in the Philadelphia Ordnance District* (Philadelphia: A. Pomerantz and Company, 1921), 374.

[5] *Ibid.*, 311–312.

[6] It was not only America's immense industrial base that led the Allies to depend on a neutral nation an ocean away. The United States also had favorable access to raw materials. The Allies needed huge quantities of Chilean nitrates that were a key ingredient in all explosives. Fortunately, the opening of the Panama Canal in 1914 greatly facilitated shipments to the East Coast. Earlier shipments that had to make the long journey around the tip of South America could go to Europe as easily as to America. (Williams Haynes, *American Chemical Industry* (New York: D.Von Noststrand, 1954), vol. 2, 63–65.). The United States itself was also a source of critical materials that were unavailable or in short supply in Europe. American cotton was combined with nitric acid, made from Chilean nitrates, and sulphuric acid, made from Texas sulphur, to produce smokeless powder, the propellant charge for small arms and big guns. (*Ibid.*, vol.3, 29–33). US copper production, which had grown quite large from supplying the burgeoning electrical industry, was also critical to the Allied war effort. In 1916, three-quarters of the copper used in British munitions factories came from America. Generally, it was a few large firms — American Smelting and Refining, in the case of copper — that supplied most of the munitions purchased by the Allies. (Reutter, *op. cit.* note 3, 124–125.)

[7] Alfred D. Chandler, Jr., *The Visible Hand: The Managerial Revolution in American Business* (Cambridge, MA: Harvard University Press, 1977), Part I.

[8] John Kenly Smith, Jr., 'The Scientific Tradition in American Industrial Research', *Technology and Culture*, 31 (1), (1990), 123–131.

[9] Thomas J. Misa, *A Nation of Steel: The Making of Modern America, 1865–1925* (Baltimore: Johns Hopkins, 1995), ch. 3.

[10] David A. Hounshell and John Kenly Smith, Jr., *Science and Corporate Strategy: DuPont R & D, 1902–1980* (New York: Cambridge University Press, 1988), ch. 1.

[11] W.J. Reader, *Imperial Chemical Industries: A History* (London: Oxford University Press, 1970), vol. I 214–215.
[12] Benedict Crowell and Robert Forrest Wilson, *The Armies of Industry: I. Our Nation's Manufacture of Munitions for a World In Arms, 1917–1918* (New Haven: Yale University Press, 1921), 27–34.
[13] R. Norris Shreve, *The Chemical Process Industries* (New York: McGraw Hill, 1956), 453–454.
[14] 'Production of Ordnance Material by the DuPont Company during the European War', *op. cit.* note 2.
[15] Crowell and Wilson, *op. cit.* note 12, 157–161.
[16] Arthur Pine Van Gelder and Hugo Schlatter, *History of the Explosives Industry in the United States* (New York: Columbia University Press, 1927), 945–954.
[17] *Ibid.*, 164–166.
[18] Haynes, *op. cit.* note 6, vol. 3, 196–200.
[19] *Ibid.*, vol. 2, 464.
[20] These calculations were made from the list of all DuPont munitions contracts in Hagley Museum and Library, LMSS 10A, Box 383.
[21] On the smokeless powder process, see Shreve, *op. cit.* note 13, ch. 22 and Edward C. Worden, *Nitrocellulose Industry* (New York: Van Nostrand, 1911), vol. 2, ch. 18.
[22] At the beginning of the war DuPont was selling powder to the government for 53c per pound. The Old Hickory contract had a target price of 44.5c per pound.
[23] Hagley Museum and Library, LMSS 10A, Box 383, Brown to the Executive Committee, 26 October 1914.
[24] Worden, *op. cit.* note 21, 898–901.
[25] *Ibid.*, 902–912 and Shreve, *op. cit.* note 13, ch. 22.
[26] Shreve, *op. cit.* note 13, 455–6.
[27] Hounshell and Smith, *op. cit.* note 10, 41, 48, 76–77.
[28] Alfred D. Chandler, Jr. and Stephen Salsbury, *Pierre S. du Pont and the Making of the Modern Corporation* (New York: Harper & Row, 1971), 364.
[29] Brown to the Executive Committee, *op. cit.* note 23.
[30] Chandler and Salsbury, *op. cit.*, note 28, 367–370.
[31] Hagley Museum and Library, LMSS 10A, Box 383.
[32] On Hopewell, see Hagley Museum and Library, Accession 1662, Box 29, Ramsay to du Pont, 5 February 1916; and , Hagley Museum and Library, Accession 1793, Pierce to Patterson, 'Hopewell Works', 18 January 1917. See also other material in Accession 1793 concerning the fire.
[33] Chandler and Salsbury, *op. cit.* note 28, 370.
[34] Hagley Museum and Library, LMSS 10A, Box 383. The shell figures are in Crowell, *op. cit.* note 1, 37.
[35] Chandler and Salsbury, *op. cit.* note 28, 380.
[36] Hagley Museum and Library, LMSS 10A, Box 383.
[37] *Ibid.*
[38] Hagley Museum and Library, Accession 1793, 'Old Hickory'.
[39] Chandler and Salsbury, *op. cit.* note 28, 400–402.
[40] Crowell, *op. cit.* note 1, 14–15.
[41] The War Industries Board was established by President Wilson in 1916 and given broad powers to mobilize the American economy for war. Headed by financier Bernard Baruch, the Board generally used persuasion rather than coercion to achieve its goals. (Haynes, *op. cit.* note 6, 42–52.)
[42] Chandler and Salsbury, *op. cit.* note 28, 403–409.
[43] *Ibid.*, 412–425.
[44] *Ibid.*, 423.
[45] See various accounts in Hagley Museum and Library, Accession 1793.
[46] *Ibid.*
[47] Hagley Museum and Library, Accession 1793, 'Old Hickory'.
[48] For DuPont's development, see Hounshell and Smith, *op. cit.* note 10.

# SCIENCE AND THE MILITARY: THE KAISER WILHELM FOUNDATION FOR MILITARY-TECHNICAL SCIENCE*

INTRODUCTION

Since antiquity, and especially since the scientific revolution of the 17th century, scientists in Europe have been closely involved with the application of technical knowledge to military purposes. However, with the early twentieth century, this collaboration took on a new dimension. Whereas earlier, it generally fell to the military to apply empirical knowledge offered by science, now, scientists sought military applications for their discoveries, and entered into new institutional arrangements with the military.

In Germany, the history of this interaction remains largely unexplored.[1] The first institutionalized ties between science and the military in Germany can be traced to the First World War. The years between 1900 and 1914 saw a frantic arms buildup between the major powers. After Germany's ambitious naval programme was de facto scuttled in 1911, attention turned to the Prussian army. Training able-bodied men was the first and foremost goal of the chief of the General Staff, Helmuth Graf von Moltke.[2] In his view, traditional personnel-intensive warfare was to feature in the future, as in the past, and the role of matériel-and technology-oriented armaments would remain on a modest scale. The imperial government's technical and economic preparations for war were also limited.[3] A military-technical academy to train engineering officers was established in Potsdam in 1903, but only a few officers attended, and these were limited to the transportation branch, the corps of engineers, and the technical institutes of the army.[4]

However, the Prussian army did have two agencies responsible for procurement and testing – the Inspektion der technischen Institute der Infanterie (Inspectorate for the Technical Institutes of the Infantry) and the Inspektion der technischen Institute der Artillerie (Inspectorate for the Technological Institutes of the Artillery). The former encompassed the munitions factory and the Infanterie-Konstruktionsbüro (Infantry Construction Bureau) in Spandau, together with the arms factories in Danzig, Spandau and Erfurt. The latter supervised the Artillerie-Konstruktionsbüro (Artillery Construction Bureau, created in 1894) and the cannon foundry in Spandau; the artillery workshops in Danzig, Spandau, Lippstadt, and Straßburg; the shell

---

*This is a substantially revised and shortened version, without the documentary appendix, of my paper, 'Wissenschaft und Militär: Die Kaiser Wilhelm Stiftung für kriegstechnische Wissenschaft', *Militärgeschichtliche Mitteilungen*, 44 (1991), 73–120. I thank the editors for the excellent translation and for focusing the text.

factories and fuse laboratories in Spandau and Siegburg; and the state gunpowder factories in Hanau and Spandau. Aside from this, the Prussian Feldzeugmeisterei (Quartermaster's Office) had in Berlin a civilian-run Militärversuchsamt (Military Testing Office) founded in 1890 as an explosives testing office. The states of Bavaria and Saxony owned similar facilities, but Württemberg did not. The Imperial Navy possessed a few technical facilities, under the Inspektion der Marineartillerie and the Inspektion des Torpedo- und Minenwesens (Inspectorates of Naval Artillery and of Torpedoes and Mines).[5]

To date, there have been no comprehensive studies of these technical facilities or their influence upon the development of weapons. It seems, however, that their task was not so much scientific research and development, as weapon-specific testing and approval of equipment manufactured by the private- and state-owned arms industry. On the other hand, the private arms suppliers, such as Fried. Krupp, had not only scientific laboratories, but also ranges for testing weapons and munitions. Despite the importance of military technology, the scientific facilities of the army enjoyed no special encouragement. The army depended on the scientific and productive capabilities of private industry.[6]

Moreover, neither the military nor the government attempted to coordinate and direct economic preparations for war. As the need for industrialization became apparent, these shortcomings quickly came to light. In August 1914, Wichard von Moellendorff and Walther Rathenau drew attention to the lack of economic preparations, and attempted an adhoc solution by the creation of a Kriegs-Rohstoff-Abteilung (War Raw Materials Department, in the War Ministry).[7] No comparable solution presented itself for lack of technological war preparation. Since the General Staff called for manpower to offset the numerical advantage of the enemy, nearly all able-bodied men, including many scientists, were drafted. No one considered whether their scientific qualifications could be better utilized. At the same time, many scientists saw it as completely natural that they should serve the Fatherland in the trenches with weapons in hand, showing their allegiance to the Reich (as set out in the famous October manifesto of the 93 intellectuals, 'An die Kulturwelt' [To the Civilized World]).[8] In consequence, universities and research laboratories were quickly deserted, even though the army soon needed technical solutions. Such tasks were assigned according to personal contacts, without coordination with the armaments industry.[9]

Some scientists responded to these tasks with almost unlimited willingness. Pacifists like Albert Einstein were the exception. Some senior researchers, such as Fritz Haber and Emil Fischer, began organizing their research to meet wartime needs. For the first time, the military, industrialists, and scientists worked together on problems of the war economy, creating such new organizations as the Kommission zur Beschaffung von Kokereiprodukten (Commission for the Procurement of Coking Products), the Salpeterkommission (Nitrate Commission), the Nährstoffausschuß (Foodstuffs Committee), the Kriegsausschuß für Ersatzfutter (War Committee for Substitute Fodder), the Kriegsausschuß für pflanzliche und tierische Öle und Fette (War Committee for Vegetable and Animal Oils and Fats), and many others.[10]

Although many of these measures were initiated in the opening months of the war, it took some time to grasp that modern warfare had entered a new dimension. The personnel-intensive warfare of the first two years gave way to the substitution of machines for men. From the summer of 1916 on, this change found its expression in the Hindenburg programme, the Auxiliary Service Law, and the establishment of the War Office. With this transition to matériel-intensive, technological warfare came the first systematic militarization of science in Germany. With this, it became necessary to coordinate military and contract research. Even before Hindenburg and Ludendorff were appointed to the Oberste Heeresleitung (supreme command) of the army on 28 August 1916, there were plans for the systematic application of science to armaments. In this process, the Kaiser Wilhelm Foundation for Military-Technical Science played a leading role.[11]

### ORIGIN OF THE KAISER WILHELM STIFTUNG FÜR KRIEGSTECHNISCHE WISSENSCHAFT (KAISER WILHELM FOUNDATION FOR MILITARY-TECHNICAL SCIENCE)

During the second year of the war, the government undertook greater efforts to involve scientists and engineers in the solution of military problems. These approaches were not initiated by the military, however, but rather by civilians, who had seen the technological deficiencies in German mobilization.[12] A suggestion by Albrecht Schmidt, who led the Laboratory for Patent Affairs of the Farbwerke (Dye Works) in Höchst, was evidently an important factor in the establishment of the new Kaiser Wilhelm Foundation for Military-Technical Science (KWKW).[13]

In late 1915 or early 1916, Schmidt, a cousin of Friedrich Schmidt-Ott,[14] ministerial director and later Prussian Kultusminister (Minister of Education), contacted Fritz Haber,[15] whose Kaiser Wilhelm Institute (henceforth: KWI) for Physical Chemistry and Electrochemistry had from late 1914 been converted into a centre for chemical warfare research.[16] Schmidt proposed the creation of a foundation with an endowment of 250,000 marks 'to reward persons who have rendered a scientific or technical service to the war effort and who "can use it".'[17] Schmidt envisaged that Captain Haber would manage the foundation and propose suitable individuals for honours. The documents do not show whether Schmidt was thinking of himself, but he did belong to the circle of possible candidates. He had developed a tear gas, as well as a method to produce artificial fog (so-called 'Höchst fog'), which was used in the Battle of Jutland on 31 May 1916.

The immediate impetus for Schmidt's suggestion might have been the government's reluctance to award military medals and distinctions for service in the arms industry. Indeed, it was not until 1917 that Schmidt was honoured for his military contributions, by the granting of the title of professor and a few medals. It seems that even in later years, Schmidt never realized that his suggestion had led to the establishment of the KWKW. Of course, this is not surprising, as his original proposal for monetary awards was overtaken within a year.[18]

In response to Schmidt's proposal, Haber consulted the Prussian Minister of War, Adolf Wild von Hohenborn, as well as Schmidt-Ott, who was responsible for scientific foundations in the Kultusministerium (Ministry of Education). Haber considered that neither he, nor his Institute, but only the Kaiser-Wilhelm-Gesellschaft zur Förderung der Wissenschaften (KWG = Kaiser Wilhelm Society for the Advancement of the Sciences)[19] was qualified 'to accept such an endowment and to award [honors] with the necessary authority, if indeed any should do so other than the Kaiser or the Minister of Education'.[20] In a letter to Schmidt-Ott,[21] Haber developed Schmidt's original blueprint, since he, too, believed that the Iron Cross was too sparingly distributed for noncombatant services. He proposed that an 'artistically designed, embossed iron medal', together with a monetary award, should be bestowed upon deserving civilians as well as military personnel for 'important' means of warfare, either scientific or technical. To forestall self-nomination, and to facilitate objective appraisal, Haber drew up a list of criteria for selection. To ensure the requisite expertise, both the War Ministry and the Ministry of Education were to participate. However, in view of the rivalry between the two ministries, Haber suggested putting the Foundation under the KWG.[22]

Schmidt-Ott evidently approved of the idea to honour military technical services, but opposed giving the task to the KWG, on grounds of military secrecy. Concern about the KWG's discretion seems premature, as von Harnack had not yet delivered his famous critique of wartime 'profiteering,' which was to produce long-lasting animosity in business circles, especially among munitions manufacturers.[23] Besides, the KWG could have created a special commission, whose members were sworn to secrecy. Schmidt-Ott did not advance the more pointed criticism that the KWG lacked experts to evaluate military innovations and inventions. As a rule, the directors of the KWIs were scientists and not technical people or engineers.[24]

What Schmidt-Ott really wanted to achieve was a modified 'Schmidt-Haber' proposal to set up regular faculty positions for engineers at the Königlich Preußischen Akademie der Wissenschaften zu Berlin (Royal Prussian Academy of Sciences in Berlin). At the turn of the century, Friedrich Althoff, the 'uncrowned king of Prussian science policy',[25] had tried in vain to establish an academy of technical sciences in Prussia and to secure membership for engineers and technical scientists in the Berlin Academy. Similar efforts in Bonn, Göttingen, and Heidelberg also failed.[26] In 1916, Schmidt-Ott hoped that Wilhelm II could expand the Berlin Academy by creating three positions for 'technical members', who would then create a 'technical commission' for awarding honours.[27] However, following talks with Haber, Schmidt-Ott dropped this plan. The Academy was predominantly humanities-oriented, and unwilling to acknowledge the technical disciplines as suitably academic activities.

Haber must have informed him of the resistance he would meet. In any case, it seemed unwise to establish new positions through the Kaiser. The earlier expansion of the Academy by three seats for the directors of the first KWIs in Berlin-Dahlem had been seen as an encroachment upon the Academy's autonomy. In 1911, this generated a great deal of animosity against the KWG. The Academy's own historian and member Harnack, then president of the KWG, was unexpectedly not elected

Permanent Secretary of the Academy. This time, resistance from Academy members was likely to endanger the whole project.[28]

On 18 May 1916, following a meeting of the Military Commission of the Prussian Academy, Haber and Schmidt developed an idea to both honour innovation and create incentives. As both believed that science and technology would gain increasing importance in the preparation and conduct of war, they wanted not just a temporary wartime facility, but a permanent organization. However, they did not wish to found a new research institute, but instead to promote military-technological research at existing institutes under the authority of the Prussian Ministry of Education.[29]

One possible model was the KWI for Physical Research, proposed in 1914 by Fritz Haber, Walther Nernst, Max Planck, Heinrich Rubens and Emil Warburg. This was intended to facilitate collaboration among physicists at German universities and to put materials at their disposal for large-scale research projects.[30] Comparable British endeavours had been reported in the British weekly, Nature, and in the *Chemiker-Zeitung*. On 3 June 1916, fourteen days after the conversation between Haber and Schmidt-Ott, an article by the chemist Dr.-Ing. Bruno Wäser appeared on the front page of the *Chemiker-Zeitung*, entitled 'Über die Schaffung einer Zentralstelle für technische und wissenschaftliche Forschung (On the Creation of a Central Agency for Technical and Scientific Research).'[31] Although this essay could not have inspired the earlier discussion,[32] it did refer to much earlier British endeavours probably known to Haber and possibly also to Schmidt-Ott. Wäser also took up the ideas of a Berlin patent lawyer, Alard du Bois-Reymond, which in turn derived from comments by the English journalist and writer, H. G. Wells. Given the Allied blockade, Wäser assumed

> that Germany and the powers economically close to and affiliated with it will find themselves converted to some degree of economic autarky after the war; [hence] an organization for the cultivation of technological and scientific research can only be beneficial.
>
> Such a central agency, either publicly or privately funded, could be created on the model of the Imperial Patent Office. In the style of the war committees, it would be staffed with sworn, technically trained officials and divided into subcommittees for giving out information, raw materials usage, and ideas for inventions. The patent office could also serve as a model in regard to the careful and extended registration process involved. After the central agency is fully developed, it would consist of a number of expert committees, perhaps for architecture, civil engineering, mechanical engineering, electrical engineering, chemistry, metallurgy, textiles, aviation, general sciences, physics, objects of everyday use, etc. Individual classes would have to be subordinated to this scheme, somewhat like patent classes.

Since the question of financing would be critical to its founding an agency, Wäser proposed that a 'combination of the great trade associations could themselves create such a central agency.'[33]

Alard du Bois-Reymond had been less definite in his proposals, which in 1915 he published in a slim booklet entitled, *Vom Deutschen Michel*.[34] A patent lawyer, familiar with the problems of converting inventions into technology, he was skeptical of Wells' proposals for a 'Ministry of Inventions'.

> Led with discretion, it can help develop something worthwhile. But it cannot produce surprises of decisive importance in this war, or even modifications of tactics. What lasting benefits it may create will unfold only afterward, in peacetime, and will then be just as much use to us as to the English, assuming that in this war we do not unlearn how to learn from our neighbors.[35]

Du Bois-Reymond neither made concrete suggestions like Wäser, nor did he name Wells as his source. It seems unlikely that his essay was even known to Haber or Schmidt-Ott, but Haber could have seen the same report in *Nature* as du Bois-Reymond on British scientific developments which, despite the war, still found its way into German libraries. In a surprisingly open manner, *Nature* had reported on plans in Britain. One could read how, in 1915, ideas that had been tossed around since the beginning of the war were being consolidated into a centralized guidance of science, presaging the establishment of a new Department of Scientific and Industrial Research (DSIR) in 1916.[36]

During the first months of the war, the British government – unlike the German – had gone to unprecedented lengths to encourage collaboration between science and industry.[37] New facilities and research committees were brought into existence, and a host of incentive programmes and organizations were created. Nevertheless, as *Nature* pointed out – as late as the autumn of 1915, there were still many scientists 'whose energies and expert knowledge are not being effectively used.'[38]

Following initiatives by the Royal Society and the Chemical Society, but also owing much to ideas circulating in the Ministry of Education since 1909, the DSIR became a model for corresponding institutions in other Anglo-Saxon countries. An initial memorandum dates from May 1915, another from June 1915, and a third – on the conception, responsibilities, and financing of the new organization – from July 1915. The DSIR was to support basic scientific and industrial research, coordinate research programmes, and intensify cooperation between science and industry. It was established with a fund, The Imperial Trust for the Encouragement of Scientific and Industrial Research, appropriated by Parliament for a five-year period in the uncommonly large amount of £1 million.[39] Surviving records in Germany do not show to what extent Haber or Schmidt-Ott were aware of these considerations. The establishment of the DSIR, and the later establishment of the National Research Council (NRC) in the USA, as well as complementary organisations in Canada, South Africa, India and Australia,[40] show that there were efforts to intensify research in all industrial nations, and that Germany did not hold a dominant position in the promotion of science for the purposes of war.

With respect to accommodating and financing their new research organization, Haber and Schmidt-Ott looked away from the KWI for Physical Research, and followed, probably without realizing it, British example. In the case of the KWI, there was a building to serve as a lecture hall, library, and archive building, but actual experiments were performed 'in the laboratories of the researchers involved'. Its budget of 75,000 marks for individual projects had to come in equal shares from the KWG, the Koppel Foundation, and the government of Prussia.[41] However, the construction of a building and the financing of military-technical research through the KWG and the government were not feasible. Rather, Schmidt's proposal to establish a foundation seemed the ideal form of organization.

In order to support a large number of projects, a foundation would need much more capital than Schmidt's proposed endowment of 250,000 marks. At an interest rate of 5%, this would yield only 12,500 marks per year. However, since neither the government nor the KWG would add significantly to an endowment, money would have to be found another way. A time-consuming public fund drive was out of the question.[42] In the end, Schmidt-Ott and Haber must have hit upon the idea not to solicit anonymous donors known to Schmidt, with total pledges of 250,000 marks, but rather to ask the Jewish banker and industrialist, Leopold Koppel. Koppel had endowed, among other things, Haber's KWI for Physical Chemistry and Electrochemistry.

Very little is known about Koppel,[43] but he certainly did not do badly by the war.[44] The inventions of Haber's Institute (gas masks, filters, etc.) took industrial form through his Gesellschaft für Verwertung chemischer Produkte [Corporation for the Utilization of Chemical Products]. Finding Haber's proposal interesting, Koppel was at once ready 'to offer the Kaiser and the Army a facility endowed with two million [marks of] war loans.'[45] Haber did not fully inform Koppel about his preliminary conversations with Schmidt-Ott and Schmidt. Instead, he gave the impression that they had just discussed problems stemming from the shift of Haber's Institute from pure research to chemical warfare, so that Koppel could later claim the idea of founding a military-technical foundation as his own.[46]

Within just three weeks – between the end of May and mid-June 1916 – further talks between Koppel, Haber, Nernst, Emil Fischer, and Schmidt-Ott led to the drafting of a set of statutes for a 'Kaiser Wilhelm Stiftung für Naturwissenschaft und Technik im Heere (Kaiser Wilhelm Foundation for Science and Engineering in the Army)'. The main problem for Haber and Schmidt-Ott was that of guaranteeing 'freedom for science'. Transferring a scientist from a facility supported by the Ministry of Education to a military-run institute would be seen as restricting that freedom. Given his experience with the military administration of his institute since the beginning of 1916, Haber had reason to reject the military model, and to advocate instead the establishment of permanent commissions in which science and the military faced each other as equals. For this reason, the authors of the statutes sought to involve the military by representation, while protecting the autonomy of science. That is, the military would not be allowed to influence the scientific direction of the organization, and would be restricted to purely representative duties.

To accomplish this, the draft statutes proposed two managing bodies, the 'Kuratorium (Board)' and the somewhat disparagingly-entitled 'Verwaltung (Administration)'. The Minister of War was to be Chairman of the Board, which would contain an officer of the General Staff as managing director as well as six officers and 'scholars'. According to the statutes, the Board had the duty 'to carry out' tasks 'which serve to fulfill the goals of the Foundation'.[47] This meant ceremonial and official functions, since all 'business matters' (i.e., the approval of individual projects) would be entrusted to another body, the 'Administration'. Within the Administration, a voting majority of scientists was ensured, as it consisted of one member each from the Berlin Academy and the Leopold Koppel Foundation, together with the President of the Kaiser Wilhelm Gesellschaft. The Board's managing director, an officer, was only 'attached' to this body.

The actual work of the Foundation would proceed through so-called 'Senaten' (senates) or Fachausschüssen (expert committees), chaired by the appropriate scientific members of the board. The right to nominate scientific members of the Board belonged to the Berlin Academy, which in any case often advised the Ministry of Education on scientific matters. Senate chairmen could choose the civilian members of their senates, as they thought fit, and there was no limit on the number. The military agencies assigned the military members, whose participation required the approval of the respective chairmen.

The division of the Foundation into board, administration, and expert committees theoretically protected the autonomy of science, since the military had a majority only on the board. Postwar revision of the statutes shows that the split into two managing bodies was designed solely to reduce the military's influence on the Foundation's scientific work. After the exclusion of the military from the Foundation in 1920, these statutes called for only one managing body (the board).[48] However, this structure raised fears that the War Ministry would withhold its endorsement. Koppel was particularly worried, and sent a note to the Chief of the Civil Cabinet, Rudolf von Valentini, asking for his opinion before taking further steps.[49] Probably, he had the intention of influencing the Minister by obtaining a preemptive positive statement from the Kaiser, accepting the Foundation.[50]

Schmidt-Ott informed von Valentini about certain details of the plan, and mentioned the lingering question of whether the War Ministry would approve it. With the sentence, 'it is necessary that His Majesty reign with full benignity over the enterprise in order to assure the cheerful cooperation of the members of the scientific commissions,'[51] Schmidt-Ott sought to ensure that the Kaiser had prior knowledge. This, however, von Valentini rejected, since it seemed to him more than doubtful 'that the draft statutes would receive the full approval of the military administration'. He recognized that 'the focus of the work and the significance naturally lies in the six expert committees, whose chairmanships are reserved to the scientific members of the Board, and whose members are appointed without consulting the military'. Von Valentini therefore suggested that 'Perhaps one could find a way to involve the military Board members in putting together the expert committees, so as to avoid even a semblance of disrespect towards the participating officers'.[52]

Despite von Valentini's doubts, on 4 July 1916, Koppel took the proposition to the Minister of War, Adolf Wild von Hohenborn, who in turn informed the Kaiser.

Only three weeks later, Koppel was told that approval awaited only the formal endorsement of the agencies involved.[53] This unusually quick and unexpectedly positive response from the Minister of War can probably be attributed to the talks that took place between Haber and the Minister in the spring of 1916, and to the changing military situation on the Western Front. Technology and science were becoming increasingly important to the war effort.

The complete overhaul of the War Ministry that took place in the summer of 1916 delayed discussion of the new foundation's statutes until the end of October. However, contrary to expectations, the military made minimal changes. The new organisation's name was changed to the Kaiser Wilhelm Stiftung für kriegstechnische Wissenschaft, or the 'Kaiser Wilhelm Foundation for Military [literally: War]-Technical Science', since the Navy also wanted to make use of it. The initials 'KWKW' were adopted to abbreviate the awkward title. The Board was to have three additional military members, which led Haber and Schmidt-Ott to insist that the Kaiser also add a member of his choice. They hoped the Kaiser's supplementary delegate, the Civil Cabinet Chief von Valentini, would give them a sufficient counterweight to the numerical majority of the military.[54]

Following changes dealing with secrecy oaths and voting procedures, the Academy held its first discussion of the statutes of the KWKW on 26 October 1916. This also marked the first time that Haber took part in a session of the full Academy (although he had been a member for almost five years). He may have feared that the project would be rejected. But it was not. After talks with the Ministry of Education, the Academy, the KWG, the Koppel Foundation, and the War Ministry on 28 October, and followed by a final edition of the statutes by the Academy on 13 November 1916, Koppel sent his proposal to the new Minister of War, Hermann von Stein. Seven days later, the Ministry of War forwarded the request to the Kaiser, who granted his royal license on 17 December 1916.[55]

In January 1917, the Ministers of War and of Education countersigned the license. This was repeated in late summer, using the date 24 January 1917, since both the Imperial Office of the Interior and the Ministry of Justice insisted on countersigning. Even before these bureaucratic hurdles were overcome, the Berlin Academy had indicated its willingness to collaborate. Its secretary, the classical philologist Herman Diels, was elected the KWKW's administrative director, and its administrative office was established in the rooms of the Academy at 38, Unter den Linden.[56] Its scientific board members – i.e., the chairs of the expert committees – were elected by the Academy, and received the Kaiser's confirmation.

Reflecting the original name, the 'Kaiser Wilhelm Foundation for Science and Engineering in the Army', the KWKW consisted of three scientific and three technological expert committees. The scientific committees were chaired by Emil Fischer (Chemische Rohstoffe der Munitionserzeugung und für die Betriebsstoffe [Chemical Raw Materials for Munitions Production and Fuels]); Fritz Haber (Chemische Kampfstoffe [Military Chemicals; i.e. gunpowder, explosives, gas warfare agents, etc.]), and Walther Nernst (Physik, umfassend Ballistik, Telephonie, Telegraphie, Ziel- und Entfernungsbestimmung, Meßwesen u. dgl.

[Physics, including ballistics, telephony, telegraphy, target selection and ranging, measurement, etc.]), whereas the three technological senates were chaired by Alois Riedler (Maschinelle und verkehrstechnische Hilfsmittel der Kriegführung [Mechanical and Transport-technological support for warfare]); Heinrich Müller-Breslau (Luftfahrt [Aviation]); and Friedrich Wüst (Metallgewinnung und Metallbearbeitung [Extraction and Processing of Metals]).[57]

The Academy had a hard time selecting the senate chairman for aviation. They eventually chose Professor Heinrich Müller-Breslau, a fellow of the Academy, internationally known for his *Theorie und Berechnung der Eisernen Bogenbrücken* (*Theory and Calculation of Iron Arch Bridges, 1880*), and later famous for his plans for the Berlin Cathedral, but whose limited credentials in aviation related primarily to the construction of Zeppelins. There were other difficulties. The Foundation was, for instance, not entirely disinterested. Its deputy director, Major Klotz, was a son-in-law of Leopold Koppel; and Alois Riedler demonstrated a regrettable lack of scientific objectivity by openly promoting his own combustion turbines through the KWKW.[58]

The selection of senate chairmen, and the biased composition of their expert committees, disadvantaged scientific 'schools' having no connections in Berlin. This was no doubt one reason why Richard Zsigmondy (professor of inorganic chemistry at Göttingen), Friedrich Krüger (professor of physics at Danzig), and Max Bodenstein (professor of physical and electrochemistry at Hannover), addressed an alternative proposal to their professional colleagues in early February 1917. In collaboration with the seventy year-old chemist and Nobel Prize winner, Otto Wallach, the three wished to establish instead an 'academic' Vermittlungsstelle für technisch-wissenschaftliche Untersuchungen (Clearinghouse for Technical and Scientific Research), the purpose of which was to 'accept missions from the army administration and industry, transmitting them to individual [university] institutes for action.'[59]

Another reason grew from the fact that the KWKW did not directly deal with industrial armaments; instead, it took its instructions solely from the Ministry of War.

Zsigmondy, Krüger, and Bodenstein wanted to create an agency, in collaboration with the Deutscher Verband technisch-wissenschaftlicher Vereine (DVT, or German Federation of Technical and Scientific Associations, founded in September 1916),[60] 'which would mediate between industry, in particular those medium and smaller factories poorly equipped with testing facilities, and the institutes of higher learning'.[61]

It is no longer possible to determine who first had the idea for the Clearinghouse for Technical and Scientific Research – whether it was the university professors or the governing board of the DVT, which resolved in 1917 to create such a mediating facility within its business office.[62] An essay by Bruno Wäser, entitled, 'Über die Schaffung einer Zentralstelle für technische und wissenschaftliche Forschung' (On the Creation of a Central Agency for Technical and Scientific Research) (3 June 1916), could also have played a part; as could reports of the initiative of thirty-six

British scientists who published a memorandum in February 1916 leading to the establishment of a 'Neglect of Science Committee', under the leadership of the distinguished physicist, Lord Rayleigh.[63]

Despite – or perhaps because of – agreements with Fischer, Nernst, and Haber, a temporary parallel organization emerged. However, next to nothing is known about the efforts of the Clearinghouse, even though it had 150 professors at its disposal until the autumn of 1917 – a considerably larger number of scientific contacts than the KWKW.[64] Scientific journals mention only two reports prepared by the DVT for the Imperial Treasury, 'the first regarding the taxation of energy, namely of coal, gas, and electricity and their byproducts, the other about the calculation of horsepower for the motor vehicle tax'.[65] There is no surviving evidence of other works facilitated by the Clearinghouse.[66] The situation is somewhat better for the KWKW.

## THE WAR WORK OF THE KWKW

It is nearly impossible to describe fully the KWKW's achievements during the war, as the Heeresarchiv (Army Archive) in Potsdam was destroyed in the Second World War, and the KWKW's archives were lost. Surviving sources afford a rudimentary picture, and a planned report, which the chemist Alfred Stock was to edit, was evidently not printed before the end of the war.[67] What we do have on individual senates, members, and activities, leads us to conclude that no scientist approached by the Foundation ever turned down an appointment to a expert committee, even though the work was unpaid, and each committee had only an 'interim' allocation of 7,500 marks per year, evidently for travel expenses (tests were supposed to be carried out with military resources). But taking part did have its advantages, including the opportunity to retrieve co-workers from military service.[68]

The autonomy of the KWKW, promised by the statutes, was never properly realised. Although the administration was supposed to handle all legal affairs, the Minister of War, as chairman of the board, took over this role. At one stage, Emil Fischer, chairman of Expert Committee One, asserted that the War Ministry had even annexed the KWKW. The senate chairmen responded by organizing an assembly not provided for in the statutes, to protect their independence from the military. Committee chairmen also resisted attempts by industry to put their own men on the senates.[69]

Despite these difficulties, the KWKW began research in the spring of 1917. At first, the Army and Navy, via the War Ministry, transmitted lists of tasks to the individual senates. Later, however, even subordinate military offices directly contacted senate chairmen whom they happened to know, without regard to areas of expertise. From this, and from indications that reports were to go directly to the War Minister, one can infer a great deal of inefficiency. This is also likely to have been true of the military-technical efforts of the individual Bundesstaaten (states) of the Reich. While the senate chairmen asked for assistance from colleagues throughout the Reich, the Prussian War Ministry did not even inform the War Ministries of the other states about the existence of the KWKW until two weeks after its establishment was announced in the Prussian *Armee-Verordnungsblatt* (Army Orders Gazette) on 16 March 1917. Yet, the Prussian War Ministry had begun informing all university scientific and technical chairs in the Reich by late January.[70]

The KWKW's researchers appear not to have achieved spectacular successes. Bureaucratic obstructions may have been partly responsible, and, as du Bois-Reymond had foreseen in 1916, the time was too short for the technical realization of some innovations. Some works later published under its name, including Stephen Löffler and Alois Riedler's 'Verbrennungsturbine und Brennstoffwirtschaft' ('Combustion Turbines and Fuel Management)' and Franz Fischer's 'Bericht über den wissenschaftlichen und technischen Stand der Gewinnung von Tieftemperaturteer aus Steinkohle' ('Report on the Scientific and Technological State of Extraction of Low-Temperature Tar from Bituminous Coal'), were not in fact initiated by the KWKW.[71] However, studies done for the KWKW by Professor Prandtl and Dr. Grammel may well have led to the earliest application of the gyroscope to aviation. Further research may trace other postwar inventions and innovations to the wartime work of the KWKW.[72]

With regard to chemical warfare, the KWKW did not satisfy the Prussian War Ministry. Instead, the Ministry wanted a permanent facility with its own laboratories, to ensure the continuity of research beyond the end of the war, 'so that the present superiority of our armaments will not be lost.'[73] At the beginning of 1917, the War Ministry asked the Ministry of Education about such a facility. This was probably at the suggestion of Haber, whose KWI for Physical Chemistry and Electrochemistry had taken over war gas studies during the war, and who wished to see his institute return to pure research. Asked for his opinion, Haber proposed a new 'KWI for Applied Physical Chemistry and Biochemistry', whose function would be 'not for solving military problems but for treating their scientific foundations, as a research station of applied science.' Such an institute would not be under military control, but should come under the KWG. Haber believed that 'the new institute's activities would benefit science and peacetime industry no less than national defense, its initiator.' Because plans for the new institute were not realized during the war, Haber's own institute retained some of these functions, and thus did not confine itself to pure basic research in the postwar era.[74]

## POSTWAR DEVELOPMENTS AND THE DISSOLUTION OF THE KWKW

It was the wish of its founders that the KWKW would continue after the war ended. At the beginning of 1918, senate chairmen considered future military responsibilities in a transitional or peacetime economy. Predictably, however, the immediate post-war period saw its activities sharply decline. In January 1919, Emil Fischer announced that 'war work naturally has no further point, and...the KWKW would do best to vanish as soon as possible.'[75]

Haber, however, kept alive his interest in the relationship between science, technology, and war, and managed to bring the Foundation back to life. In May

1919, he spoke with Schmidt-Ott about 'what role the Kaiser Wilhelm Foundation for Military-Technical Sciences [sic] is destined to play in the coming decades'. His inspiration came from a British source, the Anniversary Address (30 November 1918) of the President of the Royal Society of London, delivered that year by the physicist, Sir J. J. Thomson. Haber later claimed that by a passage inserted in the annual report of the KWI for Physical Chemistry and Electrochemistry, he also managed to revive Leopold Koppel's interest in the KWKW.[76]

In early January 1920, Koppel asked the Reichswehrministerium (Ministry of National Defense) of the Weimar Republic to approve a comprehensive change in the mission and organization of the KWKW. In light of the strict limitations set by the Treaty of Versailles, Koppel wished to dissolve the KWKW's previous ties to the Ministry of War, and to affiliate it with the Prussian Academy of Science. The statutes were to be revised, so that 'henceforth the mission of the Foundation will be defined as the advancement of the public good through the collaboration of the best scientific powers in the land with the Reich government'. Naturally, the earlier military-oriented work would be used 'for the purposes of peace.'[77]

On 19 January, Defense Minister Gustav Noske announced his agreement with the proposed changes, expressed his thanks for the work of the KWKW, and held out the prospect of future support through the Army Weapons Office (Heereswaffenamt). The Berlin Academy also agreed to the proposed transformation of the Foundation 'into an institute for the advancement of the public good through the collaboration of the best scientific powers', as well as the plan 'to attach the Foundation directly to the Academy of Sciences'; and appointed 'Diels, Haber, Seckel, and Nernst to assist Koppel in drawing up the new statutes.' Subsequently, Professor Müller-Breslau, who had chaired the KWKW's aviation committee, was also made a commission member.

Within two months, Haber had revised the old statutes in accordance with the Treaty of Versailles. The Reichsministerium des Innern [Reich Ministry of the Interior] took the place of the Prussian War Ministry and on 9 March 1920, the 'fathers' of the Foundation—Haber, Schmidt-Ott and Koppel—signed the new statutes. The Foundation changed its name to the Kaiser Wilhelm-Stiftung für technische Wissenschaft (KWTW = Kaiser Wilhelm Foundation for Technical Science). According to its statutes, its purpose was 'to handle patriotic missions of the technical-scientific sort in connection with the administrative offices of the Reich and state government'. Accordingly, its structure was simplified and the six expert committees were given new names. The Committee for Chemical Raw and Process Materials for Munitions Production became the Committee of Organic Chemistry and Biochemistry, and the Committee for Military Chemicals was redesignated the Committee for Physical and Inorganic Chemistry. The Expert Committee for Physics kept its name, although without its earlier subtitle ('including ballistics, telephony, ranging, etc.'). The Committee for Mechanical and Transport-technological Support for Warfare was redefined to Mechanical Engineering, while the Committee for Aviation was expanded to become a Committee for Transportation. The Committee for Extraction and Processing of Metals became the Committee for Metallurgy. The

Academy was granted new rights by the new statutes: it could now nominate the members of individual committees. The Foundation was not, however, formally tied to the Academy.[78]

Appropriately, given the new democratic spirit, the new committee chairmen proposed in 1920 were not appointed. Instead, in the summer of 1922, the Berlin Academy appointed the chairmen, as follows:

Expert Committee for Organic Chemistry and Biochemistry: Wilhelm Schlenk
Expert Committee for Physical and Inorganic Chemistry: Walther Nernst
Expert Committee for Mechanical Engineering: Walter Reichel
Expert Committee for Transportation: Gustav Kemmann

There were a few false starts. For example, the candidate designated for the Expert Committee for Physics, Heinrich Rubens, died before taking office. And the Academy made no decision about the chairman of the Expert Committee for Metallurgy, as it awaited the appointment of a new director at the KWI for Iron Research in Düsseldorf, whose former chief, Friedrich Wüst, had led the corresponding expert committee of the KWKW.[79]

Overall, however, the KWTW's most important problem was its financial insecurity. The new statutes of 1920 provided for the Foundation to apply for subsidies from the Federal and state governments, but provided for no public subsidies. Given its meager finances, the Foundation's activities were severely limited. At one point, it concentrated on glider technology—and for the Rhön Glider Competition of 22 August 1923, set as first prize a bond in the Compagnie Hispano-Americano de Electricidad, worth a few hundred gold marks. But the situation was parlous. Lacking funds, in 1925 the members of the board decided to dissolve the Foundation, and to bequeath its remaining assets – totaling 1,224 RM – to Professor Theodor von Kármán, director of the Aerodynamic Institute of the Aachen College of Technology,[80] for tests on aircraft.[81] That proved to be the last act in the foundation's short life.

The dissolution of the Foundation marked a failure in the struggle to achieve academic status for the technical sciences in Germany. Not until the coming of the Third Reich, in 1935, was there to be a technical academy – and this was Hermann Göring's Deutsche Akademie der Luftfahrtforschung (German Academy of Aviation Research). However, despite its closure, the KWKW did set an important precedent, and in its structure, became a forerunner of today's Deutsche Forschungsgemeinschaft (DFG = German Research Association). Its six expert committees represent a prototype of the DFG'S administrative system.[82] According to Friedrich Schmidt-Ott, the KWKW can also be credited with catalysing plans for the KWI für Eisenforschung (KWI for Iron Research), established in Düsseldorf in 1917.[83]

Moreover, despite its short existence, the KWKW fundamentally changed the military's relationship with science. Although the Prussian War Ministry's attitude cannot be documented, the military's proposal of 1917, as elaborated by Haber, for a 'KWI for Applied Physical Chemistry and Biochemistry,' within the framework of the Kaiser Wilhelm Gesellschaft and under the responsibility of the Ministry of Education, reveals a changed attitude. This signified that the Army was prepared to finance basic research on problems of military technology without insisting upon direct control. During the Weimar Republic, this new KWI could not be realized, and nothing came of plans to establish independent institutes for military research. Nevertheless, the military had come to recognize the usefulness of research for arms development, and it appears that, during the Weimar period, the Heereswaffenamt (Army Weapons Agency) began secretly to finance research at the KWIs, universities, and other facilities during the 1920s and 1930s.[84]

Leaving aside the 'militarized' scientist Fritz Haber, one can conclude that the German scientists in the KWKW responded rather unenthusiastically to the military and its wishes. In some cases with considerable success, they did follow up on various projects begun in the pervasive nationalistic, patriotic spirit of the first half of the war (e.g., the development of chemical warfare agents). Yet as far as one can judge from the limited sources available, their approach to such projects remained highly 'academic' and little oriented to applications. They did not further develop the tank or anti-tank weapons, nor did they invent any other new weapons systems, as Archimedes was said to have done during the siege of Syracuse. And once the First World War was over, with the immediate foreign threat gone, most scientists quickly lost interest in cooperating with the military, especially after the financial incentive completely disappeared.

Since the end of the Cold War, German sources once thought lost have begun to emerge from previously restricted collections. It is to be hoped that subsequent research will further illuminate the relationships, sketched in the foregoing paper, between science, engineering, and the military in Germany during and after the First World War.

NOTES

[1] A useful survey of research is Everett Mendelsohn, Merritt Roe Smith, and Peter Weingart (eds.), *Science, Technology and the Military, Sociology of the Sciences: A Yearbook*, Vol. XII (Dordrecht: Kluwer Academic Publishers, 2 vols, 1988). For the National Socialist period, the first comprehensive treatment of the subject was Karl-Heinz Ludwig, *Technik und Ingenieure im Dritten Reich*, TB-Ausgabe (Düsseldorf: Droste Verlag, 1974). See also Herbert Mehrtens and Steffen Richter (eds.), *Naturwissenschaft, Technik und NS-Ideologie* (Frankfurt: Suhrkamp, 1980).

[2] Stig Förster, *Der doppelte Militarismus: Die deutsche Heeresrüstungspolitik zwischen Status-quo-Sicherung und Aggression 1890–1913* (Stuttgart: F. Steiner Verlag Wiesbaden, 1985); Michael Geyer, *Deutsche Rüstungspolitik 1860–1980* (Frankfurt: Suhrkamp, 1984).

[3] See, e.g., Lothar Burchardt, *Friedenswirtschaft und Kriegsvorsorge. Deutschlands wirtschaftliche Rüstungsbestrebungen vor 1914* (Boppard am Rhein: H. Boldt, 1968).

[4] Hans-Joachim Wefeld, *Ingenieure aus Berlin. 300 Jahre technisches Schulwesen* (Berlin: Haude & Spener, 1988), 288–296; Hans Ebert and Hermann-J. Rupieper, 'Technische Wissenschaft und nationalsozialistische Rüstungspolitik: Die wehrtechnische Fakultät der TH Berlin 1933–1945', in Reinhard Rürup (ed.), *Wissenschaft und Gesellschaft. Beiträge zur Geschichte der Technischen Universität Berlin 1879–1979* (Berlin, Heidelberg, New York: Springer, 1979), Bd. 1, 469.

[5] Edgar Graf von Matuschka, 'Organisationsgeschichte des Heeres 1890 bis 1918', *Deutsche Militärgeschichte in sechs Bänden 1648–1939*, Bd. 3, Abschnitt V (München-Herrsching: M. Pawlak, 1983), pp. 186 f. Reichsarchiv (Hg.), *Der Weltkrieg 1914–1918: Kriegsrüstung und Kriegswirtschaft*, Bd. 1 (Berlin: Mittler, 1930), pp. 391–393.

⁶ Although the Reichsarchiv claimed that 'at the outbreak of the war, the military usage of technology in the German army [was] the most advanced in comparison to all other armies' (*Weltkrieg, op. cit.* note 5, 292), it also admitted, '(o)f course the military often only hesitantly utilized modern technological achievements; mostly there was intensive testing before taking any steps to apply a promising innovation. This hesitant approach—especially in the first decade of the century, in which the technological needs of the army grew rapidly—came partly from a reluctance, conditioned by the nature of the army, to change traditional views, and partly from a certain animosity among the troops against the incursion of a technological element into the art of war, hitherto determined essentially by moral forces [!]' (*ibid.*).

⁷ Lothar Burchardt, 'Walther Rathenau und die Anfänge der deutschen Rohstoffbewirtschaftung im Ersten Weltkrieg', *Tradition*, 15 (1970), 169–196; Burchardt, 'Eine neue Quelle zu den Anfängen der Kriegswirtschaft in Deutschland, 1914', *Tradition*, 16 (1971), 72–92.

⁸ With the loss of the documents of the Prussian War Ministry, one cannot offer more precise details on conscription policies. It should be noted, however, that certain scientists were not drafted or were released after brief service – for example the inventor of coal liquefaction and later Nobel Prize winner, Friedrich Bergius, who wanted to develop his technology for industrial use. See Rasch, 'Zur Vorgeschichte der Kohlenverflüssigung bis 1945', *Energie in der Geschichte*, 11ᵗʰ Symposium of the International Cooperation in History of Technology Committee, Hg. Verein Deutscher Ingenieure (Düsseldorf: VDI, Bereich Technikgeschichte, 1984), 470 f.). Another example was Franz Fischer, director of the Kaiser Wilhelm Institute (henceforth: KWI) for Coal Research, who was assigned by the Ministry of Agriculture to conduct work in nitrogen manufacture. See Rasch, *Geschichte des Kaiser-Wilhelm-Instituts für Kohlenforschung 1913–1943* (Weinheim: VCH, 1989), 65–67). See also Bernhard vom Brocke, 'Wissenschaft und Militarismus: Der Aufruf der 93 "An die Kulturwelt!" und der Zusammenbruch der internationalen Gelehrtenrepublik im Ersten Weltkrieg', in William M. Calder III, Helmut Flashar, and Theodor Lindken, *Wilamowitz nach 50 Jahren* (Darmstadt: Wissenschaftliche Buchgesellschaft, 1985), 649–719; vom Brocke, 'Wissenschaft versus Militarismus: Nicolai, Einstein und die "Biologie des Krieges"', *Annali dell' Instituto storico italo-germanico in Trento* (Bologna: Società editrice il Mulino, 1985), vol. 10, 405–510.

⁹ For examples of the disorganized use of science, see Rasch, *op.cit.* note 8, 72, 63–66.

¹⁰ On Emil Fischer, see Gerald D. Feldman, 'A German Scientist between Illusion and Reality: Emil Fischer, 1909–1919', in Imanuel Geiss and Bernd Jürgen Wendt (eds.), *Deutschland in der Weltpolitik des 19. und 20. Jahrhunderts: Fritz Fischer zum 65. Geburtstag* (Düsseldorf, Bertelsmann Universitätsverlag, 1973), 341–362. The history of these commissions and committees has not yet been seriously studied. These institutions should not be confused with the analogous war corporations, for example, for chemicals or leather (Kriegschemikalien AG, Kriegsleder AG). On these see Alfred Müller, *Die Kriegsrohstoffbewirtschaftung 1914–1918 im Dienst des deutschen Monopolkapitals* (Berlin [DDR]: Akademie-Verlag, 1955); Regina Roth, *Staat und Wirtschaft im Ersten Weltkrieg. Kriegsgesellschaften als kriegswirtschaftliche Steuerungsinstrumente* (Berlin: Duncker & Humblot, 1997).

¹¹ L. F. Haber, *The Chemical Industry 1900–1930: International Growth and Technological Change* (Oxford: Clarendon Presss, 1971), 223; Günter Wendel, *Die Kaiser-Wilhelm-Gesellschaft 1911–1914. Zur Anatomie einer imperialistischen Forschungsgesellschaft* (Berlin [DDR]: Akademie-Verlag, 1975), Document 128, 357 f.; Lothar Burchardt, 'Standespolitik, Sachverstand und Gemeinwohl: Technisch-wissenschaftliche Gemeinschaftsarbeit 1890 bis 1918', in Karl-Heinz Ludwig, Wolfgang König (eds.), *Technik, Ingenieure und Gesellschaft. Geschichte des VDI 1856–1981* (Düsseldorf: VDI-Verlag, 1981), 211; Burchardt, 'Die Förderung schulischer Ausbildung und wissenschaftlicher Forschung durch deutsche Unternehmen bis 1918', in Hans Pohl (*ed.*), *Wirtschaft, Schule und Universität. Die Förderung schulischer Ausbildung und wissenschaftlicher Forschung durch deutsche Unternehmen seit dem 19. Jahrhundert* (published as *Zeitschrift für Unternehmensgeschichte*, 29 (1983), 34 f.; Maria Osietzki, *Wissenschaftsorganisation und Restauration* (Cologne, Vienna: Böhlau, 1984), 56; Hubert Laitko *et al.*, *Wissenschaft in Berlin. Von den Anfängen bis zum Neubeginn nach 1945* (Berlin [DDR]: Dietz Verlag, 1987), 386; Bernhard vom Brocke, 'Die Kaiser-Wilhelm-Gesellschaft im Kaiserreich: Vorgeschichte, Gründung und Entwicklung bis zum Ausbruch des Ersten Weltkriegs', in Rudolf Vierhaus and Bernhard vom Brocke (eds.), *Forschung im Spannungsfeld von Politik und Gesellschaft. Geschichte und Struktur der Kaiser-Wilhelm-/Max-Planck-Gesellschaft* (Stuttgart:

Deutsche Verlags-Anstalt, 1990), 102, note 158; Jeffrey Allan Johnson, *The Kaiser's Chemists. Science and Modernization in Imperial Germany* (Chapel Hill: University of North Carolina Press, 1990), 188, 195; J. A. Johnson and Roy M. MacLeod, 'War Work and Scientific Self-Image: Pursuing Comparative Perspectives on German and Allied Scientistis in the Great War', in Rüdiger vom Bruch and Brigitte Kaderas (eds.), *Wissenschaften und Wissenschaftspolitik: Bestandsaufnahme zu Formationen, Brücken und Kontinuitäten im Deutschland des 20. Jahrhunderts* (Stuttgart: F. Steiner, 2002), 169–179; Dietrich Stoltzenberg's *Fritz Haber: Chemiker, Nobelpreisträger, Deutscher, Jude. Eine Biographie* (Weinheim, New York, Basel, Cambridge, Tokyo: VCH, 1994), 301–309, discusses the KWKW on only two pages, without taking notice of the work it produced, and presents a misleading view of its origins and goals.

[12] See e.g., Archiv zur Geschichte der Max-Planck-Gesellschaft, Berlin [henceforth: Arch.MPG], I A 1–642. Karl Schaum (Physical Chemical Institute of the University of Giessen) to Adolf von Harnack (28 November 1916).

[13] Albrecht Schmidt (3 July 1864–27 May 1945) studied chemistry, 1883–1887, in Darmstadt, Heidelberg, and Straßburg (Ph.D. diss., 1887). He was an assistant at the Chemical Institute, Straßburg, 1887–1888, and headed the 'Scientific Chemical Inventing Laboratory' of the Chemische Fabrik auf Aktien vorm. E. Schering, Berlin, 1888–1898. He made various inventions, including an antiseptic and a gas disinfectant and moved to the Farbwerke Höchst on 1 September 1898, to set up a patent laboratory there. He developed the sulfur dyes and worked on indigo. He became Prokurist in 1910, participating from 1912 in directors' meetings; during the War, he developed an artificial fog and an eye irritant gas; became deputy board member in 1916 and received the title of professor in 1917. He became a member of the board of directors of IG Farben in 1925; and retired in 1931. Some 2,300 patents can be traced to him. For this information, we thank the former Company Archive of Hoechst AG, Frankfurt (today, HistoCom GmbH, Frankfurt-Höchst,). See also Jens Ulrich Heine, *Verstand & Schicksal: Die Männer der I.G. Farbenindustrie AG (1925–1945) in 161 Kurzbiographien* (Weinheim, New York, Basel, Cambridge: VCH, 1990), 126–128, with varying dates in some cases.

[14] Friedrich Schmidt (-Ott) (4 June 1860–28 April 1956) a leading government official and science organizer (from 1920 hyphenated his name with the maiden name of his wife, so that he is called Schmidt-Ott, although he used the name Schmidt during the period discussed). In 1895, he was appointed Vortragender Rat in the Ministry of Education; in 1898, Department Director for Universities, Science, and Art Affairs; in 1908, Ministerial Director for Art and Science; Prussian Minister of Education, August 1917-November 1918. He was the cofounder of the Emergency Association (Notgemeinschaft) of German Science (later DFG), in 1920, and was its first president to 1934. See his autobiography, *Erlebtes und Erstrebtes 1860–1950* (Wiesbaden: F. Steiner, 1952); see also Wolfgang Treue, 'Friedrich Schmidt-Ott', in Wolfgang Treue and Karlfried Gründer (eds.), *Wissenschaftspolitik in Berlin: Minister, Beamte, Ratgeber* (Berlin: Colloquium Verlag, 1987), 235–250; W. Treue, 'Neue Wege der Forschung, ihrer Organisation und Förderung: Friedrich Schmidt-Ott (4.6.1860–24.4.1956)', *Berichte zur Wissenschaftsgeschichte*, 12 (1989), 229–238. On his family, see Heine *op. cit.* note 13, 239–241.

[15] Fritz Haber (9 December 1868–29 January 1934) completed his Habilitation at the Karlsruhe Technische Hochschule in 1896, becoming professor of physical and electrochemistry, 1906–1911. He was director of the KWI for Physical Chemistry and Electrochemistry in Berlin, 1911–1933 and received the Nobel Prize in 1918 for his 1909 method of synthesizing ammonia from nitrogen and hydrogen under high pressure (the Haber-Bosch Process). He resigned as director in 1933 following anti-Semitic legislation of the Third Reich (Gesetz zur Wiederherstellung des Berufsbeamtentums). See Anne Becker, Kurt A. Becker, Jochen H. Block, 'Fritz Haber', in Wilhelm Treue and Gerhard Hildebrandt (eds.), *Naturwissenschaftler (Berlinische Lebensbilder*, 1), (Berlin: Colloquium Verlag, 1987), 167–181; Stoltzenberg *op. cit.* note 11; Margit Szöllösi-Janze, *Fritz Haber 1868 bis 1934: Eine Biographie* (Munich: C. H. Beck, 1998).

[16] In 1915, Haber joined the gas warfare experiments of the army, and in 1916, assumed technical responsibility for gas warfare in the War Ministry's Chemical Department, which he led. The KWI for Physical Chemistry and Electrochemistry, which employed 2,150 people by the end of the war, was responsible for carrying out chemical weapons experiments. See Michael Engel, *Geschichte Dahlems* (Berlin: Verlag A. Spitz, 984), 135 f.. Cf. '...*im Frieden der Menschheit, im Kriege dem Vaterlande*...':

*75 Jahre Fritz-Haber-Institut der Max-Planck-Gesellschaft. Bemerkungen zu Geschichte und Gegenwart* (Berlin: A. Sieckmann, 1986).

[17] Geheimes Staatsarchiv Preußischer Kulturbesitz, Berlin, Bestand Merseburg (henceforth, GStAPK-M), Rep. 92 NL Schmidt-Ott B XIII, 4, 33–35, Haber to Schmidt-Ott, 2 April 1916.

[18] According to the *Reichshandbuch der Deutschen Gesellschaft*, (Berlin: Deutscher Wirtschaftsverlag, 1930), vol. 2, 1654, Albrecht Schmidt was awarded the following orders and medals: Iron Cross on a black and white Band, the Order of Kaiser Franz Josef, the Officer's Cross with War Distinction, and the Turkish Crescent. Albrecht Schmidt wrote a multi-volume, comprehensive 'Erinnerungen', the typescript of which is today in HistoCom GmbH, Frankfurt-Höchst. In the five volumes on 'Der Krieg 1914–1918', there is no reference to the KWKW; however, Schmidt mentions in the account of the year 1914 (178 ff.) that his first connection with Haber was an unproductive discussion on chemical warfare, which 'led to a very fruitful collaboration between us in the ensuing gas war, about one to one and a half years later.' I thank Herr Rudolf Volz, of the former Hoechst AG archive, for his assistance.

[19] On 11 January 1911, the Kaiser-Wilhelm-Gesellschaft zur Förderung der Wissenschaften (Kaiser Wilhelm Society for the Advancement of Science, henceforth: KWG) was founded. It was intended to organize, coordinate, and finance, with private funds, scientific research outside the universities. See Manfred Rasch, 'Thesen zur Preußische Wissenschaftspolitik gegen Ende des Wilhelminischen Zeitalters', *Berichte zur Wissenschaftsgeschichte*, 12 (1989), 240–252. By 1914, the KWG had several research institutes, including the KWI for Chemistry (1911), the KWI for Physical Chemistry and Electrochemistry (both opened in 1912); the KWI for Coal Research (1914), the KWI for Arbeitsphysiologie (Labor Physiology, 1913), the KWI for Biology (1915), the KWI for Experimentelle Therapie (Experimental Therapy, 1913); the KWI for Hirnforschung (Brain Research, 1919), and the KWI for Physikalische Forschung (Physical Research, 1917). See Vierhaus and vom Brocke (eds.), *op. cit.* note 11.

[20] GStAPK-M *op. cit.* note 17.

[21] Johnson *op. cit.* note 11, 245, incorrectly dates this letter to 2 March 1916.

[22] GStAPK-M *op. cit.* note 17.

[23] Adolf von Harnack, 'An der Schwelle des dritten Kriegsjahrs: Rede am 1. August 1916 in Berlin gehalten', *Aus der Friedens- und Kriegsarbeit* (Gießen: A. Töpelmann, 1916), 331–348, especially 341–342.

[24] A clear exception is Ludwig Prandtl, whose Aerodynamische Versuchsanstalt (Aerodynamic Research Institute) occupied an ambiguous position between the KWG, the University of Göttingen, and the Göttinger Vereinigung zur Förderung der angewandten Physik und Mathematik (Göttingen Association for the Advancement of Applied Physics and Mathematics).

[25] Friedrich Althoff (1839–1908) was a Prussian civil servant and from 1897, ministerial director and head of Education Department. For Althoff's role within the Prussian Kultusministerium, see Lothar Burchardt, 'Adolf von Harnack', *Wissenschaftspolitik in Berlin*, *op. cit.* note 14, 220; Bernhard vom Brocke, 'Von der Wissenschaftsverwaltung zur Wissenschaftspolitik: Friedrich Althoff (19.2.1839–20.10.1908)', *Berichte zur Wissenschaftsgeschichte*, 11 (1988), 1–26.

[26] Karl-Heinz Manegold, *Universität, Technische Hochschule und Industrie: Ein Beitrag zur Emanzipation der Technik im 19. Jahrhundert unter besonderer Berücksichtigung der Bestrebungen Felix Kleins* (Berlin: Duncker and Humblot, 1970), 215–221; Gisela Buchheim, 'Eine Denkschrift von Alois Riedler, die Errichtung einer Akademie der technischen Wissenschaften betreffend (1899)', *NTM-Schriftenreihe zur Geschichte der Naturwissenschaften, Technik und Medizin*, 24 (1987), 121–124. On efforts outside Berlin, see Herbert Lepper, *Die Einheit der Wissenschaften: Der gescheiterte Versuch der Gründung einer 'Rheinisch-Westfälischen Akademie der Wissenschaften' in den Jahren 1907 bis 1910* (= Abhandlungen der Rheinisch-Westfälischen Akademie der Wissenschaften, Bd. 75) (Opladen: Westdeutscher Verlag, 1987). During the Kaiserreich, only two engineers became regular members of the Berlin Academy. Even during the Weimar Republic, the proportion of engineers remained small. See Wolfgang Schlicker, *Die Berliner Akademie der Wissenschaft in der Zeit des Imperialismus*, Bd. 2, Teil 2: *Von der Großen Sozialistischen Oktoberrevolution bis 1933* (Berlin [DDR]: Akademie-Verlag, 1975), 53. See also Wolfgang J. Mommsen, 'Wissenschaft, Krieg und die Berliner Akademie der Wissenschaften: Die Preußische Akademie der Wissenschaften in den beiden Weltkriegen', in Wolfram

Fischer with Rainer Hohlfeld and Peter Nötzoldt, *Die Preußische Akademie der Wissenschaften zu Berlin 1914–1945* (Berlin: Akademie Verlag, 2000), 3–23, on 11, which overlooks the goal of integrating the technical sciences.

[27] GStAPK-M, 2·2·1 Nr. 32426, p. 1–2v, Schmidt-Ott to Valentini, 17 June 1916.

[28] Conrad Grau, *Die Berliner Akademie der Wissenschaften in der Zeit des Imperialismus*, Bd. 2, Teil 1: *Von den neunziger Jahren des 19. Jahrhunderts bis zur Großen Sozialistischen Oktoberrevolution* (Berlin [DDR]: Akademie-Verlag, 1975), 218–220; Wendel, *op. cit.* note 11, 125–134.

[29] The conversation between Haber and Schmidt-Ott is dated by a handwritten note on Koppel's telegram to Haber (23 May 1916), in the Schmidt-Ott Papers: (GStAPK-M, Rep. 92 NL Schmidt-Ott B XIII, 4, 32). It can also be dated by an entry of 18 May 1916 in the personal journal records of Friedrich Schmidt-Ott (I thank Herr Dr. Hans-Dietrich Schmidt-Ott, Berlin, for this reference from his father's papers). The assessment of the importance of science and technology in warfare that Leopold Koppel gives in his 'Begründung zum Vorschlag eines Satzungsentwurfs für eine 'Kaiser Wilhelm Stiftung für Naturwissenschaft und Technik im Heer', probably derives from Haber and Schmidt-Ott. For a reference to their prewar discussions of the relationship between science and the military see GStAPK-M, 2·2·1 Nr. 21290, 222–229, Haber to Schmidt-Ott, 18 September 1917 (copy); excerpted in Wendel, *op. cit.* note 11, Document no. 127, 357.

[30] Arch.MPG I 1 A-60. Minutes of the Meeting of the Senate of the KWG (21 March 1914; 'Antrag betr. Begründung eines Kaiser-Wilhelm-Instituts für physikalische Forschung', undated transcript, possibly 1916, with a typewritten note, 'Counts as preliminary statutes, by Senate decision of 6 July 1916', in *idem*, 1 A-49. Sincere thanks to Bernhard vom Brocke, Marburg, for the reference to this KWI.

[31] Bruno Wäser, 'Über die Schaffung einer Zentralstelle für technische und wissenschaftliche Forschung', *Chemiker-Zeitung*, 40 (1916), 485 – 48; *Wer ist's*, 11th ed. (Berlin: Verlag Herrmann Degener, 1928), 1630, lists a Bruno Waeser, who is probably — despite the variant spelling — the same man. Born 21 January 1888 in Magdeburg, he studied at the T.H. Berlin. He was a works chemist at the Konsolid. Alkaliwerk and the Reichsstickstoffwerk Piesteritz, and from late 1919, a self-employed consulting chemist and writer, with numerous publications in technical journals and handbooks.

[32] An alternative explanation was that Haber and Wäser knew each other, and their ideas influenced each other.

[33] Wäser, 'Schaffung', *op. cit.* note 31, 485. In connection with the Patent Office mentioned by Wäser, note that the Brennkrafttechnische Gesellschaft (Fuel Technology Association, founded 1917) was established by an official in the Reichspatentamt (Imperial Patent Office), Wilhelm Gentsch. See Rasch, *Geschichte*, *op. cit.* note 8, 93.

[34] Alard du Bois-Reymond, 'Kriegserfindungen', in *Vom Deutschen Michel* (Berlin, Kassel: Furchev-erlag, 1915), 78–90.

[35] Du Bois-Reymond, 'Kriegserfindungen', *op. cit.* note 34, 86.

[36] 'A Mobilisation of Chemists', *Nature*, 94 (25 February 1915), 700 f.; 'Engineering, Education, and Research', *Nature*, 95 (29 April 1915), 244–248; 'The Government and Chemical Research', *idem* (13 May 1915), 295 f.; 'A Consultive Council in Chemistry', *idem* (8 July 1915), 523 f.; 'The Promotion of Research by the State', *idem* (5 August 1915), 619 f.; 'Science in the War and after the War', *Nature*, 96 (14 October 1915), 180–185; 'Science in National Affairs', *idem* (21 Oct.ober 1915), 195–197; 'The Government Scheme for the Organisation and Development of Scientific and Industrial Research', *idem* (4 November 1915), 259 f.; 'The Organisation of Scientific Research', *idem* (17 February 1916), 692–696; 'National Aspects of Chemistry', *Nature*, 97 (20 April 1916), 171–173.

[37] A first-rate depiction of British developments can be found in Peter Alter, *Wissenschaft, Staat, Mäzene: Anfänge moderner Wissenschaftspolitik in Großbritannien 1850–1920* (Stuttgart: Klett-Cotta, 1982), transl. as *The Reluctant Patron: Science and the State in Britain, 1850–1920* (Oxford: Berg, 1987). See also as 'Die Kaiser-Wilhelm-Gesellschaft in den deutsch-britischen Wissenschaftsbeziehungen', in Vierhaus and vom Brocke, *op. cit.* note 11, 735–741. Still fundamental are Roy MacLeod and E. K. Andrews, 'The Origins of the DSIR: Reflections on Ideas and Men, 1915–1916', *Public Administration*, 48 (1970), 23–48; Roy MacLeod, Ian Varcoe, 'Scientists, Government and Organised Research in Great Britain, 1914–16: The Early History of the DSIR', *Minerva*, 8 (1970), 192–216.

[38] Cited from Alter, *op. cit.* note 37, 227.

[39] Alter, *op. cit.* note 37, 226 f.

[40] On the corresponding history of the National Research Council in the United States, see Daniel J. Kevles, *The Physicists:The History of a Scientific Community in Modern America* (New York: Knopf, 1978), 102–116, esp. 111 f.

[41] According to 'Antrag', *op. cit.* note 30.

[42] More than six months of fundraising for the KWG had produced pledges of more than ten million marks by January 1911, but the actual payments were rather slow in coming; see Lothar Burchardt, *Wissenschaftspolitik im Wilhelminischen Deutschland. Vorgeschichte, Gründung und Aufbau der Kaiser-Wilhelm-Gesellschaft zur Förderung der Wissenschaften* (Göttingen: Vandenhoeck und Ruprecht, 1975), 53–83.

[43] Leopold Koppel (20 October 1854–29 August 1933), banker and industrialist of Jewish origin, who worked his way up from minor bank clerk to independent banker (Bankhaus Koppel u. Co.). From 1901, he was a stockholder in the Deutsche Gasglühlicht - Berliner Auergesellschaft, of which he was a board member since 1897, and later its principal stockholder. The metal-filament lamp invented by Auer von Welsbach was the main source of revenue for the quickly developing company, which was heavily involved with the electrification of Berlin. In 1906, the Auergesellschaft acquired the Gesellschaft für Verwertung chemischer Produkte, which later took up the production of respirators in conjunction with Haber's Institute. See B. vom Brocke, 'Kaiser-Wilhelm-Gesellschaft, *op. cit.* note 11, 98–100. In the late summer of 1905, to finance international scientific exchanges, Friedrich Althoff and Leopold Koppel founded the 'Koppel-Stiftung zur Förderung der geistigen Beziehungen Deutschlands zum Ausland (Koppel Foundation for the Promotion of Intellectual Relations between Germany and Foreign Countries)', with a capital endowment of one million marks. Cf. GStAPK-M, Rep. 76 Vc Sekt. 1 Tit. 8 Abtlg. 8, Nr. 13, Koppelstiftung. This is apparently the same as a similarly-named foundation cited in vom Brocke, 'Wissenschaftsverwaltung' (*op. cit.* note 25), 19. The Koppel Foundation's donations to the KWG were exceeded only by those of the Krupp family and of the heirs of Fritz von Friedlaender-Fuld.

[44] By November 1915, with a big business in gas masks added to its other products, the Auergesellschaft could boast a net profit of 5.4 million marks (Szöllösi-Janze, *op. cit.* note 15, 361). Unfortunately, she does not give comparative figures for other years.

[45] GStAPK-M, 2.2.1, Nr. 32426, 3–4, Koppel to von Valentini, 16 June 1916.

[46] See the references cited in note 27 and note 29 above; on the military work of the KWI for Physical Chemistry and Electrochemistry, see Lothar Burchardt, 'Die Kaiser-Wilhelm-Gesellschaft im Ersten Weltkrieg (1914–1918)', in Vierhaus and vom Brocke (eds.), *op. cit.* note 11, 164–167.

[47] Three different rough drafts and a final version of the statutes survive. The oldest Version A ('Satzung der Kaiser Wilhelm Stiftung für Naturwissenschaft und Technik im Heere') is in Arch.MPG, I A 1–1789, 5–7, evidently a copy of the draft that Koppel sent to the War Minister (4 July 1916). A later draft, Version B, whose file number came from the War Ministry, and which represents their revision of Version A, contains handwritten annotations by Schmidt-Ott. Copy in GStAPK-M, Rep. 92 NL Schmitt-Ott B XIII, 4, 49–51.

A third, later draft, Version C, which partially incorporated the handwritten marginal notes that Schmidt-Ott had made in Version B, is in Arch.MPG, I A 1–1789, 8–10. This Version C was discussed on 18 October 1916 by representatives of the Education Ministry, the Prussian Academy, the KWG, and the Koppel Foundation, with officers of the War Ministry, as shown by the handwritten revisions, which as a rule accord with the final draft (Version D, in Arch.MPG, I A-1789, 18v–20 [undated copy]). The very last revisions were settled on by the full Academy in its meeting on November 2, 1916 and were worked into the final version (in Berlin-Brandenburgische Akademie der Wissenschaften, Akademiearchiv [henceforth: BBAWA], II - XI - 159, 15).

[48] GStAPK-M, Rep. 76 Vc, Sekt. 1 Tit. 8 Abt. 8 Stiftungssachen 17 Bd. 1, 347–350, Satzung der Kaiser Wilhelm-Stiftung für technische Wissenschaft (9 March 1920). According to Paragraph 3, 'The bodies of the Foundation are a) the board, b) the expert committees'. Various other copies of the statutes are found in BBAWA, II - XI - 159, 38 (typescript) and 42 (printed).

[49] See GStAPK-M, *op. cit.* note 27.

[50] GStAPK-M, *op. cit.* note 45, Koppel to von Valentini, 16 June 1916.

51   See GStAPK-M, *op. cit.* note 27.
52   Citations from GStAPK-M, Rep. 92 NL Schmidt-Ott. B XIII, 4, 36, Valentini to Schmidt-Ott, 24 June 1916, draft in GStAPK-M, 2·2·1 Nr. 32426, 5v–6; less detailed letter from von Valentini to Koppel, 24 June 1916: draft in *idem*, 5, copy in GStAPK-M, NL Schmidt-Ott B XIII, 4, 37–37v.
53   The loss of the Potsdam Heeresarchiv during the Second World War probably destroyed the War Ministry's documents concerning the Foundation, so that no original sources remain concerning the attitude of the Prussian War Ministry towards its establishment. The relatively better preserved holdings of the Imperial Navy in the Bundesarchiv-Militärarchiv in Freiburg/Breisgau contain no relevant documents, even though the Reichsmarineamt (Imperial Naval Office) was also involved in founding the KWKW. The dates are taken from GStAPK-M, 2·2·1 Nr. 32426, 13f., Koppel to War Minister von Stein, 13 November 1916.
54   GStAPK-M, Rep. 92 NL von Valentini, Nr. 6. Haber to von Valentini, 29 December 1916.
55   GStAPK-M, Rep. 76 Vc Sekt. 1 Tit. 8, Abt. 8 Stiftungssachen 17 Bd. 1, 27, War Ministry to Ministry of Education, Oct. 27, 1916; BBAWA, II - XI - 159, 10 (showing Haber's attendance) and 15, excerpt from minutes with revision of draft statutes of 28 October 1916 by the full Academy, 2 November 1916; GStAPK-M, 2·2·1 Nr. 32426, 13f., Koppel to von Stein, 13 November 1916 [copy]; Arch.MPG, I A 1–1789, 28, Allerhöchste Kabinetts-Ordre to War Minister von Stein, 17 December 1916, [photographic copy].
56   GStAPK-M, Rep. 76 Vc Sekt. 1 Tit. 8, Abt. 8 Stiftungssachen 17 Bd. 1, 203, Justice Ministry to Ministry of Education (21 May 1917) regarding concerns about co-signing ministerial order of 24 January 1917; *idem*, 208, Schmidt-Ott to von Trott zu Solz; *idem*, 215, royal license for establishment of the foundation (certified copy); *idem*, 35, Academy to Prussian Education Ministry, 26 October 1916; BBAWA, II - XI - 159, 15, minutes of the meeting of the full Academy, 2 November 1916; *idem*, 17–20, provision of suitable rooms, with safe, in the Academy.
57   Arch.MPG, I A 1–1789, 21–22, list of members of the board and the administration of the Kaiser Wilhelm Foundation for Military Technical Science. In GStAPK-M, Rep. 76 Vc Sekt. 1 Tit. 8, Abt. 8 Stiftungssachen 17 Bd. 1, 35, Roethe to Schmidt-Ott, 26 October 1916, the following are named on the behalf of the Academy as heads of the expert committees: Emil Fischer, Walther Nernst, Fritz Haber, Alois Riedler. In *idem*, 45, Academy to Schmidt-Ott (21 December 1916), Friedrich Wüst was named as head of the Committee for Metals. BBAWA, II - XI - 159, Bl. 22, Emil Fischer to Wilhelm Waldeyer (20 December 1916), shows that Nernst, Haber, and Fischer had previously agreed to nominate Wüst to a meeting of the full Academy, without further discussion.
58   For Müller-Breslau's appointment, see GStAPK-M, Rep. 76 Vc Sekt. 1 Tit. 8, Abt. 8 Stiftungssachen 17 Bd. 1, 64, Academy to Schmidt-Ott, 25 Jan. 1917. *Idem*, 235, Riedler to Education Minister Schmidt-Ott, 22 January 1918; copies of various writings of the War Ministry from the beginning of 1918, in *idem*, 294–296v; *idem*, 236–242, memorandum by Stephan Löffler and Alois Riedler, 'Verbrennungsturbinen und Brennstoffwirtschaft'; *idem*, 243 f, handwritten notes by Krüss about discussions with Riedler and Haber on 7 February and 29 January 1918. Critique of Riedler's project in Bancroft Library, Berkeley, Emil Fischer Papers, Outgoing Letters (carbon copies) [henceforth: E. Fischer Papers, Outgoing CC's], Emil Fischer to Fritz Haber (undated, beginning of 1918); at the expert committee chairmen's meeting, 3 October 1918, Riedler was referred with his project to the Brennkrafttechnische Gesellschaft: GStAPK-M, Rep. 76 Vc Sekt. 1 Tit. 8, Abt. 8 Stiftungssachen 17 Bd. 1, 313, handwritten note by Krüss.
59   Deutsches Museum, Handschriftenabteilung, Nachlaß Wilhelm Wien C II, typed circular memo from Richard Zsigmondy, Friedrich Krüger und Max Bodenstein (11–12–13 February 1917).
60   This umbrella organization was founded in September 1916 by the Verein Deutscher Ingenieure, the Verband Deutscher Architekten- und Ingenieurevereine, the Verein Deutscher Eisenhüttenleute, the Verband Deutscher Elektrotechniker, the Schiffbautechnischen-Gesellschaft, and the Verein Deutscher Chemiker. See Burchardt, 'Standespolitik' *op. cit.* note 11, 188 f.
61   Zsigmondy, Krüger and Bodenstein, circular memo *op. cit.* note 59.
62   *Stahl und Eisen*, 37 (1917), 384.
63   Alter, *Wissenschaft*, *op. cit.* note 37, 113.
64   GStAPK-M, Rep. 92 NL Schmidt-Ott, B XIII, 4, p. 79–81v, list of military personnel and civilians nominated to the expert committees of the KWKW.

65 *Stahl und Eisen*, 38 (1918), 79 ff.

66 No relevant documents could be found in the archives of Mannesmann, Krupp, ThyssenKrupp, Siemens, or the Verein Deutscher Eisenhüttenleute.

67 On preparations for the 1917 annual report; See GStAPK-M, Rep. 76 Vc Sekt. 1 Tit. 9, Abt. 8 Stiftungssachen 17 Bd. 1. 231 f., copy of circular memo from Nernst, 30 January 1918. Alfred Stock editor. (*idem*, Krüss's handwritten note regarding the meeting of expert committee chairmen, 2 February 1918). According to Krüss's notes of the meeting, 9 May 1918, most of the records were in manuscripts, and the deadline for submission was 15 May 1918 (*idem*, 271). At the committee chairmen's meeting, 3 October 1918, printing the report was still on the agenda (*idem*, 313, Krüss's handwritten notes). On the work of Stock, who acted for Fischer as head of Expert Committee 1 from February 1918, see Egon Wiberg, 'Alfred Stock 1876–1946', *Chemische Berichte*, 83 (1950), XX–LXXVI, here XXXVI f.

68 The chairmen no doubt contacted prospective scientific members of their senates before nominating them to the Kultusministerium. See, e.g., Emil Fischer Papers, Outgoing CC's, Emil Fischer to Carl Engler, 13 February 1917; and reply agreeing to collaborate with the KWKW, *idem*, Engler File, Engler to Fischer, 20 February 1917; see also MPI für Kohlenforschung, Archiv 01–001, Emil Fischer to Franz Fischer, 1 February 1917 about work in KWKW Expert Committee 1. The applications of committee chairmen and the affirmation of scientific members are in GStAPK-M, Rep. 76 Vc Sekt. 1 Tit. 8, Abt. 8 Stiftungssachen 17 Bd. 1. Only two limited their participation – Rieppel, owing to his being plant manager of the Bavarian Aircraft Works (*idem*, 168), and Wallichs, owing to his military duties for the Kriegsamt (War Office), at the Technische Institute der Infanterie (*idem*, 177).
On the question of travel expenses, see Emil Fischer Papers, Outgoing CC's, Emil Fischer to Carl Engler, 13 February 1917 (with a 5% capital interest rate, there should have been 100,000 marks available yearly – i.e., about 15,000 marks for every expert committee). Regarding deferments see *idem*, Emil Fischer to Kriegsamt, Kriegsarbeitsamt, 12 April 1917; and the minutes of the first meeting of Fischer's expert committee, copies in: GStAPK-M, Rep. 76 Vc Sekt. 1 Tit. 8, Abt. 8 Stiftungssachen 17 Bd. 1. 190; and in Emil Fischer Papers, Kaiser Wilhelm Stiftung für kriegstechnische Wissenschaft [henceforth:KWKW] File.

69 For Fischer's assertions, see Emil Fischer Papers, Outgoing CC's (12 February, 12 April, 2 July, 6 July, and 1 December 1917). For references to the assembly of expert committee chairmen see *idem*, Emil Fischer to Kriegs-Rohstoff-Abteilung, 2 July 1917, and to Franz Fischer, 5 November 1917. See *idem*, Emil Fischer to Geschäftsführer, KWKW (16 August 1918): 'It is not the intention of the Kaiser Wilhelm Foundation to expand by choosing men active in industry, whose contributions to the Foundation's proposals or inventions could put the military in somewhat of a predicament regarding their subsequent utilization.' Cf. GStAPK-M, Rep. 76 Vc Sekt. 1 Tit. 8, Abt. 8 Stiftungssachen 17 Bd. 1, p. 226 f., minutes of a meeting on 7 January 1918 in the War Ministry regarding the utilization of patents.

70 References to a 'long wish-list of the War Ministry', and the notice that the KWKW was wholly independent of the administrative offices of the War Minister, to whom it directly submitted its proposals, are in Emil Fischer Papers, Outgoing CC's, Emil Fischer to Carl Engler, 13 February 1917. There is a reference to 'a list of the tasks currently set by the War Ministry and the Naval Office' in *idem*, Emil Fischer to Ludwig Knorr, 21 February 1917. See *idem*, Emil Fischer to Kriegs-Rohstoff-Abteilung, 2 July 1917, informing the KRA that he will request that a task assigned to him be transferred to Professor Wüst, a specialist in metals and alloys. Wrisberg notified the Bavarian War Ministry about the Foundation, 26 March 1917, which then informed its subordinate offices; see Bayerisches Hauptstaatsarchiv, Kriegsarchiv, MKr 12282, clipping from Prussian *Armee-Verordnungsblatt*, Nr. 14, 16 March 1917, 134, and *idem*, War Ministry printed document on KWKW, addressee to be filled in, 22 January 1917, with six enclosures. Copies of this document are in the Deutsches Museum, Munich, Handschriftenabteilung, Nachlaß Wilhelm Wien C II, and Emil Fischer Papers, KWKW File. On issues of technology, procurement, and inventions, see Ernst von Wrisberg, *Wehr und Waffen 1914–1918: Erinnerungen an die Kriegsjahre im Königlich Preußischen Kriegsministerium* (Leipzig: K. F. Koehler, 1922), 114–126.

71 GStAPK-M, 2·2·1 Nr. 32426, War Ministry to the General Secretary of the KWG, 1 December 1917, with draft of 'Bestimmungen betreffend die Erfindungen der Mitglieder der Kaiser Wilhelmstiftung

[sic] für kriegstechnische Wissenschaften [sic]', typescript, 36; another copy, Arch.MPG, I A 1–1789, 34–35v; final version, 13 February 1918, in *idem*, 37–38, and in GStAPK-M, 2·2·1 Nr. 32426, 38–39. 'Gesichtspunkte für die Behandlung Geheimer Sachen' are mentioned in Emil Fischer Papers, Outgoing CC's, Emil Fischer to Franz Fischer, 25 May 1917. On Riedler's turbine work see note 58. On F. Fischer's 'Bericht über den wissenschaftlichen und technischen Stand der Gewinnung von Tieftemperaturteer aus Steinkohle' and related work, see M. Rasch, *Geschichte, op. cit.* note 8, 74–80; also GStAPK-M, Rep. 76 Vc Sekt 1 Tit. 8, Abt. 8 Stiftungssachen 17 Bd. 1, Minutes of the meeting of Expert Committee 1, 1 November 1917, 251; and Max-Planck-Institut für Kohlenforschung, Archiv 01–005, State Secretary of the Imperial Naval Office (Staatssekretär des Reichsmarineamts) to F. Fischer, 5 May 1918, inquiring about Fischer's 'Bericht'. See GStAPK-M, Rep 92 NL Schmidt-Ott, B XIII, 4, 79–81v, 'Verzeichnis für die Berufung in die Fachausschüsse der KWKW vorgeschlagenen militärischen und Zivilmitglieder' (copy, no date) for an overview of work topics mentioned in the first proposals for appointments of members.

72   For particular areas of work from the proposals for appointment of expert committee chairmen, with further documentation: GStAPK-M, Rep. 76 Vc Sekt 1 Tit. 8, Abt. 8 Stiftungssachen 17 Bd. 1 pp. 84, 86, 97, 99 f., 105 f. For glimpses into the work of Expert Committee 1, see the minutes of two meetings, 16 June and 1 November 1917, *idem*, 189–202, 247–264; also Emil Fischer Papers, KWKW File. My thanks to Herr Jobst Broelmann of the Deutsches Museum for information concerning the gyroscope.

73   Arch.MPG, I A 1–1789, pp. 26a–26c, War Ministry to Education Ministry, 13 February 1917 (copy).

74   GStAPK-M, 2·2·1 Nr. 32426, pp. 31–33, War Ministry to the Education Ministry, 13 February 1917 (copy); Haber to Schmidt-Ott, 18 September 1917, *op. cit.* note 29. Wendel's excerpt from this document *op. cit.* note 11 and note 29, deals not with cooperation between Haber's KWI for Physical Chemistry and the KWKW, but rather with intended cooperation between the KWKW and the yet to be founded KWI for Applied Physical Chemistry and Biochemistry. Further information from Dr. Dietrich J. Stoltzenberg, Hamburg (cf. *op. cit.* note 11).

75   GStAPK-M, 2·2·1 Nr. 21290, p. 278, handwritten note by Krüss on the meeting of the KWKW, 8 February [1918]. Citation from Emil Fischer Papers, Outgoing CC's, Emil Fischer to Alfred Wohl, 4 January 1919.

76   GStAPK-M, Rep. 92 NL Schmidt-Ott C 84, 38 f., Haber to Schmidt-Ott, 2 May 1919. For Koppel's revival of interest see *idem*, Haber to Schmidt-Ott, no date (circa 1920). For Thomson's address, see *Proceedings of the Royal Society of London*, Series A, XCV (1918), 253 f.

77   BBAWA, II - XI - 159, Bl. 34/1–34/2, Koppel to the Reichswehrministerium, 8 January 1920 (copy).

78   BBAWA, II - XI - 159, Bl. 36, excerpt from minutes, meeting of the full Academy, 29 January 1920; *idem*, Bl. 38 and 42, statutes (draft and final version), also in GStAPK-M, Rep. 76 Vc Sekt 1 Tit. 8, Abt. 8 Stiftungssachen 17 Bd. 1, 331–334, 347–350, 383 f.

79   *Idem*, p. 362, presentation on staffing of the six expert committees (late summer 1921); *idem*, p. 378, Academy to the Prussian Education Ministry, 28 July 1922.

80   Theodor von Kármán (11 May 1881 - 7 May 1963), studied 1898–1902 in Budapest, and 1906–1908 in Göttingen, Ph.D. 1908; became Privatdozent in Göttingen in 1910, and was then professor of mechanics (as successor to Arnold Sommerfeld) and aerodynamics at the Aachen T.H. (1913–1933). He became director of the Guggenheim Aeronautical Laboratory at the California Institute of Technology in Pasadena from 1933 to 1949, where, with his theory of turbulence, he fostered the development of modern variable swept-wing aircraft.

81   GStAPK-M, Rep. 76 Vc Sekt 1 Tit. 8, Abt. 8 Stiftungssachen 17, Bd. 1, 387 f., Minutes, meeting of the KWTW, 29 October 1922. Hugo Andres Krüss, later director-general of the Prussian National Library, supported gliders; for an 18-page memorandum on gliders and the possibilities of promoting them, see *idem*, 389–402, for documents concerning the Rhön Glider Competition, 22 August 1923, see *idem*, 441–445; concerning dissolution of the Foundation and use of the remaining assets, see *idem*, Bd. 2, 10, Reichsministerium des Innern to Prussian Kultusministerium, 7 January 1926.

82   Other models for the DFG's predecessor, the Emergency Association of Science founded by Haber and Schmidt-Ott in 1920, were the KWI for Physical Research, founded in 1914, and the Carnegie Institution, which was founded in 1902, with eighteeen expert committees. For references to the DFG and the Carnegie Institution, I thank Bernhard vom Brocke. On the DFG, see Kurt Zierold, *Forschungsförderung*

*in drei Epochen: Deutsche Forschungsgemeinschaft. Geschichte, Arbeitsweise, Kommentar* (Wiesbaden: F. Steiner, 1968), 48 f., 57–61, as well as Notker Hammerstein, *Die Deutsche Forschungsgemeinschaft in der Weimarer Republik und im Dritten Reich: Wissenschaftspolitik in Republik und Diktatur* (Munich: C. H. Beck, 1999), 47 ff., which unfortunately goes too little into the organization or its predecessors.

[83] Friedrich Schmidt-Ott, *Erlebtes op. cit.* note 14,.145; Franz Wever, 'Aus der Geschichte des Max-Planck-Institutes für Eisenforschung', Boris Rajewsky and Georg Schreiber (eds.) *Aus der Deutschen Forschung der letzten Dezennien. Festschrift Ernst Telschow zum 65. Geburtstag*, (Stuttgart: Thieme, 1956), 302 f. A historical study of the Düsseldorf KWI for Iron Research is still needed; see Lothar Burchardt, 'Förderung' *op. cit.* note 11, 182f., and Ulrich Marsch, *Zwischen Wissenschaft und Wirtschaft. Industrieforschung in Deutschland und Großbritannien 1880–1936* (Paderborn, Munich, Vienna, Zürich: Schöningh, 2000), 339–349.

[84] For the KWI for Applied Physical Chemistry and Biochemistry, see notes 73–74. The documents on the financing of research by the Heereswaffenamt are extremely skimpy. However, references can be found in the Bundesarchiv-Militärarchiv, Freiburg/Breisgau, RH 8/v. 919, 'Verkehr mit Industrie und Wirtschaft'. According to an inventory 'Reichsmittel für wissenschaftliche Forschungen' (21 September 1927), the Heereswaffenamt and others maintained connections to the KWI for Iron Research, the Silesian Coal Research Institute, the Mineralöl- und Braunkohlenforschungsinstitut (Mineral Oil and Lignite Research Institute) of the Charlottenburg T.H. to the Physikalisch-Technischen Reichsanstalt (PTR = National Institute for Physics and Technology), to the Chemisch-Technischen Reichsanstalt (CTR = National Institute for Chemistry and Technology), the VDI, VdEh and other facilities. The financial contributions of the Heereswaffenamt were as a rule small. E.g., on the support of the Silesian Coal Research Institute see Manfred Rasch, 'Die Montanindustrie und ihre Beziehungen zum Schlesischen Kohlenforschungsinstitut der Kaiser-Wilhelm-Gesellschaft: Ein Beitrag zu Wissenschaft und Wirtschaft in der Zwischenkriegszeit', *Technikgeschichte*, 55 (1988), 15 and note 42. According to an internal memo of the Heereswaffenamt, 'Waffenprüfungswesen' (14 November 1929), the KWIs for Strömungsforschung (Aerodynamic Research), for Silikatforschung (Silicate Research), and for Iron Research received small contributions from the military budget (Bundesarchiv-Militärarchiv, Freiburg/Breisgau, RH 8/v. 919). Unfortunately, *Rüstungsforschung im Nationalsozialismus: Organisation, Mobilisierung und Entgrenzung der Technikwissenschaften*, Hg. Helmut Maier (Göttingen: Wallstein, 2002), does not deal with the historical predecessors.

PATRICE BRET

# MANAGING CHEMICAL EXPERTISE: THE LABORATORIES OF THE FRENCH ARTILLERY AND THE SERVICE DES POUDRES

## INTRODUCTION

In comparing the situation of chemistry in France and the other major nations before and during the Great War, it must not be forgotten that France was not furnished on the same scale.[1] With only a few hundreds of chemists involved, the phenomenon of 'chemical' mobilization in France appears at first sight to have been far less significant than its counterparts in Germany, Britain, or the USA.[2] Nevertheless, from Charles Moureu's militant retrospective in 1920, to Olivier Lepick's account in the late 1990s, the history of French chemistry at war has received attention from both scientists and historians of science. Their attention has, however, focused principally upon the production and use of war gases.[3] The chemical munitions and explosives industry has also been studied,[4] but far less.[5]

To date, little is widely known about the organization and daily activity of the wartime military and civilian chemical laboratories of France. In order to open this 'black box', this paper surveys the laboratories that carried out assessment, experiments and research studies — not for the Inspection des Études et Expériences Chimiques, established in 1915 for the chemical war – but for the Inspection des Études et Expériences Techniques de l'Artillerie, created before the war and, from August 1917, known as the Inspection des Inventions, Études et Expériences Techniques de l'Artillerie. In this paper, I argue that the scarcity of chemical expertise eventually led to a division of labour among the various laboratories — in 1915, *between* civil and military institutions, and, by late 1917, *among* them as well. This form of cooperation was to prove a foretaste of the more elaborate forms of military-academic cooperation that came into existence after the war.

This essay will survey the pre-war military laboratories, and their trajectory from rivalry to cooperation. I will focus upon one of the oldest government laboratories – the Laboratoire de Chimie du Service Technique de l'Artillerie – during the last two years of the war—that is, from the second half of 1916 to the Armistice, during which time the munitions effort of France was coordinated by the newly-founded Ministère de l'Armement under Albert Thomas.[6]

## FRENCH MILITARY LABORATORIES BEFORE THE WAR

In August 1914, beside the laboratories of the military factories, there were half a dozen others doing chemical work for the three services of the French military (see map, Figure 1). Most of these laboratories belonged to the Artillery. At the Dépôt

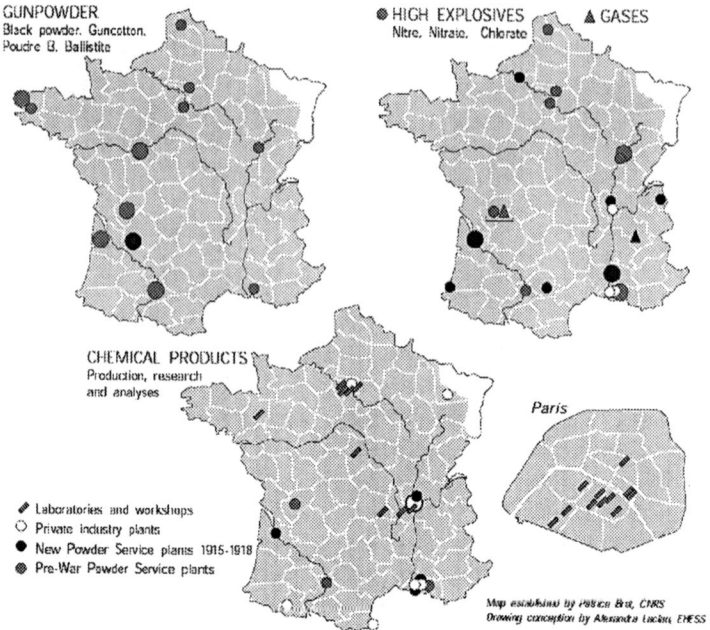

Figure 1. Chemical Explosives Production Sites in France, 1914–1918

*Source*: Louis Vannetzel, 'Le Service des Poudres', *Croix de Guerre*, special issue, (1961)

Central de l'Artillerie — the Artillery headquarters located at Place Saint-Thomas d'Aquin in Paris — stood the Laboratoire de Chimie du Service Technique de l'Artillerie (LSTA), founded in 1818.[7] As might be expected, this laboratory was especially strong in metallurgy. Outside Paris, was the Laboratoire de la Commission des Poudres de Versailles, a far younger agency. Initially set up to test gunpowder supplied to the Artillery, this devoted itself to research on propellants from 1896. Two other artillery laboratories worked on gunpowder. Not far from Paris, Le Bouchet was one of eleven powder plants, and the only one not run by the Service des Poudres et Salpêtres. Le Bouchet had carried out research on propellants for the Artillery since the Service became independent in 1875. There, during the 1870s and 1880s, Colonel (later General) Castan worked on the manufacture of large grain powder, adapted from the Belgian Wetteren powder, for the French navy's biggest guns. There also, during the 1890s and 1900s, Captains Lepidi and Schwartz worked on the stabilization of the first smokeless powder, 'poudre B', the mainstay of the French army and navy.

In 1870, just before the Franco-Prussian War, the École Centrale de Pyrotechnie, founded in 1824 at Metz, was transferred to Bourges. Metz was thought to be too close to the German border, and had actually become part of Germany following the war. The expertise of the Ecole, which related principally to fuses and rockets, also encompassed propellants and explosives.[8] However, the main laboratory for propellants and explosives belonged to the Service des Poudres. In 1887, when the Dépôt Central des Poudres et Salpêtres was installed in new buildings at Quai Henri IV in Paris, the institution took the name of the Laboratoire Central des Poudres et Salpêtres (LCP).[9] This change of name emphasized the growing importance of research and testing.[10] Its previous buildings were part of the old Paris Arsenal, which had housed the laboratories where Lavoisier worked from 1776 to 1792 and Gay-Lussac, from 1818 to 1850. There, Berzelius was a visitor and the young Liebig studied alongside Gay-Lussac. More than any other military service, the Service des Poudres developed a research tradition that drew upon extensive cooperation between its officers and the membership of the Académie des Sciences.

During the French Revolution and after Lavoisier's resignation, the management of gunpowder production was served by a galaxy of chemical talent, including Antoine-François Fourcroy, Louis-Bernard Guyton (who set up the first 'national laboratory' in 1794), Claude-Louis Berthollet, and Jean-Antoine Chaptal. Between the Restoration and the end of the Second Empire, Louis-Joseph Gay-Lussac and Jules Pelouze were the experts, but from 1870 to his death in 1907, Berthelot was the leading French chemist working on powder and explosives. As such, he was a member of the Comité Consultatif des Poudres et Salpêtres (1873) and the Conseil Scientifique of the Navy Central Laboratory (1881). He was also president of the Commission des Substances Explosives (1878), which also ran experiments at the Sevran-Livry powder plant from 1886.

The engineers of the Service des Poudres also turned out to be good chemists. Educated in mathematics, physics, chemistry and industrial drawing, they were trained to build and run mills. Two of them – Émile Sarrau and Paul Vieille – entered the Académie des Sciences. In 1884, the latter famously invented 'poudre B'. From 1910 onwards, young French chemists (*agents chimistes*) were allowed to enter the 'école d'application des poudres', and to become *ingénieurs des poudres* like others from the Ecole Polytechnique. The navy, too, had its own Central Laboratory. Founded in 1881, the Laboratoire Central de la Marine was housed in the buildings of the Dépôt Central des Poudres, and subsequently in those of the LCP. It conducted experiments in the Sevran-Livry powder factory near Paris, as did the Service des Poudres and the Commission des Substances Explosives. The first director of the Navy Laboratory was Hippolyte Sebert, a member of that Commission, and later a member of the Académie des Sciences. In 1912, the institution was officially reorganized as a Naval Artillery Research Laboratory.[11]

FROM RIVALRY TO COOPERATION: THE 'POUDRE B' DISPUTE

For several decades before 1914, the relationship between the principal consumers (the Artillery) and producers (the Service des Poudres) had been a source of national concern. This remained a problem despite the separation of the two Services in 1876, when the old Commissaires des Poudres were granted the title of *ingénieurs*, which better suited those who had been educated at the École Polytechnique. The tension between producers and consumers came to a head in March 1907, when an accidental explosion on the French battleship *Iéna* excited a huge public outcry. After two parliamentary committees, and a joint proposal by the War and Navy Ministers, the government created a permanent committee of civilian and military experts, known as the Commission Scientifique d'Étude des Poudres de Guerre,[12] which was chaired by a member of the Académie, the mathematician and physicist, Henri Poincaré.[13] The work of the Commission led in 1910 to the replacement of amylic alcohol, chosen and defended by Berthelot and Vieille, by diphenylamine as the stabilizer for 'poudre B'.

In fact, diphenylamine was already in use by Nobel and the Germans, and had been claimed by the Artillery and the Navy since 1896. But the case illustrated the situation of French military chemical research at the time of the 'Entente cordiale', and the Tangiers crisis — a situation characterised by the unshakeable pride of Berthelot and Vieille; by rivalries among the laboratories of the Poudres, Artillerie, and Marine; by the participation of the 'cannon merchants' (which led Henry Le Chatelier to resign from the Commission); and by the demand for civilian scientists and laboratories. In a letter to Poincaré in November 1907, Le Chatelier gave an illuminating account of the tensions he experienced:

> I shall tell you that Captain Robert [of the Versailles Laboratory] always takes me as a go-between when information is to be obtained from the Powder and Saltpetre Laboratory. He does not want to compromise himself by directly asking M. Vieille.

Le Chatelier then explained what he saw as a technical problem about a temperature regulator, which appears to have been a different instrument in the two laboratories, although it was used in parallel experiments. He concluded: 'A two-minutes talk between the persons concerned would have spared all that wasted time.'[14]

Poincaré replied that the routine activities of the laboratories of the Artillery and the Service des Poudres prevented them from undertaking long term projects. Accordingly, he and Le Chatelier worked out a plan for a new laboratory, devoted to research on propellants and explosives, and to a process of arbitration between producer and consumer establishments. The new, so-called Laboratoire Central des Poudres de Guerre was to have been independent from the various services, and responsible directly to the Minister of War. Its director would have been appointed by the President of the Republic, following a joint recommendation by the Ministers of War and the Navy. In fact, this laboratory was never created as such, although a laboratory of a similar type was set up for the pre-existing commission des Substances Explosives at Sevran-Levry.

## POWDER AND EXPLOSIVES RESEARCH: COOPERATION BETWEEN THE SERVICES

Paul Vieille's departure from the regular Laboratoire Central des Poudres (LCP) facilitated a reconciliation between the rival parties, and a recognition of the LCP as the major laboratory in the field. From the 1880s, the LCP had special links with the Navy Central Laboratory, also located in its buildings. After 1907, it used the naval artillerists' crushers, and in 1912, was asked to carry out research for the navy, which was awaiting the establishment of a new Naval Artillery Research Laboratory.[15] Meanwhile, new legislation in 1913 permitted six officers from other corps to be attached to the Service des Poudres, in order to move officers with chemical expertise into explosives research.

The most important of these 'new men' was the artillery officer Marie-Gustave-Albert Kœhler. Born in 1861 into an Alsacian family of dyers who chose to leave Alsace when it became German, Kœhler graduated from the École Polytechnique in 1881. From 1892 to 1911, he worked on the thermochemistry of explosives at the École Centrale de Pyrotechnie, becoming, eventually deputy director of the Ecole, and director of its research laboratory. As early as 1896, Kœhler received public acknowledgement for his work on melinites and cresylites, 'which denoted a deep knowledge of the chemistry of explosives'.[16] But the École Centrale de Pyrotechnie de Bourges was not the right place to do advanced research. In 1912, following a request by General Auguste Gaudin (Director of the Service des Poudres), Kœhler was seconded to the LCP in Paris. In 1913, he became engineer-in-chief of the Service. The same year, the Director of the Laboratory, engineer General Jean-Louis Barral, noted that Kœhler:

> ... is a scholar with the greatest knowledge, very conscientious and full of common sense, certainly too shy and too modest indeed. M. Marqueyrol, the chief engineer, has rather different qualities; together, they make an excellent staff for the management of chemical studies at the Central Laboratory.[17]

During the first years of the Great War, Kœhler seems to have taken command of the laboratory. He died in July 1916 – from overwork, it is said.[18]

Kœhler's colleague, powder engineer Marius-Daniel Marqueyrol, was slightly younger, and lived much longer. Born in 1871, and trained at the École Polytechnique in 1892, Marqueyrol was a better chemical analyst than his patron Paul Vieille, and recognised that the diphenylamine was better at stabilizing 'poudre B' than the amylic alcohol chosen by his chief. Thus, he agreed with the Artillery and Navy. Marqueyrol later succeeded Vieille, Barral and Kœhler at the head of the LCP, where he was assisted by two excellent young chemists, Henri Charles Antoine Muraour *chimiste de recherches* and Pierre Loriette *agent chimiste*. Coming from the dye industry, Muraour applied colorimetric reactions to control the stability of explosives, introduced a method to purify tolite, and developed explosives for aerial bombs. With these experts on its staff, the LCP became the undisputed centre in the field, and certainly challenged the Service des Poudres and the Artillery. Loriette, Muraour, and Marqueyrol published many joint papers in the *Mémorial des Poudres* (the journal of the Service des Poudres), which gave abundant evidence of their chemical capability. But they were too few to cope with the demands made by the coming of war.[19]

## MOBILIZATION AND MANPOWER SHORTAGES

Athough chemical weapons have become emblematic of the Great War, there was also a great demand for the analysis and use of everyday organic materials (oils, leathers) and metals. For these, both the Service des Poudres and the Artillery employed civilian and military chemists. At the outset, the most senior chemists — men too old for active service — were assigned to work in their own laboratories. But their students and younger colleagues were sent to the front, and many into infantry and artillery units, or to field hospitals. Maurice Javillier, for example, began the war in a provincial hospital, and was posted to the laboratory of the VIIth Army on the Belgian front, before being re-assigned to Paris, to become a member of the Commission des Etudes Chimiques de Guerre, and to begin work on the preparation of *ypérite* (mustard gas) with Gabriel Bertrand.[20] Later, others were called back to their patrons' laboratories or, like Ficheroulle in 1916, assigned to government factories like Le Bouchet. Some chemists – like André Wahl at the Société des Matières Colorantes in Saint-Denis, outside Paris – were mobilized for industry, while others (like Auguste Chagnon, from Saint-Gobain, and René Dubrisay, from the State Tobacco Manufacturing Agency) were sent from industry to the Artillery or to the Service des Poudres.

In August 1914, the staff of the Service des Poudres comprised only forty-four powder engineers — seven of whom were completing or had just completed their studies — and seventy-eight chemists (*agents chimistes*) and technical agents.[21] These were accorded military status by a law of 24 March 1914, but they were not numerous enough to manage industrial production, even in peacetime. So, the law also stated that, within six years, the strength of the Service should be increased to sixty-three engineers, forty-five chemists, and 110 technical agents.[22] These 'military chemists' were admitted after examination from the nine schools of applied chemistry in France. These included two schools in Paris (the École de Physique et de Chimie Industrielle de la Ville de Paris, and the Institut de Chimie Appliquée de la Faculté des Sciences de Paris), and seven provincial institutions.[23] Of course, not even the targets of peacetime — a rather tense peacetime — were met by the time the war broke out. In any case, chemists were needed more to manage the chemical installations of the Service, including quality control, than to carry out research. However, the new law also allowed the Service to engage up to five chemists of advanced levels, called 'research chemists' (*chimistes de recherches*), to be put in charge of studies on explosives.[24]

Civilian chemists also worked for the Artillery. Some, like most Artillery officers, received their training at the École Polytechnique. Thus, in 1908, the Polytechnicien Ernest Berger, who had shown ability as a laboratory chemist, was assigned to the Laboratoire de la Commission des Poudres de Versailles.[25] Before mobilization, the Laboratoire de Chimie de la Section Technique de l'Artillerie (LSTA) had four civilian chemists and ten military chemists, all working seven hours a day. Two additional chemists were requested, and three years later, by June 1917, the laboratory had fourteen civilian and five military chemists.[26] Although the number of studies before the war averaged only about one per month, by 1916 this had increased to eleven, and in 1917, to twenty-six per month. That is to say – by 1917, twenty-six times more studies were being done, with only one-third more staff.

For Major Nicolardot, Director of the LSTA, the wartime recruitment of chemists was a continuing problem. In December 1917, in answering a question on better ways to illuminate enemy airplanes, he took the opportunity to emphasize his lack of research staff: 'The question asked is new and it requires research. Considering the shortage of manpower and the intense work already done by the Laboratory, its Director has to be satisfied with giving [oral] information.'[27]

Both despite and because of the terms of mobilization, the military found itself competing with private industry for laboratory manpower. In April 1917, a disabled chemist of the LSTA, Feltz, left for 'a better paid position' in industry. Feltz's departure, a few months after the death of another chemist, Avon, brought Nicolardot into conflict with his senior officer and supervisor, Colonel Victor Leleu, Director of the Artillery Technical Section. Nicolardot asked Leleu to send him two chemists; including François Wandenbulcke, a chemist from Brittany, who had enlisted in Flanders. But Leleu refused, observing that another chemist in the Laboratory, Philippet, had been given non-chemical duties — mainly to draw blueprints of projectiles analyzed by the LSTA. Therefore, Leleu wrote the Minister: 'So, the question is not really one of replacing; it is a new increase intended to find a place for a professor who is doing chemistry at the front.'[28] The Director of the Laboratory curtly retorted: 'So, contrary to the Colonel Director's assertion ... the question is put with even more acuity than in April.'[29] Against Leleu's criticisms, Nicolardot reminded other authorities' supports, including the Inspectors General of the Artillery Technical Studies and Experiments (Généraux Inspecteurs des Études et Expériences Techniques de l'Artillerie), and even the Artillery Department (Direction d'Artillerie) at the Ministry.

Nicolardot represented his chief as a rigid bureaucrat whose quibbling got in the way of the Laboratory: 'Unfortunately, it seems that, contrary to the orders of M. the Minister of Armament and War Manufactured Products [Albert Thomas], and of the Under-Secretary of State for Inventions [Jean-Louis Breton], and contrary to the service welfare (quick execution, absence of errors), the only plausible motive for the Colonel Director's refusal [was that the Chemical Laboratory should not have a chemist] occupied in measuring, weighing, drawing and describing foreign munitions, operations which have nothing to do with chemistry'.[30] For Nicolardot, laboratory life had rules that included autonomy.

CHEMICAL WARFARE: THE USE OF ACADEMIC EXPERTISE

The French military began research on potential war gases as early as the autumn of 1914. This was also of special concern to Bertrand's laboratory of biological chemistry at the Institut Pasteur, and for Javillier, who was involved in the study of tear-gases and non-lethal gases, as were other academic chemists.[31] After the German chlorine gas attack on 22 April 1915, a civilian research committee — the Commission des Études Chimiques de Guerre — was created. By August

1915, under the command of General Ozil, the Direction Chimique des Matériels de Guerre supervised some twenty prominent academic chemists and pharmacists, led by Charles Moureu.[32] Under the control of General Perret's Inspection des Études et Expériences Chimiques, they were divided in two sections, one for Aggressive Warfare ('Produits Agressifs'), led by Moureu himself, and the other for 'Protection', including gas masks.

Thirteen major laboratories were involved in this programme. Their staff belonged mainly to institutions of higher education, including the Collège de France (Moureu from 1917, Marcel Delépine, André Mayer, Dr. Gley), the Sorbonne (Victor Grignard, Georges Urbain), the École Supérieure de Pharmacie (Moureu before 1917, then Paul Lebeau), the Faculté de Médecine (Alexandre Desgrez), and the Conservatoire National des Arts et Métiers (André Job).[33] Only a few of these laboratories were not actually academic institutions. These included the Institut Pasteur (Ernest Fourneau, Gabriel Bertrand), established in 1885; the Laboratoire de Toxicologie de la Préfecture de Police (E. Kohn-Abrest); and the Laboratoire Municipal de Chimie de la Ville de Paris (André Kling), established in 1878. The latter was the first to identify the composition of the German gas.

Some of these institutions were more or less related to the military before the war, and especially to the study of powder and explosives. This was the case with Kling's Municipal Laboratory, and Berthelot's laboratory of organic chemistry at the Collège de France, as well as the laboratory at the École de Physique et de Chimie Industrielle de la Ville de Paris (when Haller belonged to the Commission Scientifique d'Étude des Poudres de Guerre in 1907–1910), and the Laboratoire National d'Essais, at the Conservatoire des Arts et Métiers, to which Captain Cellerier was seconded.

Each of these senior academic chemists was given permission to have students recalled from the army, so as to set up research groups. Thus, Lebeau, at the École Supérieure de Pharmacie, came to lead a team of eighteen, including Léon Binet, a 'militarized' physician, who worked on defensive products. Sometimes, scientists worked together on the same projects, as did Lebeau and Urbain on offensive gases, including phosgene and vincennite.[34] The workload was so great that it had to be shared between civilian and military laboratories.

### DIVISION OF LABOUR: STUDIES OF GERMAN ARTILLERY MATERIALS

Thanks to the civilians taking up subjects on which the military laboratories were unfamiliar, the chemists of the Artillery and the Service des Poudres made a substantial contribution to their field, while allowing the military to concentrate on their own special expertise. Besides doing regular work for their supervising institutions, two laboratories — the LSTA and the Municipal Laboratory — devoted themselves to the analysis of German artillery munitions. This process of 'reverse engineering' — as tedious as it was useful — was a means not only of discovering technical features and effects, but also of evaluating the effectiveness of the Allied blockade on German industrial production, and therefore on the German economy.

Each month, starting from 1 August 1916, the Artillery's Technical Section collected and published information about German munitions — the 'Examen des Munitions Ennemies Provenant des Armées: Comptes Rendus Sommaires'.[35] After the fifth issue (1 December 1916), three specialized abstracts were then added to the general 'Comptes Rendus':[36] The first of these, a blue-covered fascicle, called Fascicule BR (for Bulletin de Renseignements), was an État Mensuel des Renseignements Techniques Nouveaux Provenant du Bulletin de Renseignements du Grand Quartier Général Relatifs aux Munitions Ennemies. Into this collection, came information from field observations, diplomatic sources in neutral countries, and spies in the occupied sectors of Allied countries and enemy territories. The second, with a yellow cover (under the short title Fascicule LM for Laboratoire Municipal), was formally entitled 'État Mensuel des Renseignements Nouveaux Relatifs à l'Artillerie et au Matériel Chimique de Guerre Provenant du Laboratoire Municipal de Chimie de la Ville de Paris et des Centres Médico-légaux qui en Dépendent'.[37] The third, with a green cover (Fascicule LSTA, for Laboratoire de la Section Technique de l'Artillerie), was first entitled 'État Mensuel des Renseignements Techniques Nouveaux Provenant du Laboratoire de Chimie de la Section Technique de l'Artillerie Relatifs aux Munitions Ennemies'. In June 1918, its title was revised, and it was called, Fascicule EAP ('Établissements de l'Artillerie et des Poudres'), with a page listing the twelve institutions which had shared analyses with the LSTA since January 1918, following ministerial instructions of 17 October 1917.[38]

These fascicles reveal the methods and processes used in the different institutions, as well as the kinds of technical information that the Artillery authorities decided to circulate. They also make it possible to quantify the work of the respective laboratories, especially the main chemical laboratories of Paris and the Artillery Technical Section (see Figure 2).

The Municipal Laboratory produced a quarterly average of twenty-five pieces of information, a number that diminished during the second quarter of 1917. Its proportional importance also declined. During the last quarter of 1916, it supplied three-quarters of all the information coming from laboratories; but only half in the first quarter 1917; two-fifths, during the second quarter of 1917; and less than one-third during 1918.[39] For Kling's civilian staff at the Municipal Laboratory, analyses were of minor importance compared to the studies on gases that were being developed in the laboratory.

With a slightly smaller average of twenty-two reports per quarter, Nicolardot's laboratory work on German munitions was less regular, but had nearly the opposite evolution – a long plateau from the second quarter of 1917 to the first quarter of 1918, then a fall. So, from one-quarter of all reports at the end of 1916, it rose to half in the first six months of 1917, and three-fifths (62 per cent) in the last six months of 1917, but declined to twenty-three per cent in 1918,[40] when studies

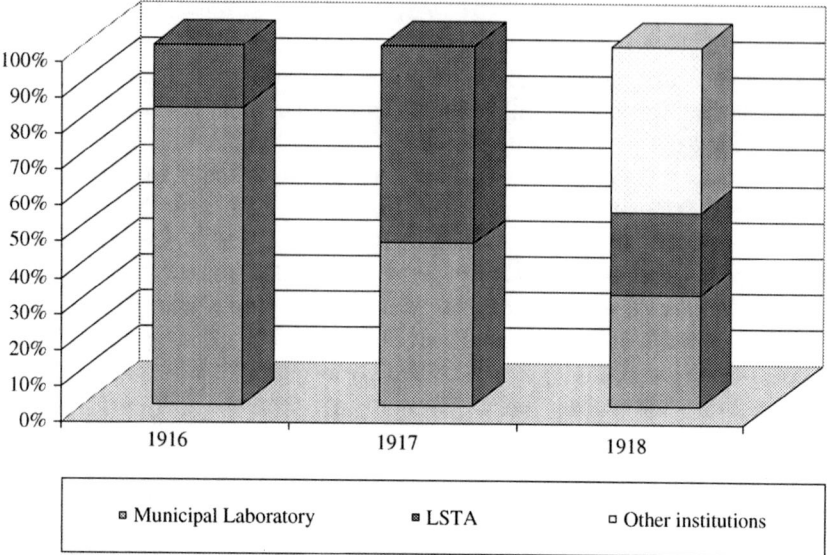

Figure 2. Studies of German Artillery Material, 1916–1918 (by establishments supervised by the Artillery Technical Section)

*Source*: "Examen des Munitions Ennemies Provenant des ArméesL Comptes Rendus Sommaires". Fascicles BR, LM, LSTA & EAP (CAA, fonds 10, ETCA, cartons 312–313).

were allocated among the other establishments of the Artillery and the Service des Poudres.

Even at the peak of the LSTA's work on German munitions, these did not represent more than one-third of its activity – only eleven of the laboratory's thirty-two reports in June 1917. The LSTA issued daily assessments that included analyses of German metals and alloys. Every quarter, at the end of the third month, it gave a synthesis of its work in each field.[41]

### THE LABORATOIRE DE LA SECTION TECHNIQUE DE L'ARTILLERIE

In 1918, the LSTA began to focus upon new fields, while many (forty-six per cent) of the routine analyses of German materials were distributed among other establishments. From most to least active were the École Centrale de Pyrotechnie, the Atelier de Chargement de Clermont-Ferrand, the Military Powder Factory at Le Bouchet, the LCP, and the Versailles Laboratory (see map, Figure 3).[42] These laboratories functioned as a network. German 105-shells were sent to the Cartoucherie de Vincennes, and gas projectiles, to the Laboratoire Municipal before being studied at the LSTA.[43]

Further cooperation took place in the concluding months of the war. While continuing its work on metals, for example, the LSTA sent its more routine analyses to the less expert laboratories of the Artillery and the Service des Poudres. For

# CHEMISTRY AT WAR IN FRANCE, 1914-1918

Figure 3. Chemical Establishments in France, 1914–1918.

*Source*: Various archival materials (CAA, SHAT...) and printed studies (Vinet *et al.*, 1919, Moureu 1920, Pique 1927, Vannetzel 1961...)

its part, it shifted to research, focusing on more complex studies for different services, including the Direction des Inventions, Études et Expériences Techniques de l'Automobile.

By June 1917, the LSTA was working on a range of industrial and commercial products from Germany, including wolfram (tungsten) from the diminishing Austrian stocks and from the factories of La Lonza, which had been reduced to treating five tons of metal a week, as contrasted with five tons every day in August 1914. The LSTA also tested wolfram from the Greek island of Thasos, which had been sent to 'Niedersehöneweide-Yohannisthal-Berlin' for mould manufacturing and furnace building.[44] It also studied low quality copra oil soap and saccharine, imported by Germany from The Netherlands, and conducted toxicological tests for evidences of morphine, codeine, colchicine, brucine, strychnine, atropine, and cocaine.[45] In addition, it made studies of fire bricks (for the Inspection of Ironworks in Tours), of 'Red Fox' Golden Virginia cigarettes (for the Economic Section) and of igniters (mercury fulminate, which was too acidic for the Artillery Technical Section), as well as on the solubility of hydrocarbons in alcohol, the metal of shells, cartridge powder for revolvers, and brass cartridges for machine guns.[46]

Sometimes, Nicolardot's reports enabled him to venture programmatic remarks. Regarding a proposal for poison bullets, by a certain Mr. Noble, he noted that the Germans were already preparing such projectiles, and that 'if we are not ready to respond to them in the same way, we must fear a dangerous moral depression as the result'.[47] To illuminate enemy airplanes, he proposed to fill rockets with ballistite as fusing powder, like modern propergols.[48] Often, Nicolardot put in requests for new instruments. For example, in 1917, he wrote:

> To be able to determine the influence of the acidity of the fulminate on its ability to detonate, it would be necessary to endow the laboratory with a small shock ram (*mouton de choc*). Then, after examination, it would be possible to compare the influence, regarding ability to detonate, of the acidity of the fulminate and of its reaction with the metal of the primer under the action of humidity.[49]

'An accurate result might not be got without using a silica resistance and operating in a vacuum,' he added. 'The current laboratory installation does not permit us to realize such a test.'[50]

Of potentially greater interest were the methodological studies requested from the LSTA by the Services. For example, the LSTA's notes for June 1917 begin with reports on quality control tests for oil sold to the artillery, and on the lubrication used on the brakes (*freins de tir*) of the 75-mm gun, built in the Atelier de Construction de Puteaux. On another occasion, the laboratory produced a method for measuring lead in brass, for the Compagnie des Forges et Aciéries de la Marine et d'Homécourt-Saint-Chamond. And on another, it derived a method for determining the boiling point of glycerines, for the Inspection of Ironworks in Lyons (Inspection des Forges de Lyon). In the latter case, the Laboratory arbitrated between the Société Française des Glycérines, in Marseilles, and the Laboratory of the Medicine Storehouse (Laboratoire d'Expertise de la Réserve des Médicaments) in the same city.[51] The next day, it completed a note on a method for grading oils.[52] When disagreement arose between two laboratories, or when results revealed abnormal compositions, analyses were repeated. Thus, a sample of perdite was re-analysed to resolve differences between the LSTA and the Laboratoire Municipal.[53]

Altogether, the number of studies and analyses steadily increased between 1916 and 1917 (see Table 1).

This table does not include analyses of leather and textile materials, which were numerous and took much time, including travel to and from provincial centres.[54]

Table 1. Studies and Analyses of the Chemical Laboratory of the Artillery Technical Section LSTA

| | Increase in Work | | |
|---|---|---|---|
| | March-May 1916 | March-May 1917 | Change |
| Studies | 33 | 78 | +45 (+136%) |
| Micrography | 22 | 66 | +44 (+200%) |
| Spectrography | 0 | 27 | +27 |
| Analyses | 948 | 963 | +15 (+1.6%) |
| | Increasing Demand | | |
| | March-May 1916 | March-May 1917 | Change |
| Studies | 35 | 121 | +86 (+246%) |
| Micrography | 25 | 73 | +48 (+192%) |
| Spectrography | 0 | 31 | +31 |
| Analyses | 1036 | 1332 | +294 (+28,6%) |
| | Incomplete Work | | |
| | March-May 1916 | March-May 1917 | Change |
| Studies | 5,7% | 35,5% | +29,8 |
| Micrography | 12% | 9,5% | −2,5 |
| Spectrography | – | 12,9% | +12,9 |
| Analyses | 8,5% | 27,7% | +19,2 |

*Source:* Nicolardot's 'Note sur la Nécessité de ne pas Diminuer le Nombre des Chimistes du Laboratoire'. CAA, LSTA, Note 583, 10 September 1917, 2.

However, it does show how data can mask a real increase in activity. In December 1917, Nicolardot observed that:

> The number of measuring out operations (*dosages*) and research studies still remains the same for these analyses. M. Philippet, with the help of the assistant chemist (*aide-chimiste*) Merlet, is able to do an even greater number of them, while drawing up the blueprints (without any description), which are of the greatest necessity, in order to give an accurate and brief report of the results obtained during unloading and dismantling enemy machines. The Minister has approved their necessity.

Nicolardot underlined the fact that the Laboratory's studies were becoming 'more important and therefore longer' than they had been a few months earlier. At the end of the summer of 1917, four studies were completed: (1) on steel patches (lopins), which emitted ammonia, and which occupied two chemists for an entire month; (2) on fireworks, which had given off spontaneous fires; (3) on means to prevent automobile radiators from freezing; and (4) on oil for brakes. Each of these studies required several hundred operations, measurements and tests.[55] And for all these, there were not enough chemists!

The sharing out of studies among the laboratories, and the division of labour in accordance with expertise, was the only way to satisfy increasing demand with a shortage of skilled scientists. A ministerial order of 17 October 1917 ordered a new distribution of work, but by this time, the rate of new analyses was already diminishing, following an order by the Director of the Artillery Technical Section, who hoped to keep up chemical activity without increasing chemical staff. In effect, Nicolardot got his wish – that 'a certain number of current analyses were to be shared out among the laboratories of the various establishments [of the Artillery], so as to let the Laboratory play its true role: a laboratory of arbitration, a laboratory of [research] studies'.[56]

## CONCLUSION

In early 20th-century France, those in charge of managing military research steadfastly claimed to be liberating the nation's laboratories from routine work, so as to enable them to devote themselves to research and to arbitrage between other establishments, such as the quality control laboratories. However, when in 1907 — in peacetime — the famous mathematician Henri Poincaré and his colleague, the chemist and engineer Henry Le Chatelier, confronted the military establishment, they failed to establish a laboratory dedicated to research. Ten years later, in 1917, a mere Artillery officer Major Nicolardot, partly succeeded where his predecessor, Captain Ducru, had failed in 1914.

The war required every French chemist to be at his most efficient, whether in civilian or military work. During the mobilization of 1914–1915, an initial division of labour took place, by which academic chemists were entrusted with research, while military laboratories were put in charge of assessment. During the reorganization of 1917, the dynamics of research finally became part of daily military work, since tests required the most expert skills available. Given the scarcity of chemical expertise, routine assessments were increasingly delegated to less expert institutions.

This impulse towards more intensive research in military laboratories came not only from the government and the military. Through less visible ways, it was also pushed by the logic of laboratory work. After the war, the Poincaré-Le Chatelier and Ducru-Nicolardot projects were finally realized, when the military powder factory at Le Bouchet was converted into a huge research centre for chemical warfare, where it continued under the supervision of academic chemists — first, Raymond Cornubert, later, Paul Lebeau — and a joint committee, the Commission des Études et Expériences Chimiques. From 1922 to 1940, the Centre d'Études du Bouchet housed several laboratories and workshops.[57] From a total work force of only 121 in 1935 — when the centre was transferred from the Artillery to the Service des Poudres — Le Bouchet grew to 290 on the eve of the Second World War — when mobilization increased its staff to 472.[58] Thus, the military-academic chemical network of the Great War presaged the first modern military research centre in France. Spurred by new fields of research, bred but barely developed in the haste of war, and not fully organized until peacetime, this came to play a major part in preparations for a future war.

## NOTES

[1] This has been properly shown for the dyestuffs industry and chemical education. See Ernst Homburg, 'The Emergence of Research Laboratories in the Dyestuffs Industry, 1870–1900', *British Journal for*

*the History of Science*, 25 (84), (1992), 91–111; and André Grelon, 'Les universités et la formation des ingénieurs en France (1870–1914)', *Formation Emploi*, n° 27-28 (1989), 65–88.

[2] Respectively, 200, 9000, 4500 and 2500 people. See Roy MacLeod, 'Chemistry for King and Kaiser: Revisiting Chemical Enterprise and the European War', in Anthony S. Travis *et al.* (eds.), *Determinants in the Evolution of the European Chemical Industry, 1900–1939: New Technologies, Political Frameworks, Markets and Companies* (Dordrecht: Kluwer, 1998), 25–49.

[3] Charles Moureu, *La Chimie et la Guerre. Science et Avenir* (Paris: Masson, 1920); Olivier Lepick, *La Grande Guerre Chimique, 1914–1918* (Paris: Presses Universitaires de France, 1998).

[4] Their increase was due to intensive work in the powder plants, both on regular propellants (gunpowder, *poudre B*), explosives (melinite, cresylite, tolite) and basic components (guncotton), or on new products, and to the creation of new factories: Bergerac, for manufacturing guncotton and *poudre B*; Sorgues, for manufacturing melinite, dinitrophenol (DNP) and melted mixtures (DD and MMN); Pont-de-Claix, for filling shells with 'special products' (opacite, yperite, phosgene, alcunite, sternite, arsines). See map, Figure 1.

[5] 'Le Service des Poudres Pendant la Guerre de 1914–1918', *Revue Historique de l'Armée*, 3, (1958); Louis Vannetzel, 'Le Service des Poudres', *Croix de Guerre*, special issue, (1961); André Corvisier (ed.), *Histoire Militaire de la France*, esp. vol. 3; Guy Pedroncini (ed.), *1870–1940* (Paris: Presses Universitaires de France, 1997).

[6] This paper is part of a personal research programme on 'The Laboratory and War in France, 1760–1955', related to the CRHST-CNRS programme. I am grateful to Mme Basque and M. Odin (Centre d'archives de l'armement, Châtellerault) and to Mlle Billoux and M. Brunet (École polytechnique, Palaiseau) for their help.

[7] The buildings housed also the *Atelier de Précision*, created in 1794, and the *Cabinet de Physique* or *Laboratoire d'Électricité*, also founded in 1818. See J. Challéat, *L'Artillerie de Terre en France Pendant un Siècle: Histoire Technique 1816–1919* (Paris: Ch. Lavauzelle, 1933, 2 vols.); *De l'Atelier de Précision à l'Établissement Technique Central de l'Armement, 200 Ans d'Histoire, Juin 1794–Juin 1994* (Montrouge: ETCA, 1995).

[8] For the *École Centrale de Pyrotechnie*, and the French tradition of research laboratories up to 1914–1918, see Patrice Bret, *L'État, l'Armée, la Science: L'Invention de la Recherche Publique en France, 1763–1830* (Rennes: Presses Universitaires de Rennes/Fondation Carnot, 2002). See also Robert Gencey, 'L'École Centrale de Pyrotechnie', *Instrumentation, Expérimentation et Expertise des Matériaux Énergétiques (Poudres, Explosifs et Pyrotechnie) du XVI$^e$ Siècle à nos Jours* (Actes des Troisièmes Journées Paul Vieille, Cité des Sciences et de l'Industrie, 19–20 October 2000), Paris, CRHST-A3P-CEA/DAM, 2001, 277–84.

[9] The obsolete name 'saltpetre' was abolished on 20 May 1914 in all the institutions of the Service.

[10] The LCP also housed the school (*École d'Application des Poudres et Salpêtres*), where future engineers were trained after the *École Polytechnique*.

[11] On this institution, see Patrick Suzzoni, *De Sully à l'ECN Paris* (Paris: DCN, 1997).

[12] On this affair, see Patrice Bret, 'La Guerre des Laboratoires: Poincaré, Le Chatelier et la Commission Scientifique d'Étude des Poudres de Guerre (1907–1908)', in *Actes du Colloque Henry Le Chatelier, Nancy, 15 Avril 2002* (Paris: Éditions CTHS, forthcoming).

[13] As the established expert on gunpowder, Berthelot had been expected to chair this committee, but died a few days before its creation.

[14] Archives de l'Académie des Sciences, Archives Le Chatelier, Le Chatelier to Poincaré, 2 November 1907: 'Pour vous donner par un exemple l'état d'esprit de quelques membres de la commission, je vous dirai que le capitaine Robert me prend toujours comme intermédiaire quand il y a à obtenir un renseignement du laboratoire des Poudres et Salpêtres, ne voulant pas se compromettre en le demandant directement à M. Vieille. Je m'occupe ainsi en ce moment de la question des régulateurs de température. Les Versaillais déclarent qu'ils ne peuvent pas régler à 25° près la température, bien qu'ils aient commandé au fournisseur de M. Vieille, un de ses appareils. Or le dit fournisseur chez lequel j'ai été pour élucider cette question m'a dit que le modèle fourni par lui à Versailles est celui que M. Vieille emploie pour les températures de 40° et qu'il ne peut aucunement convenir pour 110°. Deux minutes de conversation entre les intéressés auraient évité tout ce temps perdu.'

[15] Created by decree, 8 January 1912. See Vannetzel, *op. cit.* note 5, 176.

[16] Service Historique de l'Armée de Terre (Vincennes) (afterwards, SHAT), Kœhler file (6Ye 106050),'Travail d'avancement de 1911'.

[17] SHAT, Kœhler file,'Travail d'avancement de 1913'.

[18] Vannetzel, *op. cit.* note 5, 115. Kœhler died in Lyons, at the hospital founded by the brothers Lumière, his brother's brothers-in-law. See SHAT, Kœhler file, 22 July 1916.

[19] Marqueyrol later became Inspecteur général. He left the Service des Poudres in 1925, moved to the laboratory of a railways company, and died in 1957.

[20] The laboratory of the VIIth Army was set up in the former Customs Laboratory in Dunkirk. See Anne-Claire Déré, 'Maurice Javillier et Gabriel Bertrand, Deux Chimistes Pastoriens dans l'Institut en Guerre', in D. Aubin and P. Bret (eds.), 'Le Sabre et l'Éprouvette: L'Invention d'une Science de Guerre, 1914–1939', *14–18. Aujourd'hui. Today. Heute*, 6 (2003), 75–87.

[21] And 7720 workers. Vannetzel, *op. cit.* note 5, 118; René Pique, *La Poudre Noire et le Service des Poudres* (Paris: Société des Publications Colloïdales, 1927), 162.

[22] Besides twenty-five accountants and 329 technical under-agents, and with a total strength of 120,000 people in 1918. Vannetzel, *op. cit.* note 5, 117 and 161.

[23] *Institut de Chimie Appliquée de Toulouse, École de Chimie Industrielle de Lyon, Institut de Chimie Appliquée de Lille, Institut Chimique de la Faculté des Sciences de Montpellier, École de Chimie Appliquée de Bordeaux, École de Chimie Appliquée de Nancy, École de Chimie Appliquée de Mulhouse.* The latter one, though now in Germany, had kept the French language. Other provincial institutions were added in the 1920s. *Ibid.*, 161.

[24] *Ibid.*, 160–161. In fact, Muraour was the only one in place in 1918 (*Annuaire du Corps des Ingénieurs Militaires et des Corps d'Agents Militaires et de Sous-agents techniques Militaires des Poudres. Arrêté au 3 Mars 1918*).

[25] See Ernest Berger, *Sur la Décomposition et la Stabilisation de la Poudre B, Conférence faite devant la Société chimique de France le 16 Mars 1912* (Paris: Société Chimique de France, 1912).

[26] Centre d'Archives de l'Armement (Châtellerault) (afterwards, CAA), Fonds 10, Carton 508, Laboratoire de Chimie de la Section Technique de l'Artillerie (LSTA), Note 583, 10 September 1917, 3–4.

[27] CAA, LSTA, Note 671, 5 December 1917, 1–2.

[28] Note 583, 10 September 1917, 5.

[29] *Ibid.*, 5–6.

[30] *Ibid.*, 6.

[31] Déré, *op. cit.* note 20. Kling and Florentin worked on other materials. See Lepick, *op. cit.* note 3, 54–55.

[32] See E. Vinet, 'La Guerre des Gaz et les Travaux des Services Techniques Français', *Chimie et Industrie*, 2, (11–12), (1919), 1377–1415; Moureu, *op. cit.* note 3. Only recently have historians begun to pay attention to their work. See, for example, Déré, *op. cit.* note 19.

[33] Some of these scientists also met in the third (Chemistry) Commission de la Défense Nationale of the Académie des Science. See Archives de l'Académie des Sciences, DG 41.

[34] Vincennite was based on $AsCl_3$, with or without $SnCl^4$.

[35] See Ministerial Orders (OM) No. 16608-2/3, 3 February 1916 and No. 111246-2/3, 19 July 1916.

[36] Following the ministerial dispatch, No. 168 197-2/3, 20 October 1916.

[37] The popular name, Fascicule LM, came to be printed on the front cover after the first few issues.

[38] Ministerial dispatch, No. 154.298-2/3. The new full title was: *État Mensuel des Renseignements Techniques Nouveaux Relatifs aux Munitions de l'Ennemi, Examinées par les Établissements du Service de l'Artillerie et du Service des Poudres. École Centrale de Pyrotechnique.—Section Technique de l'Artillerie.—Atelier de Chargement de Clermont-Ferrand.—Ateliers de Construction de Bourges, Lyon, Puteaux et Rennes.—Manufacture d'Armes de St Etienne.—Commission d'Expériences de Bourges./Laboratoire Central des Poudres.—Poudrerie Militaire du Bouchet.—Commission des Poudres de Guerre de Versailles.*

[39] In spite of an increase from the first quarter (26%) to the last quarter (37%) of 1918.

[40] Average rate for 1918, but only fourteen in the second half of the year.

[41] CAA, LSTA, Carton 508: last note of every quarter, from 31 March 1917 (Note 435).

⁴² For 1916, we see Municipal Laboratory (74.42%), LSTA (15.58%); 1917, Municipal Laboratory (45.11%), LSTA (54.89%); and for 1918, Municipal Laboratory (30.92%), LSTA (23.12%), *École Centrale de Pyrotechnie* (15.03%), *Atelier de Chargement de Clermont-Ferrand* (10.69%), Le Bouchet (10.69%), LCP (3.47%), and the Versailles Laboratory (3.18%).
⁴³ CAA, LSTA, Notes 528 (26 June 1917) and 529 (27 June 1917).
⁴⁴ The slag was reduced to powder and heated with a 'compressed air *Méker* burner, on a thin plate of platinum' up to 1150 °C, then temperature was pushed by enrichment of oxygen until the platinum melting. Nicolardot wrote: 'An accurate result might not be got without using a silica resistance and operating in a vacuum. The current laboratory installation does not permit us to realize such a test.' (CAA, LSTA, Note 519, 18 June 1917).
⁴⁵ CAA, LSTA, Note 522, 20 June 1917. Such analyses were also done by the Laboratoire de Toxicologie de la Préfecture de Police.
⁴⁶ CAA, LSTA, Notes 503, 508, 515, 517, 518, 521, 523.
⁴⁷ CAA, LSTA, Note 671, 5 December 1917, 4–5.
⁴⁸ *Ibid.*, 3–4.
⁴⁹ CAA, LSTA, Note 515, 15 June 1917.
⁵⁰ CAA, LSTA, Note 519, 18 June 1917
⁵¹ CAA, LSTA, Note 688, 22 December 1917.
⁵² CAA, LSTA, Note 689, 23 December 1917, following a previous Note 676 of mid-December.
⁵³ CAA, LSTA, Note 511, 10 June 1917.
⁵⁴ On 7–8 December, Nicolardot and 'M. Vourloud' visited the *Atelier de Construction de Rennes*, in Brittany, in order to analyse the operating procedures for the salvage of tallow (CAA, LSTA, Note 675, 11 December 1917).
⁵⁵ CAA, LSTA, Note 583, 5.
⁵⁶ CAA, LSTA, Note 583, 4. Leleu's order of 2 September 1917.
⁵⁷ Physiology (1923), Protection (1924), Synthesis (1925), Therapeutics (1926), Dispersion (1926), Prophylaxy (1938).
⁵⁸ The *Laboratoire de Protection* and the *Laboratoire de Synthèse* were first created in 1922. New installations were added up to 1939: a second laboratory and a mechanical workshop for the Protection Service in 1924; an *Atelier de Demi-grand* in 1927 for the Synthesis Service; an extended laboratory and a 1000 cubic metre spherical chamber for the Dispersion Service; a laboratory for the Prophylaxy Service; and a group of laboratories for the professors of the CEEC in the mobilization of 1939–1940. See Adolphe Kovache, 'Historique du Centre d'Études du Bouchet', *Sciences et Techniques de l'Armement*, 55 (4), (1981), 651–717. From 1935 to 1978, experiments *en grand* were conducted in the related centre 'B2', located in Namous, Algeria.

# THE WAR THE VICTORS LOST:
# THE DILEMMAS OF CHEMICAL DISARMAMENT, 1919–1926

INTRODUCTION

'It will be too awful', wrote Harold Nicolson from Versailles in March 1919,' if, after winning the war we are to lose the peace.'[1] As all the world knows, the Armistice in November 1918 did not herald the coming of peace. Seldom has so deliberate a settlement been framed with such idealistic intent. Yet, amidst the transformation of borders, the collapse of empires, and the transfer of colonies, Germany refused to accept either the terms of the treaty or the fact of defeat. For over eighty years, scholars have underscored the flaws of the negotiations and the failure of Versailles. Much has been written on questions of nationalism, self-determination and reparations, the distortion of the post-war balance of power, and the rise of Fascism. However, it is also important to see Versailles as an important step – however tentative – in the long road towards arms control and disarmament. In some ways, its origins and early implementation can be viewed as an experiment, whose evident failure outlines the difficulties inherent in securing adequate safeguards and compliance.

These difficulties are nowhere more clear than in the history of Allied attempts to control and monitor postwar German industry, and its vast chemical industry in particular. One month after the Armistice, Friedrich Ebert, head of the new German republic, consoled a returning German Army at the Brandenburger Tor with the words, 'No foe has overcome you…you have protected the homeland from enemy invasion'.[2] The same might have been said to the barons of Germany's industry, who ended the war with their inventories full, their factories intact, their transportation in place. The Allies gave German chemical technology a prominent place in the restrictive provisions of the Armistice, and in the Treaty of Versailles.

Yet, strangely, given its importance, this remains a neglected corner of Allied history. Whilst historians have discussed industrial disarmament in general,[3] and the history of the chemical industry in particular, the role of 'chemical disarmament' has been the Cinderella of the story.[4] Nonetheless, the attempts of Allied officials to survey and monitor the industry lies at the heart of political, military, economic and technology policy concerning military-industrial 'dual-use.'[5]

This neglect is not less surprising, for the fact of being so well known at the time. Within scarcely two months of the Armistice, another 'war' began. This was to be fought, however, not by soldiers and generals, but by inspectors and managers and, to a lesser extent, by diplomats and politicians. Its weapons were not tanks, but statistical tables, not formulae but forms; not quotas but questionnaires. Significantly, its parameters were technical, not tactical; and its ambitions were not territorial, but economic. Allied efforts to enforce the 'chemical disarmament' of Germany had two objectives – to reduce and if possible eliminate a potential source of future military aggression, and to assist Allied civilian industry in international markets.

This, however, was not easily done. At the core of the problem was the enduring character of chemical technology. An industry that could produce both civilian fertiliser and military explosives classically exemplified the dilemma of 'dual-use', a feature whose dangers were eloquently expressed by an Allied mission in 1919: 'In the future...every chemical factory must be regarded as a potential arsenal...'.[6]

This essay considers the work of the key agency – the Military Inter-Allied Control Commission (MICC) – and the German response to its efforts. Using newly released and rarely used archives, we explore why, whatever its intentions, postwar control largely failed, and in some respects proved counter-productive. The struggle on the Rhine became a war the victors lost. German interests out-manoeuvred Allied efforts, with consequences that would echo around the world for years to come.

## THE GERMAN CHEMICAL INDUSTRY IN 1918

As in Britain and France, the war forced Germany to develop an integrated national system of weapons production on an unprecedented scale. The military-industrial struggle embraced chemistry. Despite labour and materials shortages, production rose dramatically in the last two years of the war. By September 1918, German factories, both state-owned and private, had reached the goal of 12,000 tons/month of propellant set by the Hindenburg Program two years earlier. Overall, production rose from 300 tons per month in September and October 1914, to 6,000 tons per month four years later, a twenty-fold increase. Similar increases took place in the production of explosives; by the end of the war, former dyestuffs firms were producing most of the explosives, as well as most of the nitrogen compounds required for propellants. From the autumn of 1916, contracts specified that the industry's plants for the production of nitric acid and high explosives, built or expanded with subsidies from the imperial government, be maintained in 'readiness' for 10–15 years, so that in the event of future hostilities they could quickly resume military production.[7]

In 1916, the pressures of wartime precipitated industrial cooperation through the formation of the *Interessengemeinschaft* (IG), a pooling agreement among the eight pre-war dye firms that merged to form the mighty I.G. Farben in 1925. The IG was formed in the full expectation of an economic 'war after the war' that Germany would have to wage to regain control of its international markets.[8] In 1917, the IG, whose principal factories were geographically near the largest explosives and munitions plants, agreed to supply the powder cartel (including the principal prewar explosives and powder companies) with much of the sulphuric acid and all the nitrates and high explosives the cartel needed, except nitroglycerine and nitrocellulose propellants. By this agreement, a pooling contract between the IG and the powder group was to go into effect after the war. Negotiated in the expectation of a German victory (and cancelled after the Versailles Treaty), this agreement would have given the IG an effective monopoly in high explosives production, as the wartime share of the German state factories (which the IG also supplied with much of their sulphuric and nitric acid) was negligible.[9] This agreement, along with the readiness plants, marked the industry's formal integration into future as well as current military planning – an integration that would persist, despite the forced nullification of the readiness contracts and the explosives pool in the aftermath of German defeat.

Following the Armistice came disruption in transportation, shortages of resources (especially coal), the cancellation of contracts, and sharp reductions in chemical production.[10] However, industrial capacity remained intact. Clearly, it was in Germany's interest to convert wartime facilities to peacetime production as quickly as possible. The chemical industry continued to enjoy its new-found independence from overseas nitrates.[11] Moreover, many of its wartime products found peacetime uses – for example, phosgene, picric acid, and arsenic compounds were in demand as intermediates in the production of dyes, pharmaceuticals, and other commercial products. Almost immediately, this raised a question whether the Allies could deal with the contingencies of an industry whose technology was manifestly both military and civilian.

## THE ARMISTICE AND ALLIED OCCUPATION

As late as the summer of 1918, the Allies and Germany foresaw the war as continuing well into the following year; and both sides were preparing for a spring campaign in 1919 that would have combined massive aerial attacks and chemical weapons. That mercifully this did not happen was as much as a surprise as the suddenness of the Armistice.[12] On 12 November 1918, the victorious Allies set up, under Marshal Foch, an Inter-Allied Armistice Commission to consider and resolve issues that were not settled when the Armistice documents were signed. These issues included the repatriation of prisoners, the confiscation of depots and stores, and the surrender of war materials. The authority of the Commission was to reach beyond the Western Front, and to include the Eastern Front and Alsace-Lorraine. The Armistice Commission was envisaged as the main channel of communication between the Allies and the Germany until the Treaty was ratified; at which point, it was assumed that communications would resume through normal diplomatic channels.[13]

Immediately following the Armistice, British, French, Italian and Belgian zones of occupation were agreed, and military governors representing the different Allied powers were appointed to each. Following the terms of the Armistice, the Allies occupied zones of the Rhineland – all the territory west of the river, including bridgehead areas on the right bank. These contained the principal chemical plants of Germany. British Headquarters were established in Cologne, under the command of Field Marshal Lord Herbert (later Viscount) Plumer, GCB.[14]

Understandably, many uncertainties beset the occupation. Allied objectives in relation to the chemical industry were twofold– to enforce 'chemical disarmament' and to extract technological 'know-how'. These objectives, in turn, had a double rationale – to reduce and if possible eliminate a potential source of military aggression, and to help Allied industry compete in international markets. France and Britain, in particular, were determined to secure a victory in the Factory that had eluded them on the Front. However, the Allies had conflicting visions of what victory meant. If all agreed upon the problem, there was little agreement about its solution.

The Ludwigshafen and Oppau works of BASF, like the Dormagen picric acid plant of Bayer, stood on the left bank of the Rhine, and were thus in zones of French and British occupation, respectively. The Ludwigshafen and Oppau plants were occupied by the French on 6 December 1918. Other firms (Höchst near Frankfurt, Bayer in Leverkusen near Cologne) were in the French and British sectors on the right bank. Before the end of 1918, British authorities ordered the dismantling of the Bayer mustard gas plant at Leverkusen, and the confiscation of strategic chemicals to use in dye production, as part of the Armistice provisions.[15] The British plan, in their sector – to which the French took extreme exception, in theirs – was to restart German production 'in some reasonable approximation of normal output', without at the same time posing a threat to its counterparts in British industry. This was necessary because, despite (or indeed, because of) the war, Britain was 'not yet in a position to compete with German manufacture'.[16] In furtherance of this policy, the Board of Trade granted Bayer permission to export 600 tons of dyestuffs to the unoccupied regions of Germany. Some British industrialists feared these chemicals might find their way to Britain. Yet, German firms insisted that Britain had more to fear from American than German competition; and indeed, American chemical companies were showing a keen interest in capturing their rivals' international markets.

In mid-January, 1919, General Plumer – praised for swiftly restoring food supplies and suppressing strikes – established an Industrial Department at his headquarters in Cologne, with instructions to study the economic situation in the British sector. Its work had high priority, given the need to stabilize the industrial workforce against the provocations of Bolshevik agitation; to secure new markets for British goods; and to create a context for reparations payments. At the cessation of hostilities, Bayer-Leverkusen was producing 100,000 tons of heavy chemicals and 1,000 tons of dyes per month. If Germany was to meet its war debts, the argument went, these products had to be exported, hence workers must remain employed.

To the Industrial Section were assigned three recently demobilised chemists – Dr. Shore, Thomas Slater Price (wartime head of the Royal Naval Experimental Station at Stratford, who had a doctorate from Leipzig), and Dr. G.H. Knibbs (HM Factory, Langwith) – as well as Major Hamilton McCombie, Chemical Adviser to the First Army (who had a doctorate from the German university of Strasbourg), and

Major Arthur J. Allmand, Chemical Adviser to the Second Army, who had a DSc in Chemistry from Liverpool, and who had studied with Fritz Haber at Karlsruhe before the war.[17] Most spoke excellent German. In late January, Allmand visited Leverkusen, and plans for inspections were put in place.[18]

Whilst this was underway, another initiative was undertaken by the Ministry of Munitions and, in particular, by its Departments of Trench Warfare and Explosives Supply.[19] The latter, led by Lord Moulton, FRS, had been largely responsible for the 'miracle' of Britain's chemical victory.[20] In late December, the Ministry of Munitions sent a representative with instructions to establish a Technical Commission, to study German processes, and to report directly to Moulton. The Ministry knew that Germany's chemical industry was in an 'extremely strong position both financially and technically as compared with the British'; and that much of its plant resided in the British sector. During January, Marshall visited several firms – including the Königliche Geschossfabrik in Siegburg and the Mannstaedtwerke-AKT Gesellschaft and Bayer at Leverkusen – and prepared reports on processes, manpower, and outputs. It became clear that these and other firms had taken advantage of their enforced inactivity since November 1918 to convert to commercial production. In early February, Marshall was succeeded by H.S. Denny, an Australian mining chemist and engineer, and a veteran of the huge munitions complex at Gretna, who was instructed to report on a range of issues from coal supplies to inter-allied coordination.[21] In advising the Ministry, Denny outlined plans to keep the German factories going, and their workers from going hungry.[22]

To supplement these individual efforts, the Ministry of Munitions decided to survey the entire spectrum of industrial chemistry in the British sector, and to this task appointed Brigadier (later Sir Harold) Hartley, FRS, sometime Lecturer in Chemistry at Balliol College, Oxford, and in 1918, Chief Chemical Advisor to the War Office.[23] On 29 January 1919, Hartley arrived in Cologne on the first of what would be two missions to Germany.[24] His team included fellow academics and temporary officers Capt. A.C.G. Egerton and Lt. Harold Cecil Greenwood, who had also worked in Karlsruhe with Haber;[25] the ubiquitous explosives expert, William Macnab;[26] and Herbert Levinstein, director of the leading British dyestuffs firm, soon to be merged into the British Dyestuffs Corporation.[27] They were joined by F.H. Carr from the Ministry, and the chemist A.W. Tangye, who had worked out a practical method of synthetic phenol production during the desperate days of 1915. To the Mission were also attached six American officers; representing the US Army Ordnance Department and Chemical Weapons Service; four French officers, five Italians, and one Belgian. During the first half of February, the Mission inspected a number of factories, including Oppau (the sight of which Tangye called 'wonderful'). At Bayer's headquarters in Leverkusen, they interviewed Dr. Duisberg and Dr. Quincke, and inspected Bayer's processes of ammonia oxidation. On 5 February, they visited Dormagen, where they inspected the process of explosives manufacture.[28] On 26 February, Hartley submitted his report to Lord Moulton and to the Minister of Munitions.[29]

By the end of February 1919, the Ministry thus had an overview of German activity, if only in the British zone. Three days earlier, Moulton also received a report from Marshall, outlining the economic argument for letting German firms export dyestuffs from both the British and French sectors.[30] In a lengthy memorandum of 4 March 1919, Marshall urged the resumption of trade with Britain.[31] Marshall reported separately on Bayer-Leverkusen with an appendix on Schlebusch.[32] By the end of March, the Technical Commission was wound up, and its staff returned to Britain. [33] Its final report was presented on 6 June 1919.[34]

From the outset, it was clear that the Allies faced difficulties. For a start, they could not inspect the whole of Germany's chemical industry. In fact, Bayer's centre at Elberfeld, and BASF's new ammonia plant at Leuna, as well as most of the Agfa plants, were located in unoccupied Germany, as were all the State Powder Factories of Bavaria, Prussia and Saxony. It was also clear that, owing to the economic crisis, German firms needed French and British troops to maintain order.[35] However, the French and British could also confiscate dyes and other chemicals as reparations; for these, companies were to receive compensation from the Reich, but this was not enough to guarantee cooperation with Allied inspectors.

All through 1919, the British and French continued individual inspections of works in their zones. On 14 November 1919, Sir Robert Robertson, newly elected FRS (1917) and newly knighted (1918) head of the research department at the Royal Arsenal, Woolwich – a chemist distinguished for his wartime work on TNT and 'amatol' [36] – visited factories in the neighbourhood of Cologne, Dormagen, Wahn, Leverkusen and Troisdorf, and in the American zone around Coblenz. He found that the 'easy convertibility' of Germany's chemical factories to war work 'striking', as was their production of synthetic nitrates, which had enabled them to dispense with Chilean imports. His trip to Troisdorf followed an earlier visit in July to examine the process of making nitrocellulose from wood, a subject that (along with safety) attracted considerable British attention.[37]

Not surprisingly, the giant German firms had little interest in encouraging commercial espionage; and resisted in various ways, including shutting down plants so as not to reveal their technical details. On 10 December 1918, for example, BASF reported to the German Auswärtiges Amt (Foreign Office) that the French general staff had given notice that a delegation of French chemists was *en route* to their works. The company took the Armistice provisions to mean that the French could demand information about products and raw materials only, and not about production processes, manufacturing guidelines, statistics of business operations, 'and similar things that are of their nature secret'. Demanding trade secrets would, in their view, violate the terms of the Armistice, 'which prohibits any damage to private property'.[38]

In February 1919, BASF reported to the German Armistice Commission (Deutsche Waffenstillstandskommission) that on 28 January their plants in Ludwigshafen and Oppau had come under the control of two French officers, assisted by five others, all chemists and physicists. Ostensibly there to study what BASF had produced in war materials and to prevent their further manufacture by the monitoring of inputs and outputs, the French experts were, in their view, 'in fact going far beyond these tasks', looking for precise information about production methods, the construction of apparatus, composition of products. They were using all means to get it, interviewing staff, observing operations, and taking samples. Their interest reached beyond dye production, and included the Oppau nitrogen works. Protests to the French Commissioners proved unavailing, and in the expectation that nothing would come of protests to the Armistice Commission, BASF announced that it would request compensation for damages. To prevent prying eyes, Carl Bosch ordered to be shut down those operations that risked revealing processes. As it was, the company suffered substantial losses – 25–30 million Marks from Oppau alone during the first four months of 1919. Even without the threat of French espionage, it is doubtful whether BASF could have renewed full production, given its limited coal supplies.[39]

Its competitors fared little better. Höchst supplied their French occupiers, as required, with information on war-related production, but also found French control officers taking a close interest in pharmaceuticals such as Salvarsan. A director of Höchst later asserted that the officers had passed critical information to their French competitors, Poulenc Frères and Usines de Rhône.[40] Moreover, during the Hartley Mission in February 1919, Carl Duisberg, Senior Director of Bayer (whose Leverkusen plant was Hartley's first stop) accused Levinstein of spying, and afterwards claimed that he had support from Hartley, who had allegedly rebuked Levinstein on similar grounds.[41] Whatever the truth of the allegations, there was obviously mileage in playing the Allies off against each other.

## THE TREATY OF VERSAILLES AND THE DISARMAMENT DILEMMA

The occupying forces faced a dilemma. In effect, commercial, military, and political interests were at cross-purposes. If the Allies shut down German plants to achieve 'chemical disarmament', they ran the risk of driving unemployed workers into the Bolshevik camp. Moreover, Allied textile industries needed German dyes, many of which Allied industries could not satisfactorily produce, even after confiscating German patents and factories. In any case, the British were not interested in establishing a long-term presence in the Rhineland, while the French wanted to retain long-term controls over German factories. The cash-rich Americans had no interest in occupation and little in control, although they had a good deal of interest in German technology. In failing to ratify the Versailles Treaty, the United States lost the opportunity to benefit from its technology-transfer provisions (including Article 297, which gave the Allies 'free use of German patents'); however, American companies independently confiscated German patents, and took advantage of the favourable exchange rate during the accelerating German inflation, to buy up stock in German companies. For this reason, if no other, Americans wanted the German companies to remain intact.[42]

These inter-Allied differences became painfully evident during negotiations over the Treaty, which produced compromises on the issues of occupation, controls, disarmament, and technology transfer. The relevant sections of the Treaty were embraced under Part V: Military, Naval & Air Clauses, which dealt with disarmament. These provisions fell into two broad categories. First, under Section I: Military Clauses, Chapter 2: Armament, Munitions and Material, were Articles 166–172, which specified the surrender of munitions to be destroyed or rendered useless. Second, under Section IV, Article 204 formally established the instruments of regulation – namely, the Inter-Allied Commissions of Control (which included the Military Commission, along with similar bodies for air and naval controls). The duties of the Military Commission were spelled out in Article 208:

> The Military Inter-Allied Commission of Control will represent the Principal Allied and Associated Powers in dealing with the German Government in all matters concerning the execution of the military clauses. In particular, it will be its duty to receive from the German Government the notifications relating to the location of the stocks and depots of munitions, the armament of the fortified works, fortresses and forts which Germany is allowed to retain, and the location of the works or factories for the production of arms, munitions and war material and their operations.
>
> It will take delivery of the arms, munitions and war material, will select the points where such delivery is to be effected, and will supervise the works of destruction, demolition, and of rendering things useless, which are to be carried out in accordance with the present Treaty.
>
> The German Government must furnish to the Military Inter-Allied Commission of Control all such information and documents as the latter may deem necessary to ensure the complete execution of the military clauses, and in particular all legislative documents and regulations.

The authority of the Commissioners was specified in Article 205: 'they shall be entitled, as often as they think desirable, to proceed to any point in German territory, or to send sub-commissions, or to authorize one or more of their members to go, to any such point'.[43] Thus, the Allied inspectors could descend on German factories on demand, 'any time, anywhere, without the right of refusal'.

The Military Control Commission was to receive from the German government information concerning 'the nature and mode of manufacture of all explosives, toxic substances or other like chemical preparations used by them in the war or prepared by them for the purpose of being so used'. (Article 172). The Commission was to enforce Article 168, which stated that 'all other establishments for the manufacture of any war material whatever shall be closed down', beyond that necessary for the reduced peacetime needs of the small German army, as provided for in Article 169. The German government was required to provide liaison officers and assistance for the work of the commissions (Articles 206–207).[44] However, the definition of the term 'closed down', the specification of what was to be included under 'other like chemical preparations', and the distinction between factories dedicated to 'war material' and peacetime, commercial production left gray areas that could be easily exploited.

## THE MILITARY INTER-ALLIED CONTROL COMMISSION AND THE GERMAN RESPONSE

In June 1919, the Allied Council of Four discussed the establishment of military controls in detail. Their initial assumption was that these would be completed within three months of the Treaty coming into effect.[45] To administer controls, the Allies called for three Inter-Allied Commissions (military, naval and air), to be presided over by a French general, a British admiral and a British general, respectively.[46] In July 1919, the Military Inter-Allied Control Commission was set up, and the French General Nollet was appointed President.[47]

The Military Commission divided Germany into eleven districts: four British, four French, two Italian and one Belgian. Within the Commission, three sub-commissions were appointed, dealing, respectively, with effectives (manpower), fortifications, and armaments.[48] The head of the British section, and president of the Sub-Commission of Armaments, was Maj. Gen. Sir Francis Bingham, KCB, KCMG.[49] The headquarters of the Commission were in Berlin. Within the Armaments Sub-commission was a chemical section, headed by French and British officers, Colonel Henri Muraour and Dr. Hugh (later Sir Hugh) Watts.[50] They were to do the work of chemical inspection and control, on the basis of which they would formulate recommendations for the disposition of German plant.

The work of control began slowly, and amidst difficult circumstances. On 22 June, 1919, the British scuttled the captured German fleet at Scapa Flow, 'in the expectation of a renewal of hostilities'.[51] In Britain, however, industrial demobilization had begun. On 1 July 1919, responsibility for munitions (and wartime expertise concerning chemical munitions) was returned to the War Office in London.[52] However, it was not until 15–17 September that a delegation arrived at the German War Office in Berlin, setting out the terms of industrial demobilization as required by the Treaty.[53] The Germans requested that this preliminary group be limited to 2–3 officers, but Marshal Foch insisted on sending seventy, representing the five principal Allied countries.[54] Technically, the MICC did not come into existence until the ratification of the Treaty by three of the principal powers. However, on 15 September, the first on-site inspections were held, beginning a series that would continue until 1 February 1927, and ultimately involve 400 Allied officers and 100 other personnel in a total of 33,381 missions and in the inspection of approximately 7,000 plants.[55]

During his first three months in Berlin, Bingham later recalled, the British felt 'like lepers on a leper island', preparing questionnaires to be used once the Germans appointed their Liaison Department. In October 1919, the German government reluctantly appointed a committee for military liaison, which included Generals von Cramon (from headquarters in Austria), von Kessel (chief of staff of the First Army), von Biebenstein (engineers), and Strempel (artillery). On 10 November 1919, the first British questionnaires were presented to German firms.[56] In January 1920, the Treaty came into effect, and 220 Allied officers arrived in Berlin to staff the new control commissions.

The subsequent story of questionnaires, inspections, and controls, and the German response to them, is much more complicated than earlier scholars have realized. Partly in cooperation with the government, partly on their own, the major German chemical firms coordinated their responses to the Allies, exploiting weaknesses in the inspection system, playing off controllers against each other, and taking advantage of the conflicting interests of the Allies. The IG proved to be an invaluable instrument for doing all three. When, in January, the control regime of the MICC officially began, the Allies expected a relatively short operation, with most of the initial work done within three months. In fact, the control commissions remained in operation for seven years, to 1927. This was clearly a different sort of war, fought not with munitions but with munitions in mind.

The initial phase had begun in November 1919, when the Armaments Sub-Commission submitted to the German government its first questionnaire, containing a series of technical questions. To compile answers, Berlin appointed two experts, Fritz Lenze and Fritz Haber, directors, respectively, of the Propellants Section of the Prussian Militärversuchsamt, the Military Testing Office (which in April 1920 became the Chemisch-Technische Reichsanstalt or CTR, nominally the National Office for Chemistry and Technology, but in practice limited to questions concerning explosives, including plant safety);[57] and the Kaiser Wilhelm Institute for Physical Chemistry, Haber's wartime centre of gas weapons research and development.

Unfortunately, the Allies issued their memorandum before the arrival of their technical experts, Muraour and Watts, who could have made the wording more precise and effective. In fact, it lacked 'any question related to the manufacture of cellulose powder'. Moreover, the German experts tried their best to evade the 'important questions'. When the experts did arrive in January 1920, they found it necessary to clarify the questions by consulting directly with their German counterparts.[58] Haber, at least, apparently believed that he could make use of personal friendships with the Allied experts to control the amount of information released.[59]

The Germans provided most of their official answers in April 1920 (thus, technically within the time limit set by the Treaty), and the rest by July. The Germans believed that they had gone well beyond the requirements of Article 172;[60] nevertheless, in many respects, the Allies were dissatisfied with the information they were given. The Germans found, to their dismay, that they had hinted at information that the Allies found interesting (such as the location of a secret military chemical plant in Berlin-Adlershof, about which the Allies demanded to know more).[61]

In response to Allied pressure, industrial representatives consulted with the Germans Foreign Office. The Economics Ministry and the Ministry for Reconstruction [Wiederaufbau] – which was directed in 1920–21 by Walther Rathenau – decided to comply on a limited basis, conceding the Allies the purely military information they demanded in the first questionnaire, in the hope of limiting the loss of commercial data in future. However, the Ministry of National Defense [Reichswehr] took a much harder line, and tried to keep back information it deemed of significance. This attitude corresponded with the Ministry's resistance on other fronts, from the dissolution of the General Staff to the reduction of the German army.

The influence of the Reichswehr appears to have been felt more strongly by staff at the CTR, which handled explosives and propellants questions, than by Haber, who worked in chemical warfare, and who was more closely associated with industry and with the Reconstruction Ministry (in which he, for a time, headed the chemical department). Industry sought to keep discussions with the Allies at the level of technical exchanges, over which they had some control, while avoiding a policy of resistance, such as advocated by the military, which seemed likely to result in Allied intervention. Indeed, their fears materialised in the form of the so-called London Ultimatum of 5 May 1921, in which Allied governments threatened to occupy the Ruhr if the German government continued to evade compliance on a wide range of Treaty and control questions.[62]

In addition to written information, the MICC demanded samples of war materials. One of the first chemical items was a sample of guncotton, which Bingham took to Britain in March 1920 for analysis. The reassuring result was that the German material was inferior, having excessive amounts of matter that were insoluble in acetone, probably due to the emergency use of wood cellulose. 'Stability rather low', the report laconically added.[63] The MICC also obtained for analysis samples of the German paper used as a base for 'P.E.Wolle' which, with other paper-based types, accounted for 'practically all' of the nitrocellulose propellants the Germans produced during the war. These seemed to be potentially useful as 'standard for products that may be evolved here'.[64]

In the French zone, inspection of IG plants ended when the Treaty came into force in January 1920, and the French army ended its direct occupation of these works in March, but the zone remained under military occupation, and there was occasional intervention by French troops – e.g., to quell labour disturbances. For the most part, it seems that MICC inspections of IG plants went smoothly.[65] However, the extension of MICC inspections to other German chemical factories in February 1920 produced a variety of protests, which the German Foreign Ministry and the German Military Peace Commission (Heeresfriedenskommission) conveyed to the Allied authorities. In reply, General Nollet, on 6 March 1920, insisted on his 'absolute right' to make unannounced visits. The Germans disputed this, although their Military Commission dismissed the idea of protesting against unannounced visits as 'pointless', in view of the fact that firms had known for months what was coming and should have been prepared.[66]

On 10 February 1920, a commission that included Haber and representatives of the Reich Industry Group, the Chemical Industry Association, and several IG and other firms (including Troisdorf) met to discuss arrangements which the Reich Industry Group had set up to supplement liaison with the MICC. The commission was under the direction of Professor Julius Flechtheim (an expert on corporate law and director of the former Vereinigte Köln-Rottweiler Pulverfabriken, Germany's largest wartime producer of propellants. In March 1919, the company had officially renounced the production of military explosives, changing its name to

Köln-Rottweil AG and moving to the production of commercial cellulose products instead). Following the meeting, the *Reichsverband* selected industrial spokesmen to work with the military in arranging MICC factory visits in each district. The next day, a smaller group discussed the questionnaire of 31 January, which the CTR wanted the explosives firms to answer, dealing with the production of eleven explosives and related substances (including ammonium nitrate and oleum). To provide consistent information, however, it was decided to send separate questionnaires on individual chemicals to small groups of firms (the largest number was five, for TNT) who were the principal producers, and whose process could be considered 'authoritative'. Haber presided over a meeting of IG representatives in Berlin on 24 February, telling them what to expect from the Allied chemical and explosives experts, Watts and Muraour.[67]

Early in 1920, the IG firms also met to begin coordinating responses to the Allied efforts. As it was then Bayer's turn to be the managing firm for the IG, Carl Duisberg appointed one of his chemists, Dr. Wilhelm Walter (1877–1960), as the group's liaison with the Control Commission. Walter had been to the United States and spoke English well. His responsibility was to conduct the Commission's experts on their visits to IG plants, and to check the companies' responses to Allied questionnaires, so as to ensure that the information given was consistent.[68] Bingham was notified that Walter would accompany Watts and Muraour on their factory visits. The IG asked factories to notify Bayer as soon as the MICC announced a forthcoming inspection.

Initially, this process did not go as smoothly as the IG had hoped. During the first inspection visit to Bayer's works, there was friction, as Walter apparently attempted to steer the inspectors away from sensitive areas. Moreover, some IG companies resented having their information being inspected by a rival Bayer chemist. At least one company, Griesheim-Elektron (which produced nitrates and high explosives during the war) did not want to give Walter access to its documents. However, Walter asked Bayer's legal director, Kloeppel, to ask the other firms to pressure Griesheim to accept the same procedures, so as not to draw unwanted attention from the MICC experts.[69]

In the end, it appears that these differences were smoothed over, and the IG firms managed not only to avoid serious conflicts with the MICC, but also to establish fairly cordial relations with its subcommission's representatives. This they did by emphasizing the control and disarmament aspect, including the conversion of wartime production facilities to peacetime production, while at the same time resisting Allied demands for information about chemical production processes having peacetime commercial value.[70] When, in August 1920, Dr. Watts met with British ordnance officers to discuss the answers the Germans had given to the questionnaires based on Article 172, they concluded that the Germans had provided '(f)airly satisfactory replies as to explosives, less so as to propellants'. Indeed, they suggested a follow-up visit to the propellant factories would be 'desirable'.[71] (The propellant factories, it should be recalled, were not part of the IG.) A long process of negotiations, inspections, and follow-up visits still lay ahead.

A critical moment came in late 1920 and early 1921, when leading members of the British chemical industry interest groups tried to push their government towards a much stricter policy.[72] Their campaign was made conspicuous by the publication of *The Riddle of the Rhine*, by Victor Lefebure, a wartime Anglo-French liaison officer who had helped make arrangements for the earlier missions, and who then worked for the British Dyestuffs Corporation. To Lefebure, the continuing existence of the IG's vast facilities constituted a 'riddle' that the MICC had so far failed to solve. He challenged the Commission to undertake 'very drastic action' against companies that had produced 'fifty per cent of the German shell fillings, the message of their guns....' 'It is true', he added, 'that they were manufactured in synthetic dye and fertilizer plants, but the explosives were none the less violent and the poison gases none the less poisonous.'[73] His appeal was strengthened in October 1920, when a League of Nations report focused attention to the issue of chemical disarmament.[74]

For reasons that remain unclear, it was not until April 1921 that the report of Hartley's Mission of 1919 was published as a parliamentary paper; and even then, it appeared only in an abbreviated version.[75] Its appearance, however, met vocal support from British industry, including the Association of British Chemical Manufactures, already anxious to safeguard British industry, and to ensure that Germany was not taking advantage of the situation.

In January 1920, Victor Lefebure became identified with a national campaign for German chemical disarmament. 'Disarmament', he wrote, 'must attack the potential munitions factor.' Military success, the war had shown, rested 'with the country not compelled to improvise'. It had taken the Allies four costly years to achieve the level of production that German industry had still in place. In 1921, the IG could, in his view, without difficulty supply more poison gas than would be needed to fill the stock of shell allowed to Germany by the Treaty.

In fact, the weekly tonnage of explosive allowed Germany by the Treaty was only two days' actual production. In Lefebure's view, the Treaty should have embraced the importance and nature of the chemical war, and so applied its provisions to sources of production. But Articles 168 and 169, which limited the production of war material, and required war plants to be surrendered or disbanded, were not enforceable in such a way as to restrict the use of plant for the manufacture of war material. Article 170, which provided for the import and export of munitions, failed, as it was not enforceable. Article 171 prohibited poison gas, but specified no penalties. Article 172 required Germany to disclose the nature and method of manufacture, but failed because German industry could claim commercial privilege. Germany stood ready, Lefebure said, to produce 3000 tons of gas per month, and 200,000 tons of nitrogen per year. The weekly tonnage of gas produced in the IG factories were sufficient to fill one million shells, and could supply in four days more than required to fill the full stock of shells allowed by the Treaty. 'Can a country possessing these military advantages be regarded as disarmed? The whole situation is anomalous. But for its gravity', he added, 'it would be ludicrous.'[76]

In February 1921, support for Lefebure came from another industrial quarter, when Herbert Levinstein wrote Harold Hartley, complaining that the MICC lacked sufficient expertise to supervise the German industry, and that Bingham's section was allowing Germans to hide behind claims that their plants were being used for peaceful purposes – a claim the Board of Trade was compelled to allow.

Given this pressure, the War Office commissioned Hartley to return to Germany, to check on progress being made in dismantling factories producing propellant, high explosive, and gas.[77] At the same time, the British Commission continued inspections of explosives and gas factories in the Cologne area, Bayer-Leverkusen and Dormagen, and nitroglycerin plants throughout Germany.[78] A progress report on 28 February was followed by another report on 8 March by Watts, which surveyed Leverkusen, Dormagen and other factories in the area, including Wahn, Schlebusch, and Düren.[79]

Had the Allies accepted Lefebure's arguments, the logical conclusion would have been to classify all of the IG's 'dye and fertilizer plants' as military plants, which would have led to their dismantling – an obvious boon to Allied competitors, and a weapon to force key information from German manufacturers. The not-so-hidden rationale underlying British pressure on the MICC was the attempt to gain control over critical technologies, in particular those for producing synthetic ammonia, but also others. The French had taken a different approach. Even before the Treaty came into effect, the French had already used Article 168 to threaten to close BASF's ammonia plants, unless the company licensed the Haber-Bosch technology to them. Without knowing whether the French had learned enough from their inspections to build a plant on their own, BASF could not ignore their demands. During lengthy negotiations in Paris after the treaty was signed in July 1919, BASF's chairman Carl Bosch (with the understanding of the Reich government) worked out a deal with Georges Patart, director of the French Service des Poudres.[80]

After securing support from the IG, Bosch completed a licensing agreement with the French on Armistice Day, 1919. BASF gave the French government exclusive rights to its processes within French territory. A French national corporation would build a plant to BASF's specifications, with an annual capacity of 100 tons of ammonia per day in continuous production, and with access to any technological improvements over fifteen years. BASF asked for 50 million francs, but the contract gave them only five million (somewhat more if the capacity exceeded 100 tons), plus a small royalty per kilogram of nitrogen produced during the contract period. When, during the French occupation of the Ruhr in 1923, German nationalists accused the company of treason in yielding to French demands, BASF asserted that it had acted only under duress, and thus far had done nothing to help the French. Whilst true, it was the fault of French domestic politics that delayed for several years the formation of the new company.[81]

In contrast to the French deal, there is no record of BASF licensing negotiations with the British, and in any case the company issued no other Haber-Bosch licenses during the early 1920s. Hence the British sought to use the apparatus of the Treaty to force transfers of technology. Concurrent with Lefebure's offensive, the British firm

Brunner, Mond & Co. announced in May 1921 (evidently without prior negotiations with BASF) that it had obtained from the British government a license under the Treaty to use confiscated BASF patents to make synthetic ammonia. Bosch complained to Walther Rathenau, Minister for Reconstruction, but he could offer no help. Even so, the British could only get the process to work, in 1923, with information purchased from Alsatian engineers who had worked in Germany during the war.[82]

Throughout 1921 and 1922, the Control Commission responded to domestic (mainly British) pressures with new rounds of questionnaires, surveys, inspections, and instructions to the Germans for dismantling of plants. The long series of questionnaires (BASF alone received a total of twenty-five), which every chemical and explosives company in Germany was supposed to complete, requested details of all war chemicals and explosives, supported by descriptions and enumerations of the facilities and processes to produce them. One of the first and most critical questionnaires – on the production of ammonia and nitric acid for explosives – was requested in January 1921 under the provisions of Article 172. The Germans delayed in responding, on the grounds that ammonia and nitrates were raw materials or intermediates, whereas Article 172 was supposed to apply only to finished products used as explosives or chemical weapons.[83] The delays on this and most of the other questionnaires which dealt with products of dual usage, such as alcohol, acetone, and glycerine, continued until August 1921.

In June 1921, Harold Hartley arrived in Germany on his second mission, intending to interview German experts, including Haber. On his return to England, Hartley compiled a secret report for the British government. Although he did not deal with the touchy issues of ammonia, nitrates, and other raw materials and intermediates for explosives, this second visit, which coincided with another MICC request for more information, may have tipped the Germans (still under the threat of the London Ultimatum) towards a policy of co-habitation with Allied demands.[84] During July and August 1921, German civilian and military representatives met with the major chemical and explosives firms. Again they reiterated their contention that the Allies (in this case, mainly the British) had a right only to information regarding finished products used directly as weapons, and not about substances that were raw materials or intermediates.

Thus, the Germans insisted on withholding information about nitrates and other commercial products, about which the Allies could not insist without suggesting economic espionage. On the other hand, the German government accepted their Treaty obligation to deliver information about explosives and chemical agents, and to do so as promptly and fully as possible, so as to forestall possible retaliation.[85] In cases where providing information might impair the position of a company—such as Köln-Rottweil, which was trying to shift from explosives to peacetime products like artificial fibres, and in doing so wanted to apply techniques (e.g., solventless powders) developed for explosives, the government agreed to provide compensation. This was, however, an exceptional case.[86]

In fact, once having safeguarded their main commercial technologies, German chemical experts and IG firms worked in concert with the Commission, furnishing it with evidence to demonstrate the success of the Commission, so as to justify its (somewhat limited) activities to the Allied governments. It seems clear from the correspondence between the Germans and the Commission during this period that German experts like Haber and Lenze (then senior department head at the CTR), as well as corporate leaders like Carl Duisberg, felt they could work with the Allied experts and did not want them to be recalled and replaced by other, potentially more hostile men.[87] To avoid raising suspicions about things the Allies might not yet know, the Germans took care to keep all their answers to questions as consistent and as concise as possible.[88] At the same time, the Allied experts resented the pressures coming from the Lefebure camp. In response to anxious queries by Walter, they assured the Germans that they did not support Lefebure's arguments.[89] Moreover, uncomfortable with allegations of industrial espionage, Allied experts backed off from asking questions that German companies deemed sensitive, provided they were willing to cooperate by dismantling facilities developed during the war primarily for the production of explosives and chemical weapons.

In this spirit of collaboration, it became possible for firms to retain capacities they had before the war. One toxic substance that Germany was thus allowed to produce was phosgene, a toxic gas used as an intermediate in the manufacture of dyes. BASF backed its interpretation of Article 172 by a legal opinion that it applied *'only to chemical warfare agents'* (emphasis added) as such, and not to intermediates also used in dyes or fertilizers (e.g., ammonia). Thus, the Allies obtained details on BASF's processes for arsenicals, but not on the ethylene and thiodiglycol used in making mustard gas (and indigo), nor on ammonia or (apparently) nitric acid. German complaints forced the Commission's representatives to back down repeatedly – for example, by not objecting when, in 1924, German directors withheld details about a new methanol synthesis, which BASF's chemists had developed since the war at their Leuna ammonia plant.[90]

Similar issues confronted the inspection of dyestuffs operations. However, in this case, French inspectors gained such insight into BASF that its director, Carl Bosch, was forced to negotiate with Joseph Frossard, head of the Compagnie Nationale des Matières Colorantes et des Produits Chimiques (CNMC), with a view to granting France an exclusive license to operate BASF patents and to sell products in France and in French colonies overseas. Bosch contacted Duisberg of Bayer, so as to involve the entire IG group.[91] As with the Haber-Bosch technology, it seemed better to negotiate an agreement from which the company could profit, than to risk losing everything – or, as the minutes of BASF's Select Commission of the Supervisory Board put it, 'in order thereby at least to obtain an equivalent for the information that the French commissioners are obtaining by industrial espionage'.[92] The IG companies jointly negotiated an agreement assigning their patents to CNMC and sharing their know-how with the French, dividing up the market along lines sketched by BASF's directors in 1919. In return, the IG would receive 16.75 million francs over ten years, as well as half of CNMC's profits until 1965. Germany voided the agreement after the Ruhr invasion of 1923; when the French renewed negotiations after 1924, the advantage lay with the Germans.[93]

The IG offered a similar deal, including patent-sharing, to the British Dyestuffs Corporation (BDC) in November 1921, but the BDC refused until the deteriorating situation in the industry and currency stabilization led to new negotiations in November 1923. Then, only a temporary agreement could be reached, because the BDC required the consent of the British government for a permanent deal. Political changes prevented action until after 1925, when the British government withdrew from the BDC, but then negotiations were suspended when BDC merged into what became Imperial Chemical Industries (ICI) in 1926, and the German IG group became I.G. Farben.[94]

In mid-1922, the MICC requested and obtained from the IG its final sets of plant survey forms and questionnaires. In some cases it also requested follow-up information, which IG firms also supplied. The information covered explosives (nitrocellulose and nitroglycerine propellants, TNT, picric acid, and a host of other nitrogenous 'substitute' high explosives), as well as the principal chemical warfare agents.

In their responses, German firms were required to list all buildings and apparatus for producing war chemicals, indicating the status of these facilities at the outbreak of war; to say what had changed and what they had produced during the war, and to describe their status at the end of the war, and at the time of the questionnaire. Buildings for war production were to be indicated on factory layouts, supplemented by pictures to give evidence of actual demolition or conversion. The Germans had also to provide written descriptions of processes for war chemicals, accompanied by schematic diagrams. All in all, the Allies requested an amazing quantity of information that would have given a detailed picture of the industry – provided that the Germans complied. Of course, the MICC had the right to follow up with its own inspections, to confirm what the Germans had reported. Much of the documentation the main IG firms provided – or at least, that which can be checked against internal archives – appears to be accurate and complete. Nevertheless, the Germans steadfastly refused to disclose information about commercial products, or to permit inspectors to examine internal documents.[95]

In 1923, the inspections came to a precipitous end. As foreshadowed by the London Ultimatum, and in retaliation against German non-compliance in reparations, France occupied the Ruhr district. Inflation was reaching a critical point, and people were paying off prewar mortgages with postage stamps. France wanted gold. Germans responded with a combination of passive resistance and active intimidation and sabotage that made further inspections impossible. After the situation calmed down in 1924, the Inter-Allied Control Commission conducted its last, follow-up inspections (of 392 plants),[96] and began winding down, before pulling out entirely in 1927, when it yielded to a much lighter regime under the auspices of the League of Nations.

One of the last acts of the MICC was to ensure that former German state explosives factories were shut down. Most of the principal state factories had been privatized in 1919 under a general holding company called the Deutsche Werke AG, which alleged that all facilities for producing munitions were being converted to commercial uses. One of the factories that did not belong to the Deutsche Werke group was the Gnaschwitz plant in Saxony which, with a few other state plants was also privatized, and sold to other companies.[97] The MICC argued that the Board of the Deutsche Werke were handpicked by German military authorities, and thus represented not a true privatization but a kind of dummy corporation, a continuation of the old military plants by other means. Disputes ensued as the MICC tried to shut down and dismantle the Deutsche Werke. The commercial companies which had before the war produced explosives for the mining industry and sportsmen, as well as for the military, sought a return to peacetime levels of production while maintaining ties with the state. The MICC attempted to prevent all but one of these companies from producing for the military. Some stopped producing nitrocellulose, and turned to celluloid or other peacetime cellulose products; others, which had produced nitroglycerine as a mining explosive before the war, simply continued to do so, as the Treaty permitted. Henri Muraour, the French chemical representative on the MICC, was not alone in expressing frustration that, because of the industry's 'dual-use' features, it would be virtually impossible 'by peacetime measures to prevent Germany from maintaining or constituting sufficient plant capacity to assure its wartime production.'[98]

## CONCLUSION

This essay has examined the efforts of the Allies to disarm German chemical industry and to prevent its rearmament, while obtaining important technical information about chemical production during the war. We have shown that, in large part, these efforts failed. In itself, this is hardly an original conclusion. It is well known that the Allies failed to prevent German rearmament. But we can now understand why this was particularly so in the context of the chemical industry. In brief, the Military Inter-Allied Control Commission suffered four critical faults: first, it lacked infrastructure, intelligence, and powers to sanction.[99] This is not surprising, given the unprecedented nature of forced disarmament and the lack of role models available to a Control Commission. Second, it reflected a sense of political failure, in that the French and British never achieved common cause in arms control, and often worked at cross-purposes. As we have seen, the German firms magnified these differences and did their best to exploit them.

Third, the Allies suffered from a failure of principle, arising from the Treaty's intrinsic failure to resolve the 'dual-use dilemma', the 'riddle' that, long before Lefebure, Brigadier Hartley had foreseen.[100] The Military Inter-Allied Control Commission could and did order the German government to sell or demolish its state-owned military production facilities; but its authority over private industry was almost ineffective. Despite abrogating contracts and dismantling plant, as well as destroying or rendering useless for military purposes a vast quantity of explosives and propellants,[101] MICC left largely intact such companies as Bayer, BASF, Höchst, the Nobel Dynamite concern, and Köln-Rottweil. These companies together had produced more wartime explosives than the state factories, but unlike the latter, they continued to operate after reconversion to peacetime. Nor did the MICC shut down the research operations of scientists who, like Fritz Haber, continued to provide the German government with secret advice on chemical warfare.

The fourth failure of the Military Inter-Allied Control Commission was perhaps the most significant, but also the most difficult to foresee. This was its failure to recognize the *stimulating* effect of its work. The Treaty and the control system produced the unintended consequence of prolonging wartime tensions. Allied policy reinforced synergies among German firms, as well as between the firms and the government, giving industry an incentive to keep structures that might in other circumstances have been dismantled. In consequence, one could argue that the German companies prospered, not only *despite* the Commission's presence, but actually *because* of it.

Although great attention has been paid to the Allied exploitation of German technical information after 1945, the question of technology transfer after the First World War has been strangely neglected. It is not clear whether in any critical case the Control Commission obtained information of peacetime value that could not have been obtained through other means. In synthetic ammonia, for example, the French succeeded in obtaining information by collaborating with BASF, albeit by bluffing – a more effective approach than trying to carry out a threat through the Commission. However, by the mid-1920s, American, French and Italian firms developed competitive synthetic processes, and world output of synthetic ammonia was growing to saturation, regardless of the information obtained (or rather, not obtained) through the Treaty. The situation in the dye industry is similar, in that it appears that licensing and the confiscation of patents proved more effective as a means of obtaining technology than either inspections or questionnaires. Here again, the postwar period saw a rapid expansion in dyestuffs that quickly saturated the market, and stabilization came only with concentration into larger units and by the formation of international cartels and conventions between erstwhile competitors.[102]

Where the Control Commission did succeed in technology transfer, it did so in the limited field of explosives and chemical warfare agents, to which the Germans adhered in a strict interpretation of Article 172. The German government seems to have concluded that it was better to give abundant information on questions that had little relevance to peacetime products, than to share information of commercial value. In the long run, this proved to be a profitable approach, as it offered relatively little benefit to Allied competitors.[103]

The Control Commission neither dismantled German industry, as the French devoutly desired; nor did it diminish German competitiveness, as the British fervently hoped. Yet, it did delay, at least, the immediate proliferation of chemical explosives and gas weapons, and postponed their reproduction. Moreover, the very existence of the control framework perhaps contributed to an atmosphere in which – whatever weapons might be produced – a general assumption of 'no first use' (at least in Europe) could be seen to apply.

Here, the curtain falls. In a sense, the drama shifts from chemical to other weapons of mass destruction. Much to everyone's surprise, mustard and nerve gas did not overtake the world in 1939, despite massive preparations for their use, nor were they used in Europe before 1945, although the same cannot be said for China, India and northern Africa. And whatever the reservations of the current American administration, the Chemical Weapons Convention of 1993, which entered into force in 1997, has achieved ratification by 175 states-parties, representing 95% of the world's population, making it one of the most widely accepted treaties of its kind in the world.[104] Today, six nations are declared possessors of chemical weapons, and taking a lesson from the experience of 1919, they are monitored by the Organisation for the Prohibition of Chemical Weapons (OPCW) in The Hague. Under the CWC, there have been 2,200 inspections conducted in 860 of the 6000 declared military and industrial facilities that have been declared, worldwide.[105] There is a widespread norm against the use of chemical weapons.[106] Still, the world is not a safer place. At best, we hope the experience of the Great War will be revisited – not simply as a textbook case. Its lessons remain vivid and valid, and contain features that both shaped the past, and condition the future.

### ACKNOWLEDGEMENTS

This essay derives from a larger project on 'The Great War and Modern Chemistry', supported by grants from the National Science Foundation Research and the Australian Research Council, which the authors gratefully acknowledge. We also thank the Rockefeller Foundation's Bellagio Study and Conference Center for a collaborative residency during which we drafted the first version. We are indebted to the archives cited, and to the many other libraries and collections we have used in Germany, France, England and the United States, as well as to many colleagues for their help, support, and suggestions.

### NOTES

[1] Harold Nicolson, *Peacemaking* (London: Constable, 1933), diary entry for 24 March 1919, 236.
[2] Friedrich Ebert, *Schriften, Aufzeichnungen, Reden; Mit unveröffentlichten Erinnerungen aus dem Nachlass* (Dresden: C. Reissner, 1926), 127.
[3] Fred Tanner, 'Versailles: German Disarmament after World War I', in Fred Tanner (ed.), *From Versailles to Baghdad: Post-War Armament Control of Defeated States* (New York: United Nations, 1992), 5–25; Michael Salewski, *Entwaffnung und Militärkontrolle in Deutschland, 1919–1927* (Munich: Oldenbourg, 1966), 40–56, 99–108.
[4] Historians have tended to focus upon German industry *per se*. See, for example, Joseph Borkin, *The Crime and Punishment of IG Farben* (New York: Free Press, 1978) (based in part upon unverifiable

secondary sources); and Gottfried Plumpe, *Die I.G. Farbenindustrie AG: Wirtschaft, Technik und Politik, 1904–1945* (Berlin: Duncker and Humblot, 1970), 107–130.

[5] Lothar Meinzer, 'Productive Collateral' or 'Economic Sense': The BASF under French Occupation, 1919–1923', in Anthony S. Travis *et al.* (eds.), *Determinants in the Evolution of the European Chemical Industry, 1900–1939* (Dordrecht/Boston/London: Kluwer, 1998), 51–63; Werner Abelshauser *et al.*, *German Industry and Global Enterprise. BASF: The History of a Company* (New York: Cambridge University Press, 2004), esp. 183–189. The principal account in English of Germany's chemical warfare establishment is L. F. Haber, *The Poisonous Cloud: Chemical Warfare in the First World War* (Oxford: Clarendon Press, 1987), but this gives little information about the disarmament and control regime; see 287, and 304–305. Given the timeliness of these issues today, it is surprising that they are absent from Richard J. Shuster, *German Disarmament After World War I: The Diplomacy of International Arms Inspection, 1920–1931* (London: Routledge, 2006).

[6] National Archives (Kew) (hereafter NA), MUN 4/7056, *Report of the British Mission Appointed to Visit Enemy Chemical Factories in the Occupied Zone Engaged in the Production of Munitions of War* (London: Ministry of Munitions of War, Department of Explosives Supply, February, 1919), 12. A shorter version, with this quotation, was published in 1921.

[7] See Jeffrey Allan Johnson, Chapter 1, *supra*.

[8] For the IG pooling agreement see Plumpe, *op. cit.* note 4, 96–100; Abelshauser *et al.*, *op. cit.* note 5, 171–173. The eight member firms were Agfa (Aktiengesellschaft für Anilin-Fabrikation, Berlin-Lichtenberg, Treptow, and Wolfen in central Germany), BASF (Badische Aniline- und Soda-Fabrik, Ludwigshafen and Oppau on the upper Rhine in the Bavarian Palatinate), Bayer (Farbenfabriken vorm. F. Bayer, Elberfeld in Westphalia and Leverkusen on the lower Rhine near Cologne), Höchst (Farbwerke vorm. Meister, Lucius & Brüning, Höchst bei Frankfurt/Main), Cassella (near Frankfurt/Main); Kalle & Co. (Biebrich, near Wiesbaden on the Rhine), Griesheim-Elektron (Chemische Fabrik Griesheim-Elektron, Frankfurt/Main and Bitterfeld, in Central Germany), and Weiler-ter Meer (Uerdingen) on the Rhine.

[9] For the explosives contract see Hoechst Archives (Höchst a.M.) (hereafter, HA), 18/1–21: Abkommen mit Pulverfabriken.

[10] Bundesarchiv-Militärarchiv (Freiburg/Br.) (hereafter, BAMA), PH2-87, Lieferung von Feldartillerie-Munition während des Krieges 1914/1918.

[11] Margit Szöllösi-Janze, 'Losing the War but Gaining Ground: The German Chemical Industry during World War I', in J. Lesch (ed.), *The German Chemical Industry in the Twentieth Century* (Dordrecht: Kluwer, 2000), 91–121.

[12] Roy MacLeod, 'The "Arsenal" in the Strand: Australian Chemists and the British Munitions Effort, 1916–19', *Annals of Science*, 46 (1), (1989), 45–67.

[13] The Armistice Commission was dissolved on 13 January 1919, three days after the ratification of the Peace Treaty. See NA, WO 144: Inter-Allied Armistice Commission: War Diary and Despatches of Chief of British Delegation, 1918–1920.

[14] For Plumer (1857–1932) see *Dictionary of National Biography (1901–1950)*, (Oxford: Oxford University Press, 1951), 347.

[15] Bayer-Archiv (Leverkusen) (hereafter, BAL), 202–28, vol. 1 (Abwicklung ehemaliger Kriegsbetriebe: Pikrinsäure-Versand), correspondence April-November 1919.

[16] MUN 7/539. Mission to Germany, 1919: H. S. Denny - Technical Commission to Germany, Denny to McCall, 8 February 1919.

[17] For Price (1875–1949), see *Obituary Notices of Fellows of the Royal Society*, 7 (20), (November 1951), 469–474; for McCombie (1880–1962), *Journal of the Royal Institute of Chemistry*, 86 (1962), 448; for Arthur Allmand (1885–1951), see *Journal of the Royal Institute of Chemistry*, 76 (1952), 270. Drs Shore and Knibbs have not yet been traced.

[18] Churchill College, Cambridge, Archives (CCA), Hartley Papers, Box 34, File 12. Report on visits paid to Leverkusen Works, 27, 28 and 30 January 1919. See also NA Mun 7/539.

[19] A 12-volume *History of the Ministry of Munitions* was prepared and printed, but not officially published, in 1919. See Roy MacLeod, 'The Chemists Go to War: The Mobilization of Civilian Chemists and the British War Effort, 1914–1918', *Annals of Science*, 50 (1993), 455–481.

[20] See George Dewar, *The Great Munition Feat, 1914-1918* (London: Constable & Company, Ltd, 1921); Lord Moulton, *Science and War: The Rede Lecture, 1919* (Cambridge: Cambridge University Press, 1919).

[21] Henry Samuel Denny (?- 1938), *Who Was Who*, Vol. III (1929-40), (London: A&C Black, 1941), 354.

[22] MUN 7/539. Denny to McCall, 8 February 1919.

[23] Harold Hartley (1878-1972) was educated at Dulwich and Oxford, and studied chemistry with Richard Willstätter in Munich, before graduating from Oxford in 1900, and becoming a Fellow of Balliol. In 1915, he was sent to France as Chemical Adviser, Third Army. In 1917, he became Assistant Director of Gas Services at GHQ, and in 1918, transferred to the Ministry of Munitions as director of the Chemical Warfare Department. In 1919, the department was transferred to the Artillery, and he returned to Oxford. In 1921, he helped set up the Research and Development Establishment of the War Office, and served on its Chemical Warfare Board until the 1950s. See *Biog. Memoirs, Fellows of the Royal Society*, 19 (December 1973), 348-373, esp. 356-357.

[24] CCA, Hartley Papers, Box 33,'Tangye's Time-Table of the Mission in 1919'.

[25] For Alfred (later Sir Alfred) Egerton (1886-1959) see *Dictionary of National Biography (1951-60)*, (Oxford: Oxford University Press, 1971), 330-331. For Harold Cecil Greenwood (1887-1919), *Journal of the Chemical Society*, 117 (1), (1920), 462-464.

[26] Macnab (1858-1941), the most celebrated authority on explosives in Britain before the war, joined Lord Moulton's Committee on Explosives in 1914 and from 1915 to 1918 served as Technical Adviser to the Department of Explosives Supply, Ministry of Munitions. See *Journal and Proceedings of the Institute of Chemistry* (1941), Part V, 285-186; *Who Was Who, Vol. IV (1941-1950)*, (London: A&C Black, 1964), 743.

[27] For postwar British industrial realignments, see W. J. Reader, *Imperial Chemical Industries: A History* (London: Oxford University Press, 1970).

[28] See BAL 202-28 Vol. 2 (Interalliierte Kontrollkommission General Bingham) for notes on this visit from the German side.

[29] NA MUN 4/6737, H. Hartley, 'Report of the British Mission Appointed to Visit Enemy Chemical Factories in the Occupied Zone Engaged in the Production of Munitions of War' (typescript, 26 February 1919); cf. printed version, *op. cit.* note 6.

[30] MUN 7/539, H. S. Denny - Technical Commission to Germany. "General Remarks on Chemical and Dyestuff Situation" (with cover letter dated 23 February 1919).

[31] CCA, Hartley Papers, Box 33. H. Marshall, 'The Industrial Position at Cologne' (March, 1919).

[32] MUN 7/539. Mission to Germany. Farbenfabriken vorm. Fried. Bayer & Co. Leverkusen Works: General Description and Process Report (no date, ca. February 1919).

[33] MUN 7/539. Statement of Allowances, 7 April 1919.

[34] MUN 7/539. Report of Technical Commission to the Military Governor of the Occupied German Territory.

[35] See Craig Patton, *Flammable Material: German Chemical Workers in War, Revolution, and Inflation, 1914-24* (Berlin: Haude and Spener, 1998), 131, 171, 180-181, 240, 244-248.

[36] For Sir Robert Robertson, FRS (1869-1949), see *Dictionary of National Biography, 1901-1959* (Oxford: Oxford University Press, 1961), 369.

[37] NA SUPP 28/209, R.D. Report 37, C/2184, C/2162; SUPP 28/100, German: N/C [nitrocellulose] (Propellants), 15 July 1919. Later British inspections compared German conditions unfavourably in regard to protection for workers; see SUPP 28/32, German: Factories (23 September 1921).

[38] Bundesarchiv (Berlin-Lichterfelde) (hereafter, BAB), R85, Nr. 907, Bl. 327 (copy of telegram); Nr. 908, Bl. 76 (letter), BASF to Ausw. Amt, 10 December 1918.

[39] BAB R85, Nr. 912, Bl. 98, BASF to Ausw. Amt, 9 February 1919; Meinzer, *op. cit.* note 5, 54-56; Abelshauser *et al. op. cit.* note 5, 183.

[40] HA, 18/1-25: Besetzung, Angaben über frühere und jetzige Produktion (1918-19). Ernst Bäumler, *Die Rotfabriker: Familiengeschichte eines Weltunternehmens* (Munich: Piper, 1988), 259-260. See also HA 18/1-34: Erfüllung des Friedensvertrages; Interalliierte Kontroll-Kommission (1919-1923); HA 18/1-48: MIKOKO [=Militarische Interalliierte Kontroll-Kommission], 1917-1920; 18/1-49.

[41] BAL 202-28, *op. cit.* note 28.

42 On the patent question, see Haber, *op. cit.* note 5, 219–220; Kathryn Steen, 'German Chemicals and American Politics, 1919–1922', in Lesch, *op. cit.* note 11, 323–346; Steen, 'Patents, Patriotism, and 'Skilled in the Art': USA v. The Chemical Foundation, Inc., 1923–1926', *Isis*, 92 (1), (2001), 91–122.
43 Trevor N. Dupuy and Gay M. Hammerman (eds.), *A Documentary History of Arms Control and Disarmament* (New York: R. Bowker, 1973), 95.
44 *Ibid.*, 84–85, 94–95; Tanner, *op. cit.* note 3, 11–16; Salewski, *op. cit.* note 3, 40–56.
45 Salewski, *op. cit.* note 3, 40.
46 *Ibid.*, 41.
47 Salewski, *op. cit.* note 3, 44. On General Nollet (1865–1941), who led the MICC until 1924, see Charles M. E. Nollet, *Une Experience de désarmement: cinq ans de contrôle militaire en Allemagne* (Paris: Gallimard, 1932).
48 Tanner, *op. cit.* note 3, 12, 14–15.
49 Francis R. Bingham, 'Work with the Allied Commission of Control in Germany, 1919–1924', *Journal of the Royal United Services Institution*, LXIX (1924), 747–763 at 753. Major General Hon. Sir Francis Richard Bingham (1863–1935) served as Director of Artillery at the War Office between 1913 and 1916, and as a Member of the Council of the Ministry of Munitions between 1916 and 19. See *Who Was Who, Vol. III (1929–40)*, (London: A&C Black, 1940).
50 Haber, *op. cit.* note 5, 304. Hugh (later Sir Hugh) Edmund Watts (1888–1958) was educated at the University of London and the University of Zurich. Before the war, he worked in the industrial laboratories of Wright, Layman and Umney, and later, Blundell Spence and Col. During the war, he was an assistant works manager at HM Factory, Hackney Wick. After service with the Military Commission, he was appointed (1924) HM Inspector of Explosives at the Home Office, and later rose to be Chief Inspector. See *Journal of the Institute of Chemistry*, 83 (1959), 347, 451; *Who Was Who, Vol. V (1951–1960)*, (London: A&C Black, 1964), 1143. Henri-Charles-Antoine Muraour (1880-), was in 1919 Ing. en Chef des Poudres, and later Ingenieur General des Poudres. He studied in Germany before the war, and afterwards became perhaps the most distinguished explosives chemist in France. He was widely known as the author of *Poudres et Explosifs* (Paris: Presses Universitaires de France, 1947).
51 Salewski, *op. cit.* note 3, 39.
52 Haber, *op. cit.* note 5, 286.
53 SHAT, 4 N 97: C.M.I.C. Allemagne: Rapport final de la C.M.I.C. 1927, 27.
54 Bingham, *op. cit.* note 49, 748.
55 United Nations Archive, Geneva (UNA). League of Nations, Inter-Allied Military, Aeronautical and Naval Control Commissions – Inter-Allied Military Control Commissions: Germany. Col. 57, Articles 164 à 172 inclus du Traité de paix. Armament, munitions et materiel, 14–15; list in Col. 57, Annexe no. 1: repertoire alphabetique des usines ayant travaillé pour la guerre: fabrications destinées à l'armée. See Tanner, *op. cit.* note 3, 15; Salewski, *op. cit.* note 3, 135.
56 Bingham, *op. cit.* note 49, 749.
57 On Lenze (1866–1936[?]), who became director of the CTR (1923–1932), see L. Metz, 'Fritz Lenze', *Zeitschrift für das gesamte Schiess- und Sprengstoffwesen*, 27 (1932), 109–114.
58 See SHAT, 4 N 96: C.M.I.C. en Allemagne, 3d subfolder: Mission Nollet 1921. Fabrication des poudres; C.M.I.C. Etat-major. 3ème Section. Muraour, Résumé des renseignements relatifs à la fabrication des poudres récueillis jusqu'ici en Allemagne par la Commission Interalliée (Berlin, 24 August 1920), 1.
59 See BAL, 202–28, vol. 2, notes (28 February 1920) on meeting with Haber in Berlin (24 February 1920).
60 BAB R3301, Nr. 1888. Müller (Reichsmin. f. Wiederaufbau) to Goeppert (Ausw. Amt) (no date), "Für die Presse: Zu dem Artikel in der B.Z. vom 4.I.21 'Reuter über Deutschlands Vertragstreue'".
61 BAB R3101, Nr. 1630, Sitzung am 17 August 1921, betreffend Auslegung von Artikel 172 des Friedensvertrags (Berlin, 20 August 1921), 171–182, esp. 174.
62 *Ibid.*; also see Nr. 1631, 51–51RS. On the ultimatum, see Salewski, *op. cit.* note 3, 152–153, 173–174; Tanner, *op. cit.* note 3, 22–23.
63 NA SUPP 28/33, German: Guncotton, ref. to 33/1A/56 (19 March 1920).
64 NA SUPP 28/100, German: N/C [nitrocellulose] (Manufacture), reference card for 18 November 1921.

[65] For BASF, see Meinzer, *op. cit.* note 5, 55.

[66] Politisches Archiv des Auswärtigen Amts (Berlin), R33072, B4: Besichtigungen und Durchsuchungen, Bd. 1 (2.20–9.20), Nollet to v. Cramon (6 March 1920), Heeresfriedenskomm. to v. Keller, Ausw. Amt, Friedensabt, 23 March 1920.

[67] BAL, *op. cit.* note 59,'Besprechung im Verein zur Wahrung... 10. Februar 1920'; minutes (Berlin, 11 February and 28 February 1920). On Köln-Rottweil, see 'Wirtschaftliche Rundschau', *Zeitschrift für das gesamte Schiess- und Sprengstoffwesen*, 14 (1919), 163, 212. On Flechtheim (1876–1940), see Annegret Heymann, *Der Jurist Julius Flechtheim: Leben und Werk* (Köln: Heymanns, 1990).

[68] BAL 271/2.1, Walter, personnel file.

[69] BAL 202–28, Vol. 2. Walter to Kloeppel (from Höchst, 25 February 1920).

[70] BAL 202–28, Vol. 2, 'Besprechung im Verein zur Wahrung... 10. Februar 1920', (Berlin, 11 February 1920); also minutes (28 February 1920), questionnaires, correspondence with MICC, and notes of inspections.

[71] NA SUPP 28/100, German Questionnaire, reference to C/2208 (3.8.20); cf. additional references listed here, also WO 188/166 to175 (German descriptions of production processes for chemical warfare and explosives).

[72] 'Dyes as the Key to Gas Warfare: The German Surprise: Result of a Chemical Monopoly', *The Times*, 23 August 1920); 'German Poison Gas: Allies Method of Retaliation: The New Chemical Warfare', *The Times*, 31 August 1920; "Poison Gas in War: The Struggle for the Initiative: Large Scale Production', *The Times*, 28 October 1920.

[73] Victor Lefebure, *The Riddle of the Rhine Chemical Strategy in Peace and War* (London: Collins & Co, 1921), Conclusion.

[74] CCA, Hartley Papers, Box 33, AI 3A Gas shells, File 7 Addenda ca 1919–21. League of Nations, Report of the Permanent Advisory Commission for Military, Naval and Air Questions (22 October 1920); also 'Chemical Disarmament and the League of Nations. Report of Permanent Advisory Commission' [nd; 1920].

[75] *Report of British Mission to Enemy Chemical Factories*, February 1919, Cmd 1137 (London: HMSO, 1921). See also NA WO 142/273.

[76] CCA, Hartley Papers, Box 33, V. Lefebure, 'Chemical Disarmament', January 1921.

[77] Haber, *op. cit.* note 5, 7.

[78] NA WO188/114/G9, Watts to Bingham, 15 February 1921.

[79] CCA, Hartley Papers, Box 35.

[80] Tanner, *op. cit.* note 3, 168–169; Karl Holdermann, *Im Banne der Chemie: Carl Bosch* (Düsseldorf: Econ-Verlag, 1953), 169–170; Abelshauser *et al.*, *op. cit.* note 5, 184. In connection with these negotiations, BASF persuaded its IG partner firms to abrogate their pooling agreement with the explosives cartel; on the explosives pooling agreement see *op. cit.* note 9.

[81] Meinzer, *op. cit.* note 5, 60–63; Haber, *op. cit.* note 5, 237; Abelshauser *et al.*, *op. cit.* note 5, 186.

[82] Abelshauser *et al.*, *op. cit.* note 4, 186.

[83] BAB R3301, Nr. 1888, questionnaire (s. Bingham, dated 15. January 1920 [sic: 1921]), forwarded to Reichsmin. f. Wiederaufbau from Reichswehrmin., 24 January 1921.

[84] CCA, Hartley Papers, Box 31, Chemical Warfare 1915–18, Hartley, 'German Chemical Warfare Organization and Policy' (no date, but after June 1921), discussed in Haber, *op. cit.* note 5, 2–3, 7, 9, and cited *passim*.

[85] BAB R3101, Nr. 1630, minutes, meeting of 6 July 1921 in Reichsministerium für Wiederaufbau, 30–31; BAB *op. cit.* note 61, 171–172, 182.

[86] BAB R3101, Nr. 1630, minutes, meeting of 18 August 1921, concerning compensation, 168–170; Köln-Rottweil AG to Reichswirtschaftsmin. (22 August 1921), 267–270RS; report on follow-up meeting with KR on 25 August (30 August 1921), 272–276.

[87] Cf. Haber and Lenze remarks in BAB R3301, Nr. 1888, minutes of meeting (16 February 1921) in the Wiederaufbauministerium regarding Article 172.

[88] Cf. BAB R3101, Nr. 1630, 22–23, Epstein (KWI Phys. Chem.) to Reichswehrgruppenkommando I Berlin, 11 July 1921.

[89] BAL 202–28, Vol. 2, Besuch von General Bingham... (8 December 1921).

90   Abelshauser *et al.*, *op. cit.* note 5, 188–189 (emphasis in original citation).
91   This began in June 1919, while Bosch was with the German delegation to Versailles. Plumpe, *op. cit.* note 4, 122, suggests that the account given in Borkin, *op. cit.* note 4 (based on Holdermann, *op. cit.* note 80), is fanciful.
92   Cited in Abelshauser *et al.*, *op. cit.* note 5, 187.
93   Haber, *op. cit.* note 5, 275.
94   *Ibid.*, 273–275. ICI did eventually join a European dyestuffs cartel with I.G. Farben, but not until 1931.
95   Internal copies of the surveys and related information are kept in BAL, HA, and the BASF Unternehmensarchiv in Ludwigshafen. It has not yet been possible to locate a complete set of the 1922 forms that were submitted to the Control Commission or to German authorities.
96   UNA, *op. cit.* note 55, Inter-Allied Military Control Commissions: Germany. Col. 57, Articles 164 à 17214–15.
97   SHStA Dresden, Nr. (P)74955: 1918–1920. Wirtschaftsmin., Abt. f. Handel & Gewerbe, to Reichswehrbefehlsstelle (Dresden, 24 September 1919).
98   SHAT, 4 N 96: C.M.I.C. en Allemagne, Mission Nollet 1921. Nollet to Ministre de la Guerre, Paris (Berlin, 9 September 1920), confidential message forwarding preliminary report by Muraour.
99   See 'Lessons from the Disarmament of Germany after World War I', Office of Strategic Services Research and Analysis Branch Report no. 2799 (Washington, DC, 22 March 1945), esp. 18–25, which points out that the German government convicted of treason individuals who acted as informers for the MICC (others simply disappeared), whereas there is no record of a similar conviction of individuals who harassed the MICC or otherwise violated the Versailles Treaty, despite its explicit inclusion in the Weimar constitution. Our thanks to Jay Lockenour for bringing this report to our attention.
100  *Report of the British Mission*, *op. cit.* note 6.
101  Theodor Mente, 'Entwicklung des Sprengstoffwesens in Deutschland in und nach dem letzten Kriege', *Zeitschrift für das gesamte Schiess- und Sprengstoffwesen*, 24 (1929), 373–376, 423–427, on 376; Bingham, *op. cit.* note 49, 753–759.
102  Haber, *op. cit.* note 5, chs. 9–10.
103  Roy MacLeod, 'l'Entente Chimique: l'Echec de l'Avenir à le fin de la Guerre, 1918–1922', *Le Sabre et L'Eprouvette: l'Invention d'une Science de Guerre, 1914/1939*, in *14–18: Aujourd'hui*, 6 (2003), 135–153.
104  Martine Letts, Robert Mathews, Tim McCormack and Chris Moraitis, 'The Conclusion of the Chemical Weapons Convention: An Australian Perspective', *Arms Control*, 14 (3), (1993), 311–31.
105  John Howell, 'The Chemistry of Regional Security: Looking for a Better Return on the Chemical Weapons Convention', Conference on Science, Security and Safeguards, Sydney, 16 November 2005.
106  Richard M. Price, *The Chemical Weapons Taboo* (Ithaca: Cornell University Press, 1997).

# REVISITING THE 'CHEMISTS' WAR', 1914–1923: MODERN WAR, MUNITIONS, AND NATIONAL SYSTEMS

## INTRODUCTION

The Ordnance Office, faced with an unprecedented demand for gunpowder, did not receive the quantity expected. Throughout the war it made various unsuccessful attempts to obtain more cooperation from the powder makers, who themselves had considerable problems in achieving a product of a satisfactory standard and in the necessary quantities, especially when trying to balance the demands of government and private trade.[1]

This quotation sounds strikingly close to the situation in the First World War, or at least during the first year or two and, depending on the protagonist, considerably longer. It is, in fact, a description of the munitions crisis facing Britain in that 'other', much earlier 'World War', the Seven Years' War (1756–1763). My research in the history of explosives and munitions has focused on developments to the end of the nineteenth century. In reading the papers in this book, I experienced a sense of *déjà vu*. Despite enormous differences in the scientific, technological, industrial, political, and bureaucratic landscapes, the same general concerns and challenges that faced the Great War had faced the belligerent powers 150 years earlier. It is useful to consider these in sketching the modern development of munitions, and to use them as a heuristic guide in seeking comparisons and contrasts with 1914–1918. In this way, we would hope to draw out the larger implications of this book for the study of science, technology and warfare.

## MUNITIONS SUPPLY IN THE SEVEN YEARS' WAR, 1756–1763

### *State Institutional Management*

During the Seven Years' War, possibly for the first time on a global basis, European nations took upon themselves the management and supply of munitions. Both France and England, the major protagonists, found their needs not met by traditional means, whether contracting through the powder 'ferme' in France, or buying from private powder makers in England. France, in particular, faced the additional challenge of securing adequate supplies of one of the principal raw materials for gunpowder, saltpeter, the best contemporary source for which was India.

The result, over the next decades, was a history of comprehensive reform and reorganization – in England, government purchase of some private powder mills, including Faversham Mills (1759) and Waltham Abbey (1783),[2] and in France, a government-supported scientific programme of artificial saltpetre production – an 18th-century analogue of the Haber-Bosch process – centred on the formation of the *Régie des Poudres* (1775).

*State-Supported Scientific Management and Research*

'The chief support of war must, after money, now be sought in chemistry.'[3]

In both France and England, the responses to these wartime challenges involved research, education, and quality control. In France, the major figure was Antoine-Laurent Lavoisier; his English equivalent was William Congreve (the Elder).

In these ways, the crisis of the Seven Years' War (and the subsequent American and French Revolutions) thus inaugurated the modern era of munitions. These formed part of a much larger move towards rationalist and bureaucratic management, documented for France by Ken Alder in *Engineering the Revolution* and earlier by Charles Gillispie in *Science and Polity at the End of the Old Regime*. This resulted in the development, in France, at least, of what has been called the 'second scientific revolution'.

During the French Revolution, government management became even more clearly marked. Although Lavoisier was a casualty of the Revolution, his chemical colleagues, such as Berthellot and Guyton de Morveau, and the scientifically trained '*élèves des poudres*', were pressed into the service of the State to develop ways of collecting, refining and incorporating raw materials to make gunpowder quickly and efficiently, as Patrice Bret has indicated. An equally significant outcome of the Revolution was the establishment of the *École Polytechique*, the *Écoles de Genie*, and the Napoleonic requirement that heads of powder factories (that is, the *Service des Poudres*) be *polytechniciens*. I have argued elsewhere that the scientific engineering research styles instituted through these schools continued through the 19th century and climaxed in the work of Émile Sarrau and, even more importantly, Paul Vieille, the inventor of the first successful military smokeless powder, in the mid-1880s.[4]

A spin-off of the French tradition was the development of American munitions. E.I. DuPont was one of Lavoisier's *élèves des poudres* and subsequently carried his French science-based training to DuPont in the United States. Moreover, the *École Polytechnique* tradition was institutionalized at the West Point Military Academy, which produced by the 1840s what Stanley Falk has called a generation of 'soldier-technologists', including Alfred Mordecai and T.J. Rodman, the inventor of prismatic powder, the most successful form of black powder for large guns.[5]

In England, it is less easy to see institutional and intellectual continuity from Congreve through the 19th century. But the officers at the newly reconstituted Royal Gunpowder Factory were trained at the Royal Artillery School at Woolwich; and in the post-Crimean War period, the Ordnance Select Committee of the War Office instituted a series of committees comprised of scientifically trained officers and scientists to examine the entire range of armaments. These included 'explosives committees' from 1858 to 1866, from 1869 to 1881 and from 1888 to 1891 (the latter developed cordite). There were also 'guncotton committees' from 1864 to 1868, and again from 1871 to 1874.[6]

In addition, in 1855, the British government established the office of War Department Chemist, in the person of Frederick (later Sir Frederick) Augustus Abel, FRS. Abel devoted a good part of his investigative attention to guncotton: from the early 1860s, he had hoped to 'tame' it so that it could serve as a reliable smokeless propellant; he did not succeed (at least not until cordite was developed) but by 1866, had developed a way of insuring its stability by 'pulping' it to destroy most of its fibrous structure.

In fact, the first nitrocellulose smokeless powder was the result of the combined traditions of France and England.

## THE PROBLEMATIC ROLE OF CHEMISTRY IN 19TH-CENTURY MUNITIONS

The career of Abel raises the question of the role of chemistry — and chemists — in the development of munitions. Abel was one of the earliest students of the Royal College of Chemistry in London, and perhaps the only one to carve out a career with the military.

This is of great significance. As Hermann Boerhaave suggests, chemistry was looked upon in the 18th century as a promising science for the improvement of munitions. Indeed, chemists played a major role in the State reform and rationalization of munitions. But, despite the enormous growth in chemistry as an academic discipline, its role in providing a scientific basis for industrial research and the development of smokeless powder and high explosives through the fabrication of nitrocellulose, nitroglycerine, and picric acid, chemistry and chemists were, for the most part, marginal to the development of munitions for most of the 19th century.

Part of the reason was that munitions chemistry did not pose interesting research questions. The chemical reaction of gunpowder explosion was understood reasonably well by 1800, and attracted little attention. Conversely, chemical considerations gave way to physical problems in the improvements of gunpowder. For example, the suiting of black powder for the large guns from the 1860s involved considerations of size and shape of powder grain/cartridge, and powder density.

The experience of Germany illustrates the marginality of chemistry. Despite the fact that German chemistry came to dominate the world during the 19th century, there is little evidence of German chemists' interest in munitions. One of the few examples of research was carried out by Robert Bunsen and his Russian artillery student, Leon Schischkoff, and published in 1857. In this 'Theory of Gunpowder', Bunsen deployed thermochemical considerations to gain purchase on such parameters of explosion as temperature, pressure and energy. But, to the best of my knowledge, this was a one-shot for Bunsen, and he never returned to munitions. Moreover, he did not carry out this research under government or military patronage.

However, thermochemistry turned out to be the one area in which a few chemists did play a more central role. The most important was Marcellin Berthelot. In the wake of the Franco-Prussian War, Berthelot turned his recent interest in thermo-chemistry to the study of explosions of all kinds, ranging from black powder to high explosives. This resulted in his ambitious *Sur la Force des Poudres et des Matières Explosives* (1872, 1882). The historical valuation of Berthelot has not been high in recent years, but I have recently argued that he played an important role in giving structure to the work of *polytechniciens* such as Vieille.[7] France seems to have been virtually unique in the early 20th century in possessing a government-supported theoretical and experimental tradition in the study of munitions.[8] But, as Bret shows, by the eve of the Great War, this advantage was offset by the lack of trained chemists, as well as by rivalries and lack of coordination between the munitions laboratories. Nor was munitions production closely associated with industrial research. Indeed, as Jeffrey Johnson's paper shows, there was resistance among German firms to take up munitions and high explosives production.

The principal reasons for this are straightforward: munitions and explosives production is dangerous, both to physical plant and to the health of the workers. Moreover, munitions production is economically precarious. In time of war, business is good, but there has always been a problem in maintaining capacity when demand is low. This was true in the 18th century, and it was true at the end of the 19th. By the early 20th century, there was a third reason. In countries with 'mixed' State and private munitions production, there was often a degree of competition; even more important, there was a sense that military explosives needed to be under government security. Thus, the French government, in what was viewed at the time as novel, placed the procedure for making the first smokeless military powder, Vieille's *'poudre B'*, under conditions of strict secrecy.

There were some major exceptions to the rule that munitions were marginal to industrial research, as exemplified by DuPont and the Nobel-Dynamite Trust and its international cartel sequel of blasting and propellant explosives companies. But the profitable core of their business lay in civilian explosives, notably dynamite and blasting gelatine. DuPont's products profile in 1905 was as follows:

> The black powder business was old, fragmented, and declining in both sales and earnings. The dynamite business was newer, concentrated, growing and highly profitable. Smokeless powder, a firearms propellant, was still in its infancy. DuPont's smokeless powder business was aimed primarily at the military market, although some powder was sold in the sporting market. This business had just become profitable when the cousins consolidated the industry.[9]

TRANSITION: NOBEL AND THE NOBEL CARTEL

In his seminal account of the British chemical industry, W.J. Reader commented that:

> By 1907 the world explosives industry consisted of two large groups — the Anglo-German alliance and du Pont — and a third considerably smaller:

the Latin group .... Between each other, the groups had developed a very effective system of diplomacy, combined with the organization necessary for the running of their increasingly complex joint affairs. The explosives industry, in other words, had reached the peak of its pre-1914 evolution.[10]

In 1886, the Nobel-Dynamite Trust, a cartel of companies manufacturing Nobel's blasting explosives, came into being. Although international in scope, its principal axis lay between Great Britain and Germany. By the time that this cartel was organized, Alfred Nobel was well on his way to developing a major form of a smokeless propellant out of nitrocellulose and nitroglycerine, which he named 'ballistite'. Its military potential was immediately recognized, and the Trust was extended to include companies making military propellants by the General Pooling Agreement of 1889.[11]

The resultant 'Anglo-German alliance' – reflecting, once again, the principal business axis – dominated the military propellants market from 1890 until 1914. As late as 2 July 1914, the Anglo-German alliance and DuPont had in place an agreement arranging for cross-licensing of processes and patents, and of payments between them relating to specific technical items. When, less than a month later, the war began, 'the explosives makers — British, American, and German alike — were taken completely by surprise.'[12] These international cartels and networks of the explosive industry, deriving from the Nobel-Dynamite Trust, have been something like absent guests at this conference. The dominance of this international structure, and its sudden and unexpected break-up upon the outbreak of war, are crucial factors in understanding the challenges that faced the belligerents at the outset of the war, and what they did in its first year.

## THE FIRST WORLD WAR: PLUS ÇA CHANGE...

Three challenges confronted the munitions supplies system in the Seven Years' War. These were no less relevant to the First World War. They are: the creation of state management; the securing of raw materials; and the development of state supported scientific management and research.[13]

The differences between the conflicts relate to the vast differences between pre-industrial and industrialized (or partially industrialized) societies, and to the resulting differences between pre-industrial and industrialized warfare. Nevertheless, there is an over-riding commonality between the two situations: un-preparedness and the need to take unprecedented steps to ensure adequate supplies of munitions.

In respect to the first challenge, both the earlier and later eras looked to *central coordination* to scale up production. But in both cases, and certainly in 1914, this imperative was tempered by uncertainty over how long the war would last and what would be the liability of over-extension if the war might not last long.[14] For some countries, the situation was also conditioned by the existence of the Nobel cartel, which closely linked two of the principal combatants. Although the cartel was not formally dissolved until the end of 1915,[15] the British and German components quickly went their separate ways. Indeed, I would hazard the suggestion that the pre-existent cartel structure might have aided each of the national components in scaling up. The British company, Nobel Explosives of Glasgow, was the main producer of the smokeless powder, cordite, and began to expand its facilities almost immediately.[16] In Germany, as Johnson recounts, the head of the German Nobel component, Gustav Aufschläger, was almost immediately made head of the country's War Chemicals Corporation, on 30 September 1914.[17]

Germany's War Chemicals Corporation was, of course, a *State* entity designed to coordinate chemical components. Britain soon followed suit with its Ministry of Munitions, in July 1915. Once there was the prospect of a long war, requiring munitions of unprecedented scale and diversity, government coordination was inevitable. A comparative study of the different national systems of munitions production in the war is for this reason valuable. No one system was without its problems: a mixed system often produced incoherence and difficulty of coordination but, as Johnson and Brooks show for Austria and Russia respectively, a State system could be stultified. The leisurely interventions and reforms produced by the Seven Years' War, mostly taking place in the decades *after* the war had ended, were impossible. What was to be coordinated was, of course, vastly more complicated than in the 18th century, when the challenge was exclusively black powder. In 1914, 'munitions' certainly included propellant powder, but also high explosives,[18] and, somewhat later, poison gases. Moreover, coordination carried over to chemical materials that were not strictly speaking military, such as pharmaceuticals.

Another domain of munitions that required coordination was the production of chemical intermediaries. Whereas the intermediaries for black powder were naturally occurring materials, and their refinement for use was relatively unproblematic (except for saltpetre), in 1914 they involved products from other natural sources (coal) and, more important, other industries, particularly chemical dyes. Germany was far and away the leader in this field, and one of the problems facing the Allies, as Chauveau recounts, was how to compensate for their pre-war reliance upon Germany by coordinating the scaling-up of their own industries. But even in Germany, it was by no means easy to pull recalcitrant and reluctant firms into concerted action. In wrestling with the challenge of coordination, a number of papers in this volume take up corporate motivation. At least during the first two years of the war, especially in the United States, the profit motive was a dominant factor. A longer range motive was the preparation for post-war growth, as Erik Langlinay shows in relation to *Établissements Kuhlmann*.

In securing raw materials, some commonalities between the 18th century and 1914 – 18 are clear enough: Indian saltpetre and Chilean nitrates and British blockades, for instance. But the much greater complexities of the military materials, and their interrelation with component materials produced by other industrial processes, made the challenge daunting. Germany was at the end of the scale in two respects: it was effectively cut off from natural nitrates, but it had the most developed science-based industry, capable of producing synthetic ammonia and thence nitrates. Another 'raw material' worth mentioning is professional know-how.[19] Cocroft notes the reliance in England on German expertise prior to the war; this was a common feature of most European countries, and had to be replaced quickly.

Another challenge to coordination common both to the 18th century and 1914-18 was the lack of experience on the part of government, military and civilian. In the Seven Years' War, reforms were protracted. But even in the 20th century, problems sometimes took years to work out. This is brought out by Kathryn Steen, in relation to the history of US Ordnance, Patrice Bret on the laboratories of the artillery and the powder administration, and Nathan Brooks on Russia. Britain seems to have suffered less initially from this problem, but the government did have to go outside British industry to secure the American Kenneth B. Quinan from South Africa, to be the arch-designer of their scaled-up plants.

Although all of the belligerents were 'boot strapping', what they did was not simply serendipitous; it reflected traditions and styles of governmental, military, and industrial practice. This reflection is shown in the contrast between the initial reliance on private initiative in the United States and the French penchant for government committees. Another area of coordination involved the development of production sites. Factories were constructed on an unprecedented scale for making propellants, high explosives, and guncotton, and their study today involves archaeological as well as historical techniques, as Wayne Cocroft and Sebastian Kinder well demonstrate. Interestingly, they (and Smith) also show that these plants were endowed with characteristic cultural features that functioned as expressions of legitimation and national pride – building ornamentation for the Europeans; in the United States, hyperbolic rhetoric in the DuPont pamphlet on 'Old Hickory'.

Finally, there is the challenge of state management. There was little tradition of research chemists playing an active role in state munitions or research. If 1914–18 was indeed a 'chemists' war', what comes through is how this role had to be invented, much as it had been in the 18th. In many, or perhaps all, of the protagonist countries, there was suspicion or disdain on the part of the military towards the scientists. In Russia, it went so far as to preclude chemists from participating in munitions work: they were largely relegated to work on pharmaceuticals for much of the war. Yet chemistry — and chemists — were certainly needed.

But needed specifically for what? What exactly did chemists do to support the war effort? This issue receives most comprehensive and detailed treatment in the three papers devoted to the French chemical war effort: Chauveau's on the governmental coordination bureau, the Chemicals and Pharmaceuticals Office; Bret's on the military munitions laboratories; and Langlinay's on chemical industry. Also interesting is the issue of motivation: which scientists were 'self-recruited', and was their motive pure patriotism, or the opportunity for advancing the status of scientists, as Pancaldi argues?

Going beyond these cases are two other comparative issues. One concerns the *international* system. Any model for a national system will necessarily have to take into account the international and multinational networks of supply and competition between and and within companies. This last is brilliantly exemplified in Homburg's study of Royal Dutch Oil. A second and final issue that surfaces throughout these essays, and is discussed by Johnson and Macleod, is the significance of these massive and enormously complex build-ups for the future development of chemical industry and armaments for the next war. Many of the most ambitious projects of the war were ready to come on line only in its closing days, and were shut down when the war ended. Yet, did they prepare their protagonists for new industrial and military developments? Whether or not we follow Tom Hughes' doctrine of 'technological momentum', there seems a consensus that they did. But the outcome of this should be saved for the next book.

NOTES

[1] Jenny West, *Gunpowder, Government, and War in the Mid-Eighteenth Century* (London: Woodbridge, Suffolk and Rochester, NY: Boydell Press, 1991), 77–78.

[2] For an overview of these developments, see Seymour Mauskopf, 'Chemistry in the Arsenal: State Regulation and Scientific Methodology of Gunpoweder in Eighteenth-Century England and France', in Brett D. Steele and Tamera Dorland (eds.), *The Heirs of Archimedes: Technology, Science and Warfare through the Age of Enlightenment*, Jed Z. Buchwald (General Editor), Dibner Institute Studies in the History of Science and Technology (Cambridge, Mass: MIT Press, 2005), 293-330.

[3] H. Boerhaave, *Elementa Chemiae*, translated by T.L. Davis, 'Chemistry in War: An 18th-Century Viewpoint', *Army Ordnance*, 5 (30), (1925), 783. This appears to be Davis' translation.

[4] S. H. Mauskopf, 'The Experimental Study of Munitions: Scientific and Military Traditions', in Actes des Troisième Journées Scientifique Paul Vieille: Instrumentation, Experimentation et Expertise des Matérieux Énergétiques (Poudres, Explosives et Pyrotechnie) du XVIe Siècle à Nos Jours (Paris: Centre de Recherche en Histoire des Sciences et des Techniques, Association des Amis du Patrimoine Poudrier et Pyrotechnique, 2002), 11–28.

[5] The term 'soldier-technologist' comes from Stanley L. Falk, 'Soldier-Technologist: Major Alfred Mordecai and the Beginnings of Science in the United States Army' (Unpublished PhD dissertation, Georgetown University, 1959).

[6] To give a few examples of scientists on these committees: Five out of 11 members of the original Guncotton Committee, which sat from 1864 to 1868, were Fellows of the Royal Society, and included as an active member the distinguished physicist Sir George Gabriel Stokes. The chemist William Odling was a member of the Special Committee on Guncotton (1871–1874); and the chemist and physicist James Dewar and the chemist August Duprè served with Abel on the Explosives Committee that produced cordite.

[7] Mauskopf, *op. cit.* note 4, especially 18–22; S. H. Mauskopf, "Calorimeters and Crushers: The Development of Instruments for Measuring the Behavior of Military Powder," in Steven A. Walton. ed. *Instrumental in War: Science, Research and Instruments Between Knowledge and the World* (Leiden: Brill, 2005) 119–152. This deals, among other things, with the process of inventing poudre B.

[8] I think that it is significant that the chemist and physicist James Dewar was also interested in investigating such things as the temperature, pressure and energy of explosion of various explosive substances and the velocities of their explosion waves, and was conducting research on this while serving on the Explosives Committee which invented cordite. Royal Institution, James Dewar, DB1, 'General Laboratory Notes 1864–1923', DB1/5, *c*. 1888–1901. But this was Dewar's private research and I have not yet succeeded in determining the connection between Dewar's private and governmental activities.

[9] David A. Hounshell and John Kenly Smith, *Science and Corporate Strategy. Du Pont R & D, 1902–1980* (Cambridge: Cambridge University Press, 1988), 11.

[10] W.J. Reader, *Imperial Chemical Industries, A History*, 2 vols. *The Forerunners, 1870–1926* (Oxford: Oxford University Press, 1970), vol. 1, 126.

[11] The 'Anglo-German alliance' comprised the 'Explosives Groups', makers of blasting explosives (the Nobel-Dynamite Trust), and the 'Powder Group', makers of propellants. This alliance was brought into being by the General Pooling Agreement of 1889. See Reader, *op. cit.* note 10, vol. I, chs. 5, 7

and 9 for a detailed account of the Nobel cartels; for the component companies in the Anglo-German alliance (136); 151 for the component explosives firms and their connections to the armament industry c. 1905 (151); and for a diagram of the international explosives industry in 1914 (197).

[12] *Ibid.*, vol. I, 214.

[13] Subsequent to my own delineation of these challenges, I read Roy MacLeod's, 'Chemistry for King and Kaiser', in which he delineates 'an unprecedented combination' of actions in the mobilization of 1915 that are in general outline quite similar to the ones that I came up with: (1) geopolitical control over the materials; (2) the mobilization and deployment of skilled manpower; (3) the rationalization of component industries, with linkages between the supply of raw materials, the production of intermediaries, and the manufacture of munitions; and (4) clear lines of communication between applied research, chemical industry, government procurement, and military demands. MacLeod's points 3, 1 and 4 correspond more or less to my points 1, 2 and 3. I should say that, to an extent; I would be willing to include his second point about skilled workers in my list, too. See MacLeod, 'Chemistry for King and Kaiser: Revisiting Chemical Enterprise and the European War', in A.S. Travis *et al.* (eds.), *Determinants in the Evolution of the European Chemical Industry, 1900–1939* (Amsterdam: Kluwer Academics, 1998), 33.

[14] See Hew Strachan, *The First World War, Vol. I: To Arms* (Oxford: Oxford University Press, 2001), 1005–1014, for the complex and often contradictory thinking concerning the likely length of the war.

[15] See Reader, *op. cit.* note 10, vol. I, 254. For a general account of Nobel Explosives' activities in Britain during the war to the formation of Explosives Trade Ltd. in November 1918, see chapters 11 and 14. There has been remarkably little attention given to the 20th century history of the Nobel cartel in recent scholarship.

[16] Soon complemented by expansion and by purchase from the United States.

[17] This did not prevent Aufschläger from deploring to Sir Ralph Anstruther, Chair of the Trust, in 1915, the termination of 'the friendly relations which have existed between our companies for 29 years'. Quoted in Reader, *op. cit.* note 10, vol. I, 308.

[18] Shells filled with high explosives were mandated by the new positional trench warfare. See Strachan, *op. cit.* note 14, 999ff; and Roy MacLeod, 'The Chemists Go to War: The Mobilisaton of Civilian Chemists and the British War Effort, 1914–1918', *Annals of Science*, 50 (1992), 455–481.

[19] The is MacLeod's second point; see *op. cit.* note 12.

# BIBLIOGRAPHY

Abelshauser, Werner, Wolfgang von Hippel, Jeffrey Allan Johnson and Raymond G. Stokes. *German Industry and Global Enterprise. BASF: The History of a Company* (Cambridge: Cambridge University Press, 2004).

Aftalion, Fred. *A History of the International Chemical Industry*, translated by Otto Theodor Benfey (Philadelphia: University of Pennsylvania Press, 1991).

Alter, Peter. *The Reluctant Patron: Science and the State in Britain, 1850–1920* (Oxford: Berg, 1987).

Amatori, Franco and Bruno Bezza (eds.). *Montecatini, 1888–1966: Capitoli di storia di una grande impresa* (Bologna: Il Mulino, 1990), 69–148.

Associazione Chimica Industriale di Torino. *Onoranze Centenarie ad Ascanio Sobrero* (Turin: Unione Tipografico-Editrice Torinese, 1914).

*Atti del Convegno in celebrazione del centenario della morte di Ascanio Sobrero* (Turin: Accademia delle Scienze, 1989).

Ayerst, R. P. et al. 'The Role of Chemical Engineering in Providing Propellants and Explosives for the UK Armed Forces', in William F. Furter (ed.), *History of Chemical Engineering* (Washington, DC: American Chemical Society, 1980), 337–392.

Bailey, J. B. A. *Field Artillery and Firepower.* (Annapolis.: Naval Institute Press, 2nd ed. 2004).

Baruch, Bernard M. *American Industry in the War: A Report of the War Industries Board* (New York: Prentice-Hall, 1941), 165–207.

Beaton, K. *Enterprise in Oil: A History of Shell in the United States* (New York: Appleton-Century, 1957).

Beer, John J. *The Emergence of the German Dye Industry* (Urbana: University of Illinois Press, 1959).

Berger E. *Sur la Décomposition et la Stabilisation de la Poudre B. Conférence Faite Devant la Société Chimique de France le 16 mars 1912* (Paris: Société Chimique de France, 1912).

Berti, Giampietro and Piero Del Negro (eds.). *Al di qua e al di là del Piave: l'Ultimo Anno della Grande Guerra. Atti del Convegno Internazionale, Bassano del Grappa, 25–28 maggio 2000* (Milan: Angeli, 2001).

Bingham, Francis R. 'Work with the Allied Commission of Control in Germany, 1919–1924', *Journal of the Royal United Services Institution*, LXIX (1924), 747–763.

Bonino, G. B. 'Ciamician, Giacomo', in *Dizionario Biografico degli Italiani* (Rome: Istituto della Enciclopedia Italiana, 1960), 25, 118–122.

Borkin, Joseph. *The Crime and Punishment of IG Farben* (New York: Free Press, 1978)

Bowditch, Malcolm and Lesley Hayward. *A Pictorial Record of the Royal Naval Cordite Factory Holton Heath* (Wareham: Finial Publishing, 1996).

Bret, Patrice. *L'État, l'Armée, la Science. L'Invention de la Recherche Publique en France, 1763–1830* (Rennes: Presses Universitaires de Rennes/Fondation Carnot, 2002).

Bret, Patrice. 'La Guerre des Laboratoires: Le Chatelier, Poincaré et la Commission Scientifique d'Étude des Poudres de Guerre (1907–1908)', in *Actes du Colloque Henry Le Chatelier, Nancy, 15 avril 2002* (Paris: ENS Editions, forthcoming).

Buchanan, Brenda J. (ed.). *Gunpowder: The History of an International Technology* (Bath: Bath University Press, 1996).

Buchholz, Klaus. 'Verfahrenstechnik (Chemical Engineering) – Its Development, Present State and Structure,' *Social Studies of Science*, 9 (1), (1979), 33–62.

Buchinger, Marie-Luise. *Stadt Brandenburg an der Havel Teil 2: Äussere Stadtteile und eingemeindete Orte* (Worms am Rhein: Wernersche Verlagsgesellschaft, 1995).

Bud, Robert. *The Uses of Life: A History of Biotechnology* (Cambridge: Cambridge University Press, 1994), 37–45.
Burchardt, Lothar. *Friedenswirtschaft und Kriegsvorsorge: Deutschlands wirtschaftliche Rüstungsbestrebungen vor 1914* (Boppard am Rhein: H. Boldt, 1966).
Burchardt, Lothar. 'Professionalisierung oder Berufskonstruktion? Das Beispiel des Chemikers im Wilhelminischen Deutschland', *Geschichte und Gesellschaft*, 6 (1988), 326–48.
Burk, Kathleen. *Britain, America and the Sinews of War, 1914–1918* (Boston: George Allen and Unwin, 1985).
Calì, Vincenzo, Gustavo Corni, Giuseppe Ferrandi (eds.), *Gli Intellettuali e la Grande Guerra* (Bologna: Il Mulino, 2000).
Camera dei Deputati. *Relazioni della Commissione Parlamentare d'Inchiesta per le Spese di Guerra,* 2 Vols. (Rome: Camera dei Deputati, 1923).
Cappadoro, Angelo. 'Visite Agli Stabilimenti Industriali', *Giornale di Chimica Industriale ed Applicata,* 7 (2), (1920), 366–375.
Cardwell, D. S. L. *The Organisation of Science in England* (London: Heinemann, 1972).
Carls, S. D. *Louis Loucheur (1872–1931): Ingénieur, Homme d'Etat, Modernisateur de la France* (Paris: PUF, 2000).
Cartwright, A. P. *The Dynamite Company: The Story of African Explosives and Chemical Industries, Ltd.* (Capetown: Purnell, 1964).
Cayez P. *Rhône-Poulenc 1895–1975* (Paris: Armand Colin et Masson, 1988).
Cerruti, Luigi and Eugenio Torraca. 'Development of Chemistry in Italy, 1840–1910', in David Knight and Helge Kragh (eds.), *The Making of the Chemist: The Social History of Chemistry in Europe, 1789–1914* (Cambridge: Cambridge University Press, 1998), 133–162.
Challéat, J. *l'Artillerie de Terre en France pendant un Siècle: Histoire Technique, 1816–1919,* 2 vols. (Paris: Ch. Lavauzelle, 1933).
Chandler, Alfred D., Jr. and Stephen Salsbury. *Pierre S. duPont and the Making of the Modern Corporation* (New York: Harper & Row, 1971).
Chandler, Alfred D., Jr. and Stephen Salsbury. *The Visible Hand: The Managerial Revolution in American Business* (Cambridge, MA: Harvard University Press, 1977).
Charles, Daniel. *Between Genius and Genocide: The Tragedy of Fritz Haber, Father of Chemical Warfare* (London: Cape, 2005) [published by HarperCollins in the USA under the title, *Master Mind]*
Chickering, Roger and Stig Forster (eds.). *Great War, Total War: Combat and Mobilization on the Western Front, 1914–1918* (New York: Cambridge University Press, 2000).
Churchill, Winston S. *The World Crisis* (New York: Scribner's, 1931).
Ciamician, Giacomo. 'I problemi chimici del nuovo secolo', *Revista Università degli Studi di Bologna, Annuario,* (1903–4), 17–58.
Cipolla, Carlo M. *Literacy and Development in the West* (Harmondsworth: Penguin Books, 1969).
Cocroft, Wayne. *Dangerous Energy: The Archaeology of Gunpowder and Military Explosives Manufacture* (Swindon: English Heritage, 2000).
Cocroft, Wayne and Ian Leith. 'Cunard's Shellworks, Liverpool', *Archive,* 11 (1996), 53–64.
Cohendet, P. *La Chimie en Europe. Innovation, Mutations et Perspectives* (Paris: Economica, 1984).
Comparato, Frank E. *Age of Great Guns: Cannon Kings and Cannoneers who Forged the Firepower of Artillery* (Harrisburg, PA: The Stackpole Co, 1965).
Corvisier, André (ed.). *Histoire militaire de la France* (Paris: Presses Universitaires de France, 1997).
Costa, Albert B. 'Ciamician, Giacomo Luigi', in Charles C. Gillispie (ed.). *Dictionary of Scientific Biography,* (New York: Scribner, 1970–1990), vol. 3, 279–280.
Cranstone, David. 'Monuments Protection Programme: Chemical Industry' (Typescript report for English Heritage, 2002).
Crawford, Elisabeth, John L. Heilbron, and Rebecca Ullrich. *The Nobel Population, 1901–1937* (Berkeley: University of California Press, 1987).
Crowell, Benedict. *America's Munitions, 1917–1918,* 2 vols. (Washington, DC: US Government Printing Office, 1920).

Crowell, Benedict and Robert Forrest Wilson. *The Armies of Industry, I: Our Nation's Manufacture for a World in Arms, 1917–1918* (New Haven: Yale University Press, 1921).
Curtis, Harry A. (ed.). *Fixed Nitrogen* (New York: Chemical Catalogue Co., 1932).
D'Angio, Agnès. *Schneider et Compagnie et la Naissance de l'Ingénierie: Des Pratiques Internes à l'Aventure Internationale, 1836–1949* (Paris: CNRS Editions, 2000).
David-Fox, Michael. *Revolution of the Mind: Higher Learning among the Bolsheviks, 1918–1929* (Ithaca: Cornell University Press, 1999).
Daviet, J.P. *Un Destin International: La Compagnie de Saint-Gobain (1830–1939)* (Paris: Editions des Archives Contemporaines, 1988).
de Bruin, G. *Buscruytmaeckers. Ervaringen en Lotgevallen van een Merkwaardig Bedrijf in Holland* (Amsterdam: Nederlandsche Springstoffenfabriek, 1952).
de Bruin, G. *Het Geslacht Bredius: Honderdvijftig Jaren Buskruitfabricatie in Nederland, 1772–1922* (Amsterdam: Nederlandsche Springstoffenfabriek, 1947).
de Leeuw, W.C. *Kleurstoffen uit petroleum: Algemeene beschouwingen over de industrie der synthetische kleurstoffen: Rapport in opdracht der directie der Bataafsche Petroleum-Maatschappij* (Amsterdam: BPM, 1918).
De Nittis, Andrea. 'Chimica di guerra a Bologna, 1915–1918' (Tesi di Laurea, Università di Bologna, 2001).
de Vries, Joh. *Hoogovens IJmuiden: Ontstaan en Groei van een Basisindustrie* (IJmuiden: Koninklijke Nederlandsche Hoogovens en Staalfabrieken, 1968).
Del Boca, Angelo, Massimo Legnani and Mario G. Rossi. *Il regime fascista: storia e storiografia* (Rome and Bari: Laterza, 1995).
Del Negro, Piero. *Esercito, Stato, Società: Saggi di Storia Militare* (Bologna: Cappelli, 1979).
Déré, Anne-Claire. 'Maurice Javillier et Gabriel Bertrand: Deux Chimistes Pastoriens dans l'Institut en Guerre', in *Le Sabre et l'Éprouvette: l'Invention d'une Science de Guerre 1914–1939, 14–18. Aujourd'hui.*, 6 (2003), 75–88.
Dewar, George A. B. *The Great Munition Feat, 1914–1918* (London: Constable, 1921).
Di Meo, Antonio (ed.). *Storia della chimica in Italia* (Rome: Theoria, 1989).
Dosi, Giovanni. *Innovation, Organization and Economic Dynamics: Selected Essays* (Cheltenham: Edward Elgar, 2000).
Dosi, Giovanni. *Technical Change and Industrial Transformation* (New York: St. Martin's Press, 1984).
Dosi, Giovanni, Keith Pavitt, and Luc Soete (eds.), *The Economics of Technical Change and International Trade* (New York: New York University Press, 1990).
Dosi, Giovanni, Renato Giannetti, and Pierangelo Maria Toninelli (eds.), *Technology and Enterprise in a Historical Perspective* (Oxford: Oxford University Press, 1992)
Duinmayer J. and C. Groeneveld, 'Geschiedenis van het Koninklijke/Shell-Laboratorium, Amsterdam, Laboratorium der Bataafsche Petroleum Maatschappij' (Unpublished manuscript: Amsterdam, 1957).
Dyer, Davis and David B. Sicilia. *Labors of a Modern Hercules: Evolution of a Chemical Company* (Boston: Harvard Business School Press, 1990).
Ecksteins, Modris. *Rites of Spring: The Great War and the Birth of the Modern Age* (Boston: Houghton Mifflin, 1989).
Edgerton, David. *Warfare State: Britain, 1920–1970* (Cambridge: Cambridge University Press, 2006).
Egger, Rainer. 'Rüstungsindustrie in Niederösterreich und Heeresverwaltung während des Ersten Weltkrieges', *Bericht über den 16. österr. Historikertag in Krems 1984* (Vienna: Verband Österreichischer Geschichtsvereine, 1985)
Einaudi, Luigi. *La condotta economica e gli effetti sociali della guerra italiana* (Bari and New Haven: Laterza and Yale University Press, 1933).
Etzkowitz, Henry, and Loet Leydesdorff, 'The Endless Transition: A 'Triple Helix' of University-Industry-Government Relations', *Minerva*, XXVI (3), (1998), 203–218.
Etzkowitz, Henry, and Loet Leydesdorff (eds.), *Universities and the Global Knowledge Economy: A Triple Helix of University-Industry-Government Relations* (London: Cassell, 1997).
Feldman, Gerald D. *Army, Industry, and Labor in Germany, 1914–1918* (Princeton: Princeton University Press, 1966).

Ferguson, Niall. *The Pity of War: Explaining World War I* (New York: Basic Books, 1999).
Fischer, Wolfram, *WASAG: Die Geschichte eines Unternehmens, 1891–1966* (Berlin: Duncker and Humblot, 1966).
Flieger, H. *Unter der gelben Muschel: Die Geschichte der Deutschen Shell* (Düsseldorf: Verlag für Deutsche Wirtschaftsbiographien, 1961).
Forbes R.J. and D.R. O'Beirne, *The Technical Development of Royal Dutch/Shell* (Leiden: E.J. Brill, 1957).
Foreman, Stuart. 'Nitro-glycerine Washing House, South Site, Waltham Abbey Royal Gunpowder Factory, Essex', *Industrial Archaeology Review*, XXIII (2), (2001), 125–142.
Fox, M. R. *Dye-makers of Great Britain, 1856–1976: A History of Chemists, Companies, Products and Changes* (Manchester: Imperial Chemical Industries, 1987).
Freeth, Francis. 'Ammonium Nitrate and TNT in World War I', *New Scientist*, 13 (270), (1962), 157.
Freeth, Francis. 'Explosives for the First World War', *New Scientist*, 23 (402), (1964), 274–276.
Frey, M. *Der Erste Weltkrieg und die Niederlande: Ein neutrales Land im politischen und wirtschaftlichen Kalkül der Kriegsgegner* (Berlin: Akademie Verlag, 1998).
Friedman, Robert Marc. *The Politics of Excellence: Behind the Nobel Prize in Science* (New York: Times Books, 2001).
Fries, Amos A. *Chemical Warfare* (New York: McGraw Hill, 1st ed, 1921).
Fussell, Paul. *The Great War and Modern Memory* (London, New York: Oxford University Press, 1975).
Gabriëls, H. *Koninklijke Olie: De Eerste Honderd Jaar, 1890–1990* ('s-Gravenhage: Shell Internationale Petroleum Maatschappij, 1990).
Gerretson, F.C. *History of the Royal Dutch*, 4 Vols. (Leiden: E.J. Brill, 1958).
Gillet, M. *Les Charbonnages du Nord de la France au XIX Siècle* (Paris–La Haye: Mouton, 1973).
Godfrey, J. F. *Capitalism at War: Industrial Policy and Bureaucracy in the First World War in France, 1914–1918* (New York: Berg, 1987).
Goebel, Otto. *Deutsche Rohstoffwirtschaft im Weltkrieg: Einschliesslich des Hindenburg-Programms* (Stuttgart, Berlin and Leipzig: Deutsche Verlags-Anstalt, 1930).
Gooch, John. 'Italy during the First World War', in Allan R. Millett and Williamson Murray (eds.), *Military Effectiveness*, 3 vols. (Boston: Allen and Unwin, 1988), Series on Defence and Foreign Policy, Vol. 1, *The First World War*, 157–189.
Gooch, John. *Army, State, and Society in Italy, 1870–1915* (Basingstoke: Macmillan, 1989).
Goodstein, Judith R. 'The Rise and Fall of Vito Volterra's World', *Journal of the History of Ideas*, 45 (4), (1984), 607–617.
Govoni, Paola. *Un pubblico per la scienza. La divulgazione scientifica nell'Italia in formazione* (Rome: Carocci, 2002).
Griffiths, Gareth. *Women's Factory Work in World War One* (Stroud: Alan Sutton, 1991).
Guagnini, Anna. 'Life in the Slow Lane: Research and Electrical Engineering in Britain, France, and Italy, ca. 1900', in Peter Kroes and Martijn Bakker (eds.), *Technological Development and Science in the Industrial Age* (Dordrecht, NL: Kluwer, 1992), 133–153.
Guedon, Jean–Claude. 'Conceptual and Institutional Obstacles to the Emergence of Unit Operations in Europe', in William F. Furter (ed.), *History of Chemical Engineering* (Washington, DC: American Chemical Society, 1980), 45–76.
Haber, L.F. *The Chemical Industry, 1900–1930: International Growth and Technological Change* (Oxford: Clarendon Press, 1971).
Haber, L.F. *The Poisonous Cloud: Chemical Warfare in the First World War* (Oxford: Clarendon Press, 1986).
Hanslian, Rudolf (ed.). *Der Chemische Krieg* (Berlin: E. S. Mittler & Sohn, 1925, 3rd ed., 1937).
Hardie, D, and J Davidson Pratt. *A History of the Modern British Chemical Industry* (Oxford: Pergamon, 1966).

Hartcup, Guy. *The War of Invention: Scientific Developments, 1914–1918* (London: Brassey's, 1988).
Harris, J.P. *Men, Ideas and Tanks: British Military Thought and Armoured Forces, 1903–1939* (Manchester: Manchester University Press, 1995).
Hartley, (Brigadier Sir) Harold. 'A General Comparison of British and German Methods of Gas Warfare', *Journal of the Royal Artillery*, 46 (11), (1920), 492–509.
Haslam, M. *The Chilwell Story: VC Factory and Ordnance Depot* (Nottingham: The Boots Company, 1982).
Hayes, Peter. *Industry and Ideology: I.G. Farben in the Nazi Era* (Cambridge, UK: Cambridge University Press, 1987).
Haynes, William. *American Chemical Industry*, 6 vols. (New York: Van Nostrand, 1945–54).
Hendrix, P. *Henri Deterding: De Koninklijke, de Shell en de Rothschilds* ('s-Gravenhage: SDU, 1996).
Henriques, R. *Samuel Marcus, First Viscount Bearsted and Founder of the "Shell" Transport and Trading Company, 1853–1927* (London: Barrie and Rockliff, 1960),
Henriques, R. *Sir Robert Waley Cohen, 1877–1952: A Biography* (London: Secker and Warburg, 1966).
Herwig, Holger H. *The First World War: Germany and Austria-Hungary, 1914–1918* (New York and London: Arnold, 1997).
Hessell, Frederick A. *Chemistry in Warfare: Its Strategic Importance* (New York: Hastings House, 1942).
Heymann, Annegret. *Der Jurist Julius Flechtheim: Leben und Werk* (Köln: Heymanns, 1990).
Hohenberg, Paul M. *Chemicals in Western Europe 1850–1914: An Economic Study of Technical Change* (Chicago: Rand McNally, 1967).
Holdermann, Karl. *Im Banne der Chemie: Carl Bosch* (Düsseldorf: Econ-Verlag, 1953).
Homburg Ernst. 'The Emergence of Research Laboratories in the Dyestuffs Industry, 1870–1900', *British Journal for the History of Science*, 25 (84), (1992), 91–111.
Homburg, E. 'De Eerste Wereldoorlog: Samenwerking en Concentratie Binnen de Nederlandse Chemische Industrie', in J.W. Schot *et al.* (eds.), *Techniek in Nederland in de Twintigste Eeuw* (Zutphen: Walburg Pers, 2000), vol. 2, 316–331.
Homburg, E., A. Rip and J.S. Small. 'Chemici, hun kennis en de industrie', in J.W. Schot *et al.* (eds.), *Techniek in Nederland in de Twintigste Eeuw* (Zutphen: Walburg Pers, 2000), vol. 2, 298–315.
Homburg, E., J..S. Small and P.F.G. Vincken. 'Van Carbo- naar Petrochemie, 1910–1940', in J.W. Schot *et al.* (eds.), *Techniek in Nederland in de Twintigste Eeuw* (Zutphen: Walburg Pers, 2000), Vol. 2, 332–357.
Homburg, Ernst, Anthony S. Travis, and Harm G. Schröter (eds.). *The Chemical Industry in Europe, 1850–1914* (Dordrecht, NL: Kluwer, 1998),
Hounshell, David and John K. Smith. *Science and Corporate Strategy: Du Pont R&D, 1902–1980* (New York: Cambridge University Press, 1988).
Howard, Stephen. *A Century in Oil: The "Shell" Transport and Trading Company, 1897–1997* (London: Weidenfeld & Nicholson, 1997).
Hughes, Thomas Parke. 'Technological Momentum in History: Hydrogenation in Germany, 1898–1933', *Past & Present*, No. 44 (1969), 106–132.
Hughes, Thomas Parke. *Networks of Power: Electrification in Western Society, 1880–1930* (Baltimore: Johns Hopkins University Press, 1983).
Ipatieff, V. N. *The Life of a Chemist: Memoirs* (Oxford University Press: London, 1946).
Job, P. 'Notice sur la vie et les travaux de Georges Urbain (1872–1938)', *Bulletin de la Société Chimique de France*, 6, (1939), 745–766.
Johnson, Jeffrey Allan. *The Kaiser's Chemists: Science and Modernization in Imperial Germany* (Chapel Hill: University of North Carolina Press, 1990).
Johnson, Jeffrey Allan. 'Chemical Warfare in the Great War', *Minerva*, 40 (1), (2002), 93–106.
Johnson, Jeffrey Allan. 'La Mobilisation de la Recherche Industrielle allemande au service de la Guerre Chimique, 1914–1916', *Le Sabre et L'Eprouvette: l'Invention d'Une Science de Guerre, 14–18 Aujourd'hui*, (6), (2003), 89–103.
Johnson, Jeffrey A. and Roy M. MacLeod. 'War Work and Scientific Self-Image: Pursuing Comparative Perspectives on German and Allied Scientists in the Great War', in Rüdiger vom Bruch and Brigitte

Kaderas (eds.), *Wissenschaftssystem und Wissenschaftspolitik: Bestandsaufnahmen zu Formationen, Brüchen und Kontinuitäten im Deutschland des 20. Jahrhunderts* (Stuttgart: F. Steiner, 2002),169–179.

Jones, Daniel P. 'The Role of Chemists in Research on War Gases in the United States during World War I' (Unpublished PhD dissertation, University of Wisconsin, 1969).

Jones, Daniel P. 'From Military to Civilian Technology: The Introduction of Tear Gas for Civil Riot Control', *Technology and Culture*, 19 (2), (1978), 151–168.

Jones, Daniel P. 'Chemical Warfare Research during World War I: A Model of Cooperative Research', in John Parascandola and James Whorton (eds.), *Chemistry and Modern Society: Historical Essays in Honor of Aaron J. Ihde* (Washington, DC: American Chemical Society, 1983), 165–185.

Jones, H.O. and H.A. Wootton, 'The Chemical Composition of Petroleum from Borneo', *Journal of the Chemical Society. Transactions*, 91 (2), (1907), 1146–1149.

Jünger, Ernst. *The Storm of Steel*, translated by Michael Hofmann (New York: Penguin, 2004).

Kavangah, Gaynor. *Museums and the First World War* (Leicester: Leicester University Press, 1994).

Kennedy, Paul M. *The Rise of the Anglo-German Antagonism, 1860–1914* (Atlantic Highlands: Ashfield Press, 1987).

Kessler, H. *Walther Rathenau: His Life and Work* (London: G. Howe, 1929).

Kevles, Daniel J. 'George Ellery Hale, the First World War, and the Advancement of Science in America', *Isis*, 59 (4), (1968), 427–437.

Kevles, Daniel J. *The Physicists: The History of a Scientific Community in Modern America* (New York: Alfred Knopf, 1978).

Kinder, Sebastian. 'Brandenburg an der Havel Der: Industriestandort Kirchmöser von der pulverfabrik bis zum Ausbesserungswerk der Reichsbahn', *Brandenburgische Denkmalpflege*, 1 (2000), 4–16.

Klein, P.W. 'Colijn en de "Koninklijke"', in J. de Bruijn and H.J. Langeveld (eds.). *Colijn: Bouwstenen voor een Biografie* (Kampen: Kok, 1994), 105–128.

Koistinen, Paul A.C. *Mobilizing for Modern War: The Political Economy of American Warfare, 1865–1919* (Lawrence: University of Kansas Press, 1997).

Kovache, Adolphe, 'Historique du Centre d'Etudes du Bouchet', *Sciences et techniques de l'armement*, 55 (4), (1981), 651–717.

Lacaita, Carlo G. *Istruzione e sviluppo industriale in Italia, 1859–1914* (Firenze: Giunti, 1973).

Langeveld, H.J. *Dit leven van krachtig handelen: Hendrikus Colijn, 1869–1944*, Vol. 1 (Amsterdam: Balans, 1998).

Lefebure, Victor. *The Riddle of the Rhine: Chemical Strategy in Peace and War* (London: Collins, 1921).

Léger, J.-E. *Une grande entreprise dans la chimie française, Kuhlmann, 1825–1982* (Paris: Nouvelles Editions Debresse, 1982).

Lemmens, A.M.C. and G.P.J. Verbong. 'Natuurlijke en synthetische kleurstoffen in Nederland in de negentiende eeuw,' *Jaarboek voor de Geschiedenis van Bedrijf en Techniek*, 1 (1984), 256–275, esp. 269–273.

Lenselink, J. *Vuurwerk door de Eeuwen Heen* (Amsterdam: De Bataafsche Leeuw, 1991).

Lepick, Olivier. *La Grande Guerre Chimique, 1914–1918* (Paris: PUF, 1998)

Lesch, John E. (ed.). *The German Chemical Industry in the Twentieth Century* (Dordrecht, NL: Kluwer Academic Publishers, 2000).

Letts, Martine, Robert Mathews, Tim McCormack and Chris Moraitis. 'The Conclusion of the Chemical Weapons Convention: An Australian Perspective', *Arms Control*, 14 (3), (1993), 311–31.

Leydesdorff, Loet, and Henry Etzkowitz. 'Triple Helix of Innovation: Introduction', *Science and Public Policy*, XXV (6), (1998), 358–364.

Leydesdorff, Loet, and Henry Etzkowitz. 'The Triple Helix as a Model for Innovation Studies', *Science and Public Policy*, XXV (3), (1998),195–203.

Leydesdorff, Loet, and Henry Etzkowitz (eds.). *A Triple Helix of University-Industry-Government Relations: The Future Location of Research?* (New York: Science Policy Institute, State University of New York, 1998).

Levy, S. *Modern Explosives* (London: Pitman & Sons, 1920).

Levy-Leboyer, M. Interview, 'Des héritiers aux patrons', *L'Histoire, Puissances et faiblesses de la France industrielle, XIX-XXème siècles* (Paris: Seuil, 1997).
Lewes, V B. 'Modern Munitions of War', *Journal of the Royal Society of Arts*, LXIII (1915), 821–831.
Liddell Hart, B. H. *The Real War* (London: Faber, 1930).
Litvinoff, Barnet. *The Essential Chaim Weizmann* (London: Weidenfeld and Nicolson, 1982).
Mackaman, Douglas and Michael Mays (eds.), *World War I and the Cultures of Modernity* (Jackson, Miss: University Press of Mississippi, 2000).
MacLeod, Roy M.' The 'Arsenal' in the Strand: Australian Chemists and the British Munitions effort', *Annals of Science*, 46 (1), (1989), 45–67
MacLeod, Roy M. 'The Chemists go to War: The Mobilization of Civilian Chemists and the British War Effort, 1914–1918', *Annals of Science*, 50 (1993), 455–481.
MacLeod, Roy M. 'The Industrial Invasion of Britain: Mobilising Australian Munitions Workers, 1916–1919, *Journal of the Australian War Memorial*, No. 27 (1995), 37–46.
MacLeod, Roy M. 'Chemistry for King and Kaiser: Revisiting Chemical Enterprise and the European War,' in Anthony S. Travis *et al.* (eds.), *Determinants in the Evolution of the European Chemical Industry, 1900–1939: New Technologies, Political Frameworks, Markets and Companies* (Dordrecht, NL: Kluwer, 1998), 25–49.
MacLeod, Roy M. 'Secrets among Friends: The Research Information Service and the 'Special Relationship' in Allied Scientific Information and Intelligence, 1916–18', *Minerva*, 37 (4), (1999), 201–233.
MacLeod, Roy M. 'Sight and Sound on the Western Front: Surveyors, Scientists and the "Battlefield Laboratory", 1915–1918', *War and Society*, 18 (1), (2000), 23–46.
MacLeod, Roy M. 'L'Entente Chimique: l'Echec de l'Avenir à la Fin de la Guerre', *Le Sabre et L'Eprouvette: l'Invention d'Une Science de Guerre, 14–18 Aujourd'hui*, 6 (2003), 135–152.
MacLeod, Roy M. 'Towards a New Synthesis: Chemists and Chemical Industry in Europe', *Isis*, 94 (2003), 114–116.
Mai, Gunther. *Das Ende des Kaiserreichs: Politik und Kriegführung im Ersten Weltkrieg* (München: Deutsche Taschenbuch Verlag, 1987).
Marsch, Ulrich. *Zwischen Wissenschaft und Wirtschaft: Industrieforschung in Deutschland und Grossbritannien 1880–1936* (Paderborn/Munich: Ferdinand Schöningh, 2000).
Masini, P. C. 'Il Giovane Molinari', *Volontà: Rivista Anarchica Bimestrale*, 29 (1976), 469–476.
Mauskopf, Seymour H. (ed.). *Chemical Sciences in the Modern World* (Philadelphia: University of Pennsylvania Press, 1993).
Mauskopf, Seymour H. 'Calorimeters and Crushers: The Development of Instruments for Measuring the Behavior of Military Powder,' in Steven A. Walton. (ed.) *Instrumental in War: Science, Research and Instruments Between Knowledge and the World* (Leiden: Brill, 2005) 119–152.
Mauskopf, Seymour H. 'Chemistry in the Arsenal: State Regulation and Scientific Methodology of Gunpowder in Eighteenth-Century England and France, in Brett D. Steele and Tamera Dorland (eds.), *The Heirs of Archimedes: Technology, Science and Warfare through the Age of Enlightenment*, in Jed Z. Buchwald (General Editor), Dibner Institute Studies in the History of Science and Technology (Cambridge, Mass: MIT Press, 2005, 293–330.
Mauskopf, Seymour H. 'The Experimental Study of Munitions: Scientific and Military Traditions', in *Actes des Troisième Journées Scientifique Paul Vieille: Instrumentation, Expérimentation et Expertise des Matérieux Énergétiques (Poudres, Explosives et Pyrotechnie) du XVIe Siècle à Nos Jours* (Paris: Centre de Recherche en Histoire des Sciences et des Techniques, Association des Amis du Patrimoine Poudrier et Pyrotechnique, 2002), 11–28
Mazzetti, Massimo. *L'Industria Italiana nella Grande Guerra* (Rome: Stato Maggiore dell'Esercito, 1979).
Meinzer, Lothar. 'Productive Collateral' or 'Economic Sense': The BASF under French Occupation, 1919–1923', in Anthony S. Travis *et al.* (eds.), *Determinants in the Evolution of the European Chemical Industry, 1900–1939* (Dordrecht, NL/Boston/London: Kluwer, 1998), 51–63.
Mente, Theodor. 'Entwicklung des Sprengstoffwesens in Deutschland in und nach dem letzten Kriege', *Zeitschrift für das gesamte Schiess- und Sprengstoffwesen*, 24 (1929), 373–376, 423–427.

Metz, L. 'Fritz Lenze', *Zeitschrift für das gesamte Schiess- und Sprengstoffwesen*, 27 (1932), 109–114.
Ministry of Munitions. *Report of the British Mission to Enemy Chemical Factories, February 1919*. (London: HMSO, 1921).
Ministry of Munitions. *History of the Ministry of Munitions*, 12 vols. (London: HMSO, 1922).
McDowell, Charles H. 'The Work of the Chemical Section of the War Industries Board', *Journal of Industrial and Engineering Chemistry*, 10 (1918), 780–783.
McNeill, William H. *The Pursuit of Power* (Oxford: Blackwell, 1982).
Melograni, Piero. *Storia Politica della Grande Guerra 1915–1918* (Milan: Mondadori, 1998).
Miles, W. *Military Operations in France and Belgium, 1916* (London: HMSO, 1938).
Misa, Thomas J. *A Nation of Steel: The Making of Modern America, 1865–1925* (Baltimore: Johns Hopkins University Press, 1995).
Mohn, Hans. 'Ein Beitrag zur Geschichte der deutschen Pulverfabriken' (Unpublished PhD dissertation, Würzburg, 1925).
Molinari, Ettore. 'Lo Sviluppo di Alcune Grandi Industrie Chimiche in Rapporto alla Guerra', *Annali di Chimica Applicata*, IV (1917), 13–41.
Molinari, Ettore. *Treatise on General and Industrial Inorganic Chemistry*, translated by T. H. Pope (London: Churchill, 2$^{nd}$ ed., 1920).
Molinari, Ettore and Ferdinando Quartieri. *Notizie Sugli Esplodenti in Italia* (Milan: Ulrico Hoepli, 1913).
Montù, Carlo. *Storia dell' Artiglieria Italiana*, 16 vols. (Rome: Comitato per la Storia dell'Artiglieria Italiana, 1935–1955).
Morris, Peter J. T. 'The Development of Acetylene Chemistry and Synthetic Rubber by I.G. Farbenindustrie Aktiengesellschaft, 1926–1945' (Unpublished PhD dissertation, Oxford, 1982).
Morris, Peter J. T. 'Ambros, Reppe, and the Emergence of Heavy Organic Chemicals in Germany, 1925–1945', in Anthony S. Travis et al. (eds.), *Determinants in the Evolution of the European Chemical Industry, 1900–1939: New Technologies, Political Frameworks, Markets and Companies* (Dordrecht, NL: Kluwer, 1998), 89–122.
Moureu, Charles. *La Chimie et la Guerre: Science et Avenir* (Paris: Masson, 1920).
Mußmann, Olaf. 'Komplexe Geschichte: Systemtheorie, Selbstorganisation und Regionalgeschichte. Von der Papiermühle zur Pulverfabrik – ein historischer Längsschnitt der Gemeinde Bomlitz' (Unpublished PhD dissertation, University of Hannover, 1994).
Nasini, Raffaello. 'La chimica italiana nel momento attuale', *Annali di Chimica Applicata*, 5 (1916), 125–142.
Nelson, Richard R. *National Innovation Systems: A Comparative Analysis* (New York: Oxford University Press, 1993).
Nollet, Charles M. E. *Une Expérience de désarmement: cinq ans de contrôle militaire en Allemagne* (Paris: Gallimard, 1932).
O'Connell, Robert L. *Ride of the Second Horseman: The Birth and Death of War* (New York: Oxford University Press, 1995).
Offer, Avner. *The First World War: An Agrarian Interpretation* (Oxford: Clarendon Press, 1989).
Onida, Fabrizio and Gianfranco Viesti (eds.). *L'Industria della difesa: l'Italia nel quadro internazionale* (Milan: Angeli, 1996).
Pancaldi, Giuliano. 'Vito Volterra: Cosmopolitan Ideals and Nationality in the Italian Scientific Community between the Belle Époque and the First World War', *Minerva*, 31 (1), (1993), 21–37.
Paoloni, Giovanni (ed.). *Vito Volterra e il suo tempo (1860–1940)* (Rome: Accademia dei Lincei, 1990).
Partington, J. R. *The Nitrogen Industry* (New York: Van Nostrand, 1923).
Pattison, Michael. 'Scientists, Inventors and the Military in Britain, 1915–19: The Munitions Inventions Department', *Social Studies of Science*, 13 (4), (1983), 521–67.
Patton, Craig. *Flammable Material: German Chemical Workers in War, Revolution, and Inflation, 1914–24* (Berlin: Haude and Spener, 1998).
Pique, René. *La Poudre Noire et le Service des Poudres* (Paris: Société des Publications Colloïdales, 1927).
Plumpe, Gottfried. *Die I.G. Farbenindustrie AG: Wirtschaft, Technik und Politik, 1904–1945* (Berlin: Duncker and Humblot, 1990).

Prentiss, Augustin. *Chemicals in Warfare: A Treatise on Chemical Warfare* (New York: McGraw-Hill, 1937).
Price, Richard M. *The Chemical Weapons Taboo* (Ithaca: Cornell University Press, 1997).
Prior, Robin, and Trevor Wilson. *The First World War* (London: Cassell, 2001).
Prochasson, Christophe and Anne Rasmussen. *Au Nom de la Patrie – Les Intellectuels et la Première Guerre Mondiale (1910–1919)*, (Paris: Edition de la Découverte, 1996).
Pursell, Carroll. *The Military-Industrial Complex* (New York: Holt and Reinhart, 1972).
Rasch, Manfred. 'Wissenschaft und Militär: Die Kaiser-Wilhelm-Stiftung für kriegstechnische Wissenschaft', *Militärgeschichtliche Mitteilungen*, 1 (1991), 73–120.
Reader, John. *Explosives* (Harmondsworth: Penguin, 1942).
Reader, W. J. *Imperial Chemical Industries, A History*, 2 vols. (London: Oxford University Press, 1970).
Reichsarchiv. *Der Weltkrieg, 1914 bis 1918 vol. 1: Kriegsrüstung und Kriegswirtschaft* (Berlin: E.S. Mittler & Sohn, 1930).
Reid, Robert W. *Tongues of Conscience: War and the Scientist's Dilemma* (London: Constable, 1969).
Reinhardt, Carsten and Anthony S. Travis. *Heinrich Caro and the Creation of Modern Chemical Industry* (Boston: Kluwer Academic Publishers, 2000).
Reutter, Mark. *Sparrows Point: Making Steel* (New York: Summit Books, 1988).
Rhees, David J. 'The Chemists' War', *Bulletin for the History of Chemistry*, 13–14 (1992–93), 40–46.
Ricci, Aldo G. and Francesca Romana Scardaccione (eds.). *Archivio Centrale dello Stato, Ministero per le armi e munizioni. Decreti di ausiliarità* (Rome: Ministero per i Beni Culturali e Ambientali, 1991).
Rintoul, William. 'Frederic Lewis Nathan, 1861–1933', *Journal of the Chemical Society*, (1), (1934), 564–5.
Robertson, Robert. 'Some War Developments of Explosives', *Nature*, 107 (23 June 1921), 524.
Robinson, C. S. 'Kenneth Bingham Quinan, C.H., M.I.Chem.E. (1878–1948)', *The Chemical Engineer*, 206, (1966), 290–297.
Roerkohl, Anne. 'Die Lebensmittelversorgung während des Ersten Weltkrieges im Spannungsfeld kommunaler und staatlicher Maßnahmen', in Hans Jürgen Teuteberg (ed.), *Durchbruch zum modernen Massenkonsum, Lebensmittelmärkte und Lebensqualität im Städtewachstum des Industriezeitalters* (Münster: Coppenrath, 1987), 47–60.
Rohlack, Momme. *Kriegsgesellschaften (1914–1918). Arten, Rechtsformen und Funktionen in der Kriegswirtschaft des Ersten Weltkrieges* (Frankfurt/M.: Lang, 2001)
Roland, Alex. 'Technology and War: The Historiographical Revolution of the 1980s', *Technology and Culture*, 34 (1), (1993), 131–132.
Rostagno, Carlo. 'Lo Sforzo Industriale dell'Italia Nella Recente Guerra', in *Rivista d'Artiglieria e Genio*, LXV, (1926), 2074–2097.
Roth, Regina. *Staat und Wirtschaft im Ersten Weltkrieg: Kriegsgesellschaften als kriegswirtschaftliche Steuerungsinstrumente* (Berlin: Duncker & Humblot, 1997).
Rousseau, G. *Etienne Clémentel (1864–1936), Entre Idéalisme et Réalisme: Une Vie Politique (Essai Biographique)*, (Clermont Ferrand: Archives départementales du Puy-de-Dôme, 2000).
Salewski, Michael. *Entwaffnung und Militärkontrolle in Deutschland, 1919–1927* (Munich: Oldenbourg, 1966).
Sanderson, Michael. *The Universities and British Industry, 1850–1970* (London: Routledge and Kegan Paul, 1972).
Saunders, Nicholas. 'Excavating Memories: Archaeology and the Great War, 1914–2001', *Antiquity*, 76 (292), (2002), 101–8.
Saunders, Nicholas. 'The Ironic 'Culture of Shells' in the Great War and Beyond', in John Schofield, William G. Johnson, and Colleen Beck (eds.), *Matériel Culture: The Archaeology of 20th Century Conflict* (London: Routledge, 2002), 22–40.
Saunders, N.J. (ed.). *Materialities of Conflict: Anthropology and the Great War, 1914–2002* (London: Routledge, 2004).
Scardaccione, Francesca Romana (ed.). *Archivio Centrale dello Stato. Ministero per le Armi e Munizioni. Contratti* (Rome: Ministero per i Beni Culturali e Ambientali, 1995).

Schaap, D. *Geen Oorlog, Geen Munitie. De Geschiedenis van 300 Jaar Militaire Produktie* (Haarlem: Fibula-Van Dishoeck, 1979).
Schaefer, Gotthold. 'Die Bewirtschaftung von Pulver und Sprengstoff im Kriege', in Friedrich-Wilhelms-Universität zu Berlin (ed.), *Jahrbuch der Dissertationen der Philosophischen Fakultät der Friedrich-Wilhelms-Universität zu Berlin*, Dekanatsajahr 1921–22, (Berlin: Selbstverlag der Friedrich-Wilhelms-Universität zu Berlin, 1922).
Scharroo, P.W. 'Opstellen over springstoffen,' *Technisch Tijdschrift*, 1 (1914), 75–146.
Scharroo, P.W. *Springstoffen: Eigenschappen, Vervaardiging, Werking en Toepassingen, voor Militaire Vernielingen, in den Mijnbouw en in den Landbouw* (Utrecht: Bruna, 1914).
Schoen Erich. *Geschichte des Deutschen Feuerwerkswesen der Armee und Marine mit Einschluss des Zeugwesens* (Berlin: Paul, 1936);
Schroeder-Gudehus, Brigitte. 'Characteristics des Relations Scientifiques Internationales, 1870–1914', *Cahiers d'Histoire Mondiale*, 10 (1966), 161–177.
Schröter, Alfred. *Krieg-Staat-Monopol, 1914–1918* (Berlin: Akademie-Verlag,1965).
Schröter, Harm. 'Cartels as a form of Concentration in Industry: The Example of the International Dyestuffs Cartel from 1927 to 1939,' *German Yearbook on Business History 1988* (Berlin: Springer, 1990), 114–144.
Schwarte, General Max (ed.). *Die Technk im Weltkriege: Unter Mitterwerkung von 45 technischen fachwissenschaftlichen Mitarbeiten* (Berlin: E.S. Mittler, 1920).
Schweppe, J.H. (ed.). *Research aan het IJ. LBPMA 1914 – KSLA 1989: De Geschiedenis van het 'Lab Amsterdam'* (Amsterdam: Shell Research, 1989).
Seel, Wolfgang. 'Preußisch-deutsche Pulvergeschichte', Folge 2: Rauchloses Pulver, Teil 8, *Deutsches Waffen Journal*, 9 (19), (1984), 294–301.
Segreto, Luciano. *Marte e Mercurio: Industria Bellica e Sviluppo Economico in Italia, 1861–1940* (Milan: Angeli, 1997).
Shorter, J. 'Humphrey Owen Jones, FRS (1878–1912): Chemist and Mountaineer', *Notes and Records of the Royal Society*, 33 (1979), 261–277.
Shuster, Richard. *German Disarmament after World War I: The Diplomacy of International Arms Inspection, 1920–1931* (London: Routledge, 2006).
Shreve, R. Norris. *The Chemical Process Industries* (New York: McGraw Hill, 1956).
Smith, John Kenly, Jr. 'The Scientific Tradition in American Industrial Research', *Technology and Culture*, 31 (1), (1990), 123–131.
Smith, P. G. A. *The Shell that Hit Germany the Hardest* (London: Shell Marketing, 1919).
Snelders, H.A.M. 'De Scheikundige Laboratoria van de Delftse Artillerie-Inrichtingen', in H.L. Houtzager et al. (eds.), *Kruit en Krijg: Delft als Bakermat van het Prins Maurits Laboratorium TNO* (Amsterdam: Rodopi, 1988), 207–217.
Spica, Pietro. 'Il Laboratorio di Chimica Analitica dell' U.M.C.G. (Ufficio Materiale Chimico di Guerra)', *Atti del Reale Istituto Veneto di Scienze, Lettere ed Arti*, (LXXX), (1920–21), 917–1005.
Spitz, P.H. *Petrochemicals: The Rise of an Industry* (New York: John Wiley, 1988).
Steen, Kathryn. 'Wartime Catalyst and Postwar Reaction: The Making of the US Synthetic Organic Chemicals Industry, 1910–1930' (Unpublished PhD dissertation, University of Delaware, 1995).
Steen, Kathryn. 'Patents, Patriotism, and "Skilled in the Art": USA v. The Chemical Foundation, Inc., 1923–1926', *Isis*, 92 (1), (2001), 91–124.
Steger, A. 'Springstoffen en Munitie in den Modernen Oorlog te Land en ter Zee', *24e Jaarverslag 1914–1915 van het Technologisch Gezelschap te Delft*, (1915) 53–100.
Stevenson, David. *Armaments and the Coming of War: Europe, 1904–1914* (Oxford: Clarendon Press, 1996).
Stokes, R. G. *Divide and Prosper: I.G. Farben under Allied Authority, 1945–51* (Berkeley: University of California Press, 1988).
Stoltzenberg, Dietrich. *Fritz Haber: Chemiker, Nobelpreisträger, Deutscher, Jude* (Weinheim/New York: VCH, 1994).
Strachan, Hew. 'On Total War and Modern War', *International History Review*, XX (2000), 341–370.
Strachan, Hew. *The First World War: Vol. I, To Arms* (Oxford: Oxford University Press, 2001).

Stranges, Anthony. 'Friedrich Bergius and the Rise of the German Synthetic Fuel Industry,' *Isis*, 75 (279), (1984), 643–667.
Suzzoni Patrick. *De Sully à l'ECN Paris* (Paris: DCN, 1997).
Szöllösi-Janze, Margit. *Fritz Haber: 1868–1934: Eine Biographie* (Munich: C.H.Beck 1998)
Szöllösi-Janze, Margit. 'Losing the War, but Gaining Ground: The German Chemical Industry during World War I', in John E. Lesch (ed.), *The German Chemical Industry in the Twentieth Century* (Dordrecht, NL: Kluwer, 2000), 91–121.
Szöllösi-Janze, Margit. *Grossforschung in Deutschland* (Frankfurt/Main: Campus Verlag, 1990)
Tanner, Fred (ed.). *From Versailles to Baghdad: Post-War Armament Control of Defeated States* (New York: United Nations, 1992).
Teltschik, Walter. *Geschichte der Deutschen Großchemie: Entwicklung und Einfluß in Staat und Gesellschaft* (Weinheim, New York: VCH, 1992).
Thépot, A. 'Frédéric Kuhlmann: Industriel et Notable du Nord, 1803–1881', *Revue du Nord*, LXVIII (265), (avril-juin 1985).
Theissen, Andrea and Arnold Wirtgen (eds.), *Miltärstadt Spandau: Zentrum der Preussischen Waffenproduktion 1722 bis 1918* (Berlin: Brandenbürgisches Verlagshaus, 1998).
Tissier, L. 'Les Poudres et Explosifs', in Jean Gerard (ed.), *1914–1924: Dix Ans d'Efforts Industrielles et Coloniaux* (Paris: *Chimie et Industrie*, 1926), 1320–1344.
Tomassini, Luigi. *Lavoro e Guerra: La Mobilitazione Industriale Italiana, 1915–1918* (Naples: Edizioni Scientifiche Italiane, 1997).
Tomassini, Luigi. 'Le Origini', in Raffaella Simili and Giovanni Paoloni (eds.), *Per una Storia del Consiglio Nazionale delle Ricerche* (Rome and Bari: Laterza, 2 vols, 2001).
Travers, M. W. *A Life of Sir William Ramsay* (London: Arnold, 1956).
Travers, Tim. *The Killing Ground: The British Army, the Western Front, and the Emergence of Modern Warfare, 1900–1918* (London: Unwin Hyman, 1987);
Travis, Anthony S. *The High Pressure Chemists* (London: Brent Schools and Industry Project, 1984).
Travis, Anthony S. *The Rainbow Makers: The Origins of the Synthetic Dyestuffs Industry in Western Europe* (Bethlehem, Penn.: Lehigh University Press: Associated University Presses, 1993).
Travis, Anthony S. 'The Haber-Bosch Process: Exemplar of 20th Century Chemical Industry', *Chemistry & Industry*, 15 (1993), 581–5.
Travis, Anthony S. 'Modernizing Industrial Organic Chemistry: Great Britain between Two World Wars, in Anthony S. Travis, Harm G. Schröter, Ernst Homburg and Peter J.T. Morris (eds.), *Determinants in the Evolution of the European Chemical Industry, 1900–1939: New Technologies, Political Frameworks, Markets and Companies* (Dordrecht, NL: Kluwer, 1998), 171–198.
Travis, Anthony S., Harm G. Schröter, Ernst Homburg, and Peter J.T. Morris (eds.), *Determinants in the Evolution of the European Chemical Industry, 1900–1939: New Technologies, Political Frameworks, Markets and Companies* (Dordrecht, NL: Kluwer, 1998).
Trebilcock, C. 'A "Special Relationship": Government and Rearmament, and the Cordite Firms', *Economic History Review*, 2nd series, 19, (1966) 364–79.
Trumpener, Ulrich. 'The Road to Ypres: The Beginnings of Gas Warfare in World War I, *Journal of Modern History*, 47 (1975), 460–480.
van Creveld, Martin. *Supplying War: Logistics from Wallenstein to Patton* (Cambridge: Cambridge University Press, 1977).
van Creveld, Martin. *Technology and War, From 2000 BC to the Present* (New York: Free Press, 1989).
van Duin, C.F. and K. Brackmann, 'Het 'Waarom' der Moderne Brisante Nitrospringstoffen,' *Chemisch Weekblad*, 16 (1919), 501–509.
Van Gelder, Arthur Pine and Hugo Schlatter. *History of the Explosives Industry in America* (New York: Columbia University Press, 1927).
van Romburgh, P. 'Chemie en Onafhankelijkheid,', *Verslag van het Verhandelde in de Algemeene Vergadering van het Provinciaal Utrechtsch Genootschap van Kunsten en Wetenschappen* (Utrecht: Kemink, 1915), 9–37.
van Tuyll van Serooskerken, H.P. *The Netherlands and World War I: Espionage, Diplomacy and Survival* (Leiden: Brill, 2001).

Vierhaus, Rudolf and Bernhard vom Brocke (eds.). *Forschung im Spannungsfeld von Politik und Gesellschaft: Aus Anlass des 75 jährigen Bestehens der Kaiser–Wilhelm/Max–Planck Gesellschaft, 1911–1986* (Stuttgart: Deutsche Verlags-Anstalt, 1990).

Vinet, E. 'La Guerre des Gas et les Travaux des Services Chimiques Français', *Chimie et Industrie*, 2 (11–12), (1919), 1377–1415.

vom Brocke, Bernhard. 'Wissenchaft und Militarismus', in William M. Calder, III, *et al.*, *Wilamowitz nach 50 Jahren* (Darmstadt: Wissenschaftliche Buchgesellschaft, 1985), 649–719.

Wachtel, Curt. *Chemical Warfare* (Brooklyn: Chemical Publishing Co., 1941).

Wallace, Stuart. *War and the Image of Germany: British Academics, 1914–1918* (Edinburgh: John Donald, 1988).

Watts, John. *The First Fifty Years of Brunner Mond and Company* (Winnington: Brunner Mond and Company, 1923).

Weems, F. Carrington. *America and Munitions: The Work of Messrs. J.P. Morgan & Co. in the World War* (New York: privately printed, 1923).

Wegs, James R. *Die österreichische Kriegswirtschaft, 1914–1918* (Vienna: Schendl, 1979).

West, Jenny. *Gunpowder, Government, and War in the Mid-Eighteenth Century* (London: Royal Historical Society; Woodbridge, Suffolk and Rochester, NY: Boydell Press, 1991).

Wetzel, Walter. *Naturwissenschaften und chemische Industrie in Deutschland: Voraussetzungen und Mechanismen ihres Aufstiegs im 19. Jahrhundert* (Stuttgart: Franz Steiner Verlag, 1991).

Whittemore, G. F. 'World War I, Poison Gas Research and the Ideals of American Chemists', *Social Studies of Science*, 5 (2), (1975), 135–63.

Williams, William B. *Munitions Manufacture in the Philadelphia Ordnance District* (Philadelphia: A. Pomerantz and Company, 1921).

Williamson, Samuel. *The Politics of Grand Strategy: Britain and France Prepare for War, 1904–1914* (New Jersey: Ashfield Press, 1990).

Wilson, Trevor. *The Myriad Faces of War* (Cambridge: Polity Press, 1986).

Winter, J. M. *The Experience of World War I* (New York: Cambridge University Press, 1989).

Winter, Jay, Geoffrey Parker, and Mary R. Habeck, *The Great War in the Twentieth Century* (New Haven: Yale University Press, 2000).

Wittke, Gunter Michael. 'Vorgeschichte und Geschichte des Werkes für Gleisbaumechanik Brandenburg-Kirchmöser in der Zeit von 1914 bis 1932. Ein Beitrag zur Betriebsgeschichte für das Werk für Gleisbaumechanik Brandenburg-Kirchmöser' (Unpublished dissertation, Humboldt-University Berlin, 1989).

Worden, Edward C. *Nitrocellulose Industry* (New York: Van Nostrand, 1911).

Wrigley, Chris. 'The Ministry of Munitions: An Innovatory Department', in K. Burk (ed.), *War and the State: The Transformation of British Government, 1914–18* (London: Allen & Unwin, 1982).

Yerkes, R[obert] M. *The New World of Science: Its Development during the War* (New York: The Century Co., 1920).

Zabecki, David T. *Steel Wind: Colonel Georg Bruchmüller and the Birth of Modern Artillery* (Westport: Praeger, 1994).

Zagorsky, S. O. *State Control of Industry in Russia during the War, 1898–1925* (New Haven: Yale University Press:, 1928; reprinted Moscow, 1954).

# NOTES ON CONTRIBUTORS

**Patrice Bret** is the scientific head of the Department of History at the Centre des Hautes Études de l'Armement (CHEAr) of the French Ministry of Defence, and a Directeur de Recherche of the Centre National de Recherche Scientifique (CNRS), at the Centre Alexandre Koyré—Centre de Recherche d'Histoire des Sciences et des Techniques (CAK-CRHST) in the Cité des Sciences et de l'Industrie (La Villette), Paris. He has published several books in the history of chemistry, the history of military research, and the history of scientific institutions in France.

**Nathan M. Brooks** is an Associate Professor of History at New Mexico State University, Las Cruces, New Mexico. His research focuses on the history of chemistry in Russia and the Soviet Union. He spent the 2003-2004 academic year as the Edelstein International Fellow in the History of Chemical Sciences and Technology at the Chemical Heritage Foundation, Philadelphia.

**Sophie Chauveau** is an Assistant Professor at the University Lyon II, junior fellow at the Institut Universitaire de France, and a researcher at the Laboratoire de Recherches Historiques Rhône-Alpes (LARHRA). She has published *L'invention pharmaceutique. La pharmacie française entre l'État et la société au XXe siècle* (Paris, 1999), and several articles about the French pharmaceutical industry during the twentieth century. She is also co-author of a book dealing with the history of industrial property (*Des brevets et des marques, une histoire de la propriété industrielle* (Paris, 2001). Her most recent works deal with the blood organization in France since the end of WWII.

**Wayne Cocroft** is a senior archaeological investigator with English Heritage. He is the author of a national survey of the explosives industry in England, entitled *Dangerous Energy: The Archaeology of Gunpowder and Military Explosives Manufacture* (English Heritage, 2000). Recently, he has been recording the built legacy of the Cold War, and is the co-author of the *Cold War: Building for Nuclear Confrontation, 1946–1989*.

**Ernst Homburg** is a Professor of History of Science and Technology at the University of Maastricht. His he is currently writing on the history of industrial research, and on science policy. He has recently published *Groeien door kunstmest. DSM Agro 1929–2004* (Verloren: Hilversum, 2004).

**Jeffrey Allan Johnson** is a Professor of History at Villanova University. His recent work includes studies of science policymaking in Germany to 1939, of German chemical societies to 1914, and of German industrial research on chemical warfare during the First World War; and he is co-author with Werner Abelshauser and others of *German Industry and Global Enterprise. BASF: The History of a Company* (Cambridge University Press, 2004). With Roy MacLeod, he is writing a comparative history of chemical munitions in the First World War.

**Sebastian Kinder** is a Senior Lecturer in Economic Geography at the Humboldt University in Berlin. He holds a D.Phil. degree in geography from the University of Oxford. His research interests include the development of German powder plants during the First World War.

**Erik Langlinay** graduated from the Institut d'Etudes Politiques de Paris and is agrégé d'histoire. He is currently doing a doctorate on 'The French Chemical Industry, 1914–1928', at the Ecole des Hautes

Etudes en Sciences Sociales in Paris. He is also the 'attaché scientifique' of the Institut Pierre Mendès France, which hold the papers of Mendès France.

**Roy MacLeod** is Professor Emeritus of History at the University of Sydney, and a Leverhulme Visiting Professor at the University of Oxford. He has written extensively in the social history of science, medicine and technology; on military history; and on the applications of science and technology to modern warfare. In 1996, he held the Edelstein Senior International Fellowship in the History of Chemistry, and in 1998, was awarded a Humboldt Prize by the Government of Germany. He has recently held visiting appointments at Bologna, Heidelberg, Stockholm and Canberra. With Jeffrey Johnson, he is writing a comparative study of the history of chemical munitions and industrial innovation during the First World War.

**Seymour Mauskopf** is a Professor of History at Duke University in Durham, North Carolina. He was the first Edelstein International Fellow in the History of Chemical Sciences and Technology, 1988-1989, and in 2000, was the Price Fellow at the Chemical Heritage Foundation, Philadelphia. In 1998, he received the Dexter Award from the American Chemical Society for Outstanding Contributions to the History of Chemistry. His research interests include the history of chemistry and allied sciences, the history of chemical technology, focusing on munitions and explosives; and the history of parapsychology and marginal science. He has written many articles and edited two books, including *The Reception of Unconventional Science* (Westview Press, 1979) and *Chemical Sciences in the Modern World* (University of Pennsylvania Press, 1993).

**Giuliano Pancaldi** is Professor of the History of Science, and head of the International Center for the History of Universities and Science at the University of Bologna. He holds a Laurea degree in Philosophy of Science from the University of Bologna, and a D. Phil. in Modern History from the University of Oxford. His books include *Darwin in Italy. Science across Cultural Frontiers* (Indiana University Press, 1991), and *Volta: Science and Culture in the Age of Enlightenment* (Princeton University Press, 2003).

**Manfred Rasch** is head of the ThyssenKrupp Konzernarchiv and also professor of the history of science and technology as well as industrial history at the Ruhr University of Bochum, where he received his doctoral degree with a study of the Kaiser Wilhelm Institute for Coal Research. His recent books include *Technikgeschichte im Ruhrgebiet – Technikgeschichte für das Ruhrgebiet* (edited with Dietmar Bleidick) (Klartext, 2004), *August Thyssen und Hugo Stinnes: Ein Briefwechsel 1898-1922* (prepared by Vera Schmidt, edited with Gerald D. Feldman) (C. H. Beck, 2003), and *Granaten, Geschütze und Gefangene: Zur Rüstungsfertigung der Henrichshütte in Hattingen während des Ersten und Zweiten Weltkriegs* (Klartext, 2003).

**John Kenly Smith, Jr.** is Associate Professor of History at Lehigh University in Bethlehem, Pennsylvania. He is co-author with David A. Hounshell of *Science and Corporate Strategy: DuPont R&D 1902–1980* (Cambridge University Press, 1988) and numerous articles on the history of chemical technology and industrial research. He is currently working on the history of chemical catalysis.

**Kathryn Steen**, an historian of technology, is an associate professor of history at Drexel University in Philadelphia. Her interests centre on the relationship between government and industry, and on trans-national exchanges and comparisons. She is author of *Wartime Catalyst: The Making of the U.S. Synthetic Organic Chemicals Industry, 1910–1930* (University of North Carolina Press, forthcoming).

# INDEX

Abel, Sir Frederick Augustus, 249, 254
Aberdeen Proving Ground, 114
Academia, wartime (Italy), 67–70
Académie des Sciences, Paris, 146, 205, 217, 218
Acetone, 33
Aerial bombs, 36, 63, 207
Aetna Explosives Company, 105, 106, 109
Agache, E.D., 147, 150, 155, 156
Agfa (Aktiengesellschaft für Anilinfabrikation, Berlin), 4, 8, 12, 13
Air Liquide, 160
Aktien Gesellschaft Factory, Saint-Fons, 24
Amatol, 34, 35, 109
American Smelting and Refining Co, 168, 175, 177
American Synthetic Color Company, 115, 116
American University Experiment Station, 113
Ammonium nitrate, 34–36, 105, 109
Ammunition, 123, 133, 138
Anilin Syndikat, 130
Aniline, 130, 136
Architects:
  Baines, Frank, 42
  Unwin, Raymond, 42
Aromatic intermediates/aromatic substances, 124, 125, 127–130, 133–138
Artillery shells (and shell casings), xii, xiv, 2–5, 6, 8–11, 13, 15, 16, 31, 34–36, 40, 56, 63, 76–80, 84–87, 96, 105, 114, 117, 123, 148–149, 151, 168–170, 174–175, 212, 214, 233
Asiatic Petroleum Company, 126, 127
Asquith, Herbert, 36
Atelier de Chargement de Clermont-Ferrand, 212, 218, 219
Ateliers de Construction de Puteaux, 214, 218
Atlas Powder Company, 105, 110, 119
Aufschläger, Gustav, 2, 6, 252, 255

Baker, Newton, U.S. Secretary of War, 108, 175
Balik Papan, 127–129
Ballistite, 62, 214, 251
Banque de Paris et des Pays-Bas, 150, 155, 157
Barral, Jean-Louis, 207

Barthélémy, E., 150
BASF (Badische Anilin & Sodafabrik, Ludwigshafen), 4, 7–9, 12, 13, 17–19, 24, 50, 137, 224, 226, 227, 234–236, 239, 241, 242, 244, 245
Bastet, Michel, 130, 132–134, 138
Bauer, Max, 11, 14
Bayer (Farbenfabriken vorm. Friedr. Bayer & Co, Leverkusen), 4, 6, 7, 9, 11–14, 16, 224, 226, 227, 232, 234, 236, 239
Béhal, Auguste, 23, 24, 28, 29, 152
Benzene, 104, 130, 135, 137
Benzinwerke Rhenania, 127–129
Berger, Ernest, 208, 218
Berthelot, Marcellin, 205, 206, 217, 249, 250
Berthollet, Claude-Louis, 209
Bertrand, Gabriel, 208, 210, 218
Bethlehem Steel, 168, 169
Billingham, Teeside, 38, 44
Bingham, General Sir Francis, 229, 231, 232, 234
Black powder mills, 123
Blanchisserie et Teinturerie de Thaon, 150, 157
Board of Ordnance (UK), 43
Board of Trade (UK), 224
Bodenstein, Max, 188, 199
Bois-Reymond, Alard du, 183, 184, 190, 197
Bombs for aerial use, *see* aerial bombs
Borneo oil, 38, 126–132, 134, 137–139
Bosch, Carl, 7, 8, 125, 227, 234–236
Bouchet (Le), 204, 208, 212, 213, 216
Bourges, 204, 207, 213
Breton, Jean-Louis, 209
Bromine, 86
Brunner Mond & Company, 35, 36, 44, 110
Butanol, 33

Cannizzaro, Stanislao, 62
Cape Explosive Works, South Africa, 38
Carnegie Steel, 168
Cartoucherie de Vincennes, 212
Cartridges production (Italy), 63, 64
Caustic soda, 35
Cellulose, 3, 6, 13

271

Cellulose technologies, 177
Central Laboratory of the Ministry of Finance (Russia), 84, 88
Chance factory, Oldbury, Birmingham, 35
Chemical Committee of the Main Artillery Administration (GAU) (Russia), 84, 90–95
Chemical expertise, 203, 207, 216
Chemical Industry Association (Verein zur Wahrung der Interessen der Chemischen Industrie Deutschlands), 231
Chemical Section of the Supreme Chief of the Army Sanitary and Evacuation Service (SCASES) (Russia), 76, 85, 89, 92
Chemical warfare/war gases, 103, 111–118, 209
Chemicals and pharmaceuticals, 21, 23, 29, 30, 253
Chemische Fabriek Hembrug, 133
Chemische Fabriek Naarden, 136, 139
Chemisch-Technische Reichsanstalt (CTR), 230–232, 236
Chemistry in wartime, 21, 97, 254
Chemistry, pure and applied (Italy), 67
Chetwynd, Viscount, 36
Chilwell, Nottingham, 36, 37
Chilworth Gunpowder Company, Surrey, 37, 40
Chlorine, 85–87, 111, 114, 116
Chloropicrin, 86, 113, 115, 116
Chugaev, L.A., 87, 91
Ciamician, Giacomo, 62, 68–70
Civilian laboratories, 203, 206, 208, 210, 211, 241, 255
Clayton Aniline Company, 35, 135
Clémentel, Etienne, 22, 145, 152, 154, 159, 162, 165
CNR (Italian National Research Council), 68, 69
Coal tar, 137
Cochin, D., 154
Colijn, Hendrik, 131, 135, 136, 138
Collège de France, 210
Comité Consultatif de Poudres, 205
Comité des Forges, 22, 163
Comité International des Approvisionnements (International Committee for Supply), 25
Commission des Etudes Chimiques de Guerre, 150, 208, 209, 216
Commission des Substances Explosives, 205, 206
Commission for the Preparation of Asphyxiating Gases (CPAG) (Russia), 76, 86, 87, 90
Commission for the Preparation of Explosives (CPE) (Russia), 75, 79–87, 90, 91, 95
Commission Scientifique d'Etude des Poudres de Guerre, 206, 210, 217
Compagnie des Chemins de Fer du Nord, 147

Compagnie des Forges et Aciéries de la Marine-Honécourt, 150, 214
Compagnie des Matières Colorantes, 154
Compagnie des Mines d'Anzin, 147
Compagnie Nationale des Matières Colorantes et des Produits Chimiques (CNMC), 145, 153, 236, 237
Conservatoire des Arts et Métiers, 210
Contracts, 'readiness', 13, 222, 223
Cordite, 31–34
Cotelle, A., 145, 156
Council of National Defense, US (CND), 107, 119
Crozier, William, Brigadier General, 108, 120, 174, 175
Cunard's Shellworks, Liverpool, 40

Dallolio, Alfredo, 63
De Kempenaer, J.U., 132, 135
De Kok, Frits, 125, 126, 131–137, 139
De Leeuw, Willem, 133, 135–137, 139
Department of Scientific and Industrial Research (DSIR), see UK
Dépôt Central de l'Artillerie, 204
Deterding, (Sir) Henry, 125–128, 130, 131, 135–138
Deutscher Verband technisch-wissenschaftlicher Vereine (German Federation of Technical and Scientific Associations) (DVT), 188, 189
Deutsche Werke, 238
Diels, Herman, 187, 191
Diphenylamine, 108, 120, 172, 206, 207
Direction Chimique des Materiéls de Guerre, 150, 210
Direction des Industries Chimiques (Chemicals Office), 28
Dow Chemical Company, 116
Duisberg, Carl, 6, 7, 11, 14, 225, 227, 232, 236
DuPont (E.I. DuPont de Nemours and Company), xv, 104–106, 108–111, 117, 119–121, 158, 167–178, 248, 250, 251, 253, 270
Du Pont, Pierre, 173
Dutch East Indies, 126, 138
Dynamite, 62
Dye industry, 1, 3, 6, 7, 12–15, 130, 134, 136, 137, 154, 207, 239
Dyestuffs/synthetic dyes, 1, 2, 4, 121, 124, 129, 134, 136–138, 144, 146, 147, 152, 154, 157, 159, 172, 177, 216, 222, 224–226, 233, 236, 237, 239, 242, 245

# INDEX

Ebert, Friedrich, 221
Ecole Centrale de Pyrotechnie, 204, 207, 212, 217, 219, 254
Ecole de Physique et de Chimie Industrielle de la Ville de Paris, 23, 29, 152, 208, 210
Ecole Polytechnique, 147, 148, 151, 160, 205, 207, 208, 217, 248
Ecole Supérieure de Pharmacie, 30, 164, 210
Ecoles de Genie, 248
Economic mobilization, 22
Edgewood Arsenal, 114, 116–118
Edison, Thomas A., 108
Einaudi, Luigi, 65
Einstein, Albert, 180, 194
Electrochemicals, 23
English Heritage, 44
Emeryville, California, 126
Ether alcohol, 33
Explosives, 124–126, 128–130, 132, 133, 135, 137–139, 203–208, 210, 217, 222, 223, 225, 228, 230–244, 247–255
Explosives production (Italy), 61–63
Explosives, substitute (*see also*: Picric acid, TNT), 8, 10, 12

Faculté de Médecine, Paris, 30, 210
Faculté des Sciences de Paris, 30, 208, 218
Fascism, 65
Feldzeugmeisterei (Prussian Ordnance Department), 3, 5, 6, 10, 11, 17, 50, 59, 60, 180
Ferguson, Niall, xii
Fireworks laboratory, 53, 56
Fischer, Emil, 4, 7, 11, 13, 17, 180, 185, 187, 189, 190, 194, 197, 199–201
Fleurent, Ernest, 28, 30
Foch, Marshal, 223, 229
Forges de Firminy, 150
Fourcroy, Antoine-François, 205
Freeth, Dr Francis, 35
French, Field Marshall Sir John, 31
French explosives factories, 36
French war effort, 21
Frossard, Joseph, 153, 158, 160, 236

Gay-Lussac, Louis-Joseph, 205
German chemists, 38
German munitions, 47, 48, 56, 138, 211, 212
Gezamelijk Buskruidmakers van Noord-Holland, Utrecht and Zeeland, 123, 132, 134, 137, 139
Glycerine, 3, 13, 33

Gnaschwitz, 2, 13, 48-49, 238
Griesheim-Elektron, 11, 12
Grignard, Victor, 210
Griolet, G., 150, 157
Guncotton, *see also* nitrocellulose, 32, 33, 38, 47, 50, 51, 53–55, 58, 59, 62, 123, 171–173, 176, 217, 231, 243, 249, 253, 254
Gunpowder Neck Reservation, 114, 116

Haber-Bosch process, 4, 8, 16, 38, 48, 50
Haber, Fritz, 4, 17, 160, 180, 181, 183, 187, 193, 195, 196, 199, 225, 230, 239
Haber, L.F., 75
Hague Convention, 54
Hartley, Brig. Sir Harold, 225, 233–235, 238
Heeresarchiv (Army Archive), 189, 199
Heereswaffenamt (Army weapons Agency), 191, 193, 202
Hercules Powder Company, 104–106, 111, 118
High explosives, 103–111
Hindenburg Programme, 2, 14, 15, 19, 47, 50, 54, 55, 58, 181, 222
HM Factory Avonmouth, 43
HM Factory Gretna, 33, 38, 42
HM Filling Factories, 36, 41
HM National Ammonium Perchlorate Factory, 38
HM National Filling Factory No.5, Quedgley, Gloucestershire, 41
HM National Machine Gun Factory, Burton-on-Trent, Staffordshire, 43
HM Office of Works, 36, 42
Höchst (Farbwerke vorm. Meister Lucius & Brüning, Höchst a.M), 4, 7, 10–12, 18, 181, 195, 196, 224, 227, 239, 241, 244
Hopewell, Virginia, 173, 174, 176, 178
Hughes, Thomas, xiii
Hydrogenation, 125, 128

ICI (Imperial Chemical Industries), 44, 237
IEEC, *see* Inspection des Etudes et des Expériences Chimiques, 153
IG Farben, 124, 125, 136
Imperial Trust for the Encouragement of Scientific and Industrial Research, 184
Imperial War Museum, London *see also* National War Museum, 39
Industrial mobilization (Italy), 63, 64
Industrial policy, 21, 27, 29, 96, 155, 163, 260
Infanterie-Konstruktionsbüro (Infantry Construction Bureau), 179
Inspection des Etudes et Expériences Chimiques, 151, 203, 210

# INDEX

Inspection des Etudes et Expériences Techniques de l'Artillerie, 203, 209
Inspection des Forges de Lyon, 214
Inspection des Inventions, Etudes et Expériences Techniques de l'Artillerie, 203
Inspection des Inventions, Etudes et Expériences Techniques de l'Automobile, 213
Inspektion der Marineartillerie (Inspectorate of Naval Artillery, Germany), 179, 180
Inspektion der Technischen Institute der Artillerie (Inspectorate for the Technical Institutes of the Artillery, Prussia), 179
Institut de Chimie Appliquée, 208, 218
Institut Pasteur, 26, 209, 210
Interessengemeinschaft (IG), 4, 13, 222, 223, 230–234, 236, 237
Ipat'ev, V.N., 75, 79–82, 85, 86, 90–95

Jackling, Daniel C., 175
Javillier, Maurice, 208, 209, 218
Jones, Humphrey Owen, 127, 128

Kaiser Wilhelm Gesellschaft zur Förderung der Wissenschaften (Kaiser Wilhelm Society for the Advancement of the Sciences) (KWG), 182, 185, 187, 190, 196–198, 200
Kaiser Wilhelm Institute (KWI), 181, 183–185, 190–198, 201, 202, 244
Kaiser Wilhelm Institute for Physical Chemistry, 4, 230
Kaiser Wilhelm Stiftung für Kriegstechnische Wissenschaft (Kaiser Wilhelm Foundation for Military-Technical Science) (KWKW), 181, 187–193, 195, 196, 199–201
Kármán, Theodor von, 192, 201
Kerosene, 127, 128
Kitchener, Lord, 36
Knoops, Willem, 126–130, 132, 135, 138, 139
Koehler, Albert, 207
Koistinen, Paul A.C., 108
Köln-Rottweil (Vereinigte Köln-Rottweiler Pulverfabriken), 3, 10, 231, 232, 235, 239
Kommission zur Beschaffung von Kokereiprodukten (Commission for the Procurement of Coking Products), 180
Königlich Preußische Akademie der Wissenschaften zu Berlin (Royal Prussian Academy of Sciences in Berlin), 196–198
Koppel, Leopold, 185–188, 191, 197–199, 201
Kriegsamt (War Office, Prussian), 14
Kriegsausschuß für Ersatzfutter (War Committee for Substitute Fodder), 180

Kriegsausschuß für pflanzliche und tierische Öle und Fette (War Committee for Vegetable and Animal Oils and Fats), 180
Kriegschemikalien AG (KCA), 6, 7, 11, 13
Kriegsrohstoffabteilung (War Raw Materials Department) (KRA), 6, 9, 11, 14, 180, 200
Kuhlmann, Charles F., 147

Laboratoire Central de la Marine (Navy Central Laboratory), 205
Laboratoire Central des Poudres, 205, 206, 218
Laboratoire de Chimie du Service Technique de l'Artillerie, 203, 204
Laboratoire d'Expertise de la Réserve des Médicaments, 214
Laboratoire de la Commission des Poudres de Versailles, 204, 208, 218
Laboratoire de Protection, 219
Laboratoire de Synthése, 219
Laboratoire de Toxicologie de la Préfecture de Police, 210, 219
Laboratoire National d'Essais, 210
Laboratorio Pirotecnico, Bologna, 64
Lavoisier, Antoine-Laurent, 205, 248
Le Chatalier, Henri, 206, 216, 217
Lefebure, Victor (*The Riddle of the Rhine*), 233, 234, 236, 238
Lenze, Fritz, 230, 236
Lepick, Olivier, 203
Levinstein, Herbert, 225, 227, 234
Lewisite, 114, 117
Liverpool, Cunard's Shellworks, 40
Lloyd George, David, 36
Loucheur, Louis, 22, 161, 164
Loudon, Hugo, 128, 130, 131
Ludendorff, Erich, 14, 15
Lyddite *see also* picric acid, 31, 35
Lyonnaise des Eaux, 161

Macnab, William, 39, 225
Main Artillery Administration (GAU) (Russia), 77, 79, 80, 84, 90, 94
Makovetskii, A.E., 82, 85
Manpower, 83, 180, 208, 209, 225, 229, 255
Manufacture Lyonnaise des Matières Colorantes, 24
Marqueyrol, Marius-Daniel, 207, 218
Masse, René, 152, 153, 155, 159
Mechanical industries (Italy), 63
*Mémorial des Poudres*, 207
Militärversuchsamt (Military Testing Office), Prussian, 8, 180, 230

# INDEX

Military Inter-Allied Control Commission (MICC), 222, 228–235, 237–239
Military Laboratories, 203, 210, 216
Mineralölwerke Rhenania, 130
Mines de Lens, 150, 155–157
Mines de Vicoigne et Noeux, 153, 155
Ministry of Munitions (UK) – see UK Ministry Munitions, 225
Mobilization, 1, 2, 5, 6, 9, 12, 14, 15, 21, 22, 24, 26, 27, 45, 47, 49, 58, 62–65, 67, 68, 71, 72, 78, 94, 98, 103, 106–108, 112, 117–119, 145, 148, 159, 163, 167, 175, 181, 203, 208, 209, 216, 219, 229, 241, 255
Mobilization plans, 49, 58
Moellendorff, Wichard von, 180
Molinari, Ettore, 65, 66
Moltke, Helmuth Graf von, 5, 179
Mononitrotoluenes (MNT), 124, 129–133
Mordecai, Alfred, 248, 254
Morgan, J.P. & Company, 106, 108–110
Moulton, Lord (Fletcher), 35, 225, 226
Moureu, Charles, 26, 29, 164, 210, 217, 218
Müller-Breslau, Heinrich, 188, 191, 199
Munitions industry, 47, 56
Muraour, Colonel Henri Charles Antoine, 207, 229, 230, 232, 238
Mussolini, Benito, 69
Mustard gas (yperite), 43, 70, 86, 112, 113, 116, 164, 208, 224, 236

Nährstoffausschuß (Foodstuffs Committee), 180
Nathan, Sir Frederic (Controller of Propellant Supplies), 33
National Academy of Sciences (US), 108
National Aniline and Chemical Company, 116
National Research Council (US), 108, 109, 112–115, 120, 122, 184
National system (Italy), 70, 71
National War Museum, London see also Imperial War Museum, 39
Naval Consulting Board (US), 108
Nederlandsch Kleurstoffen-Fabriek, 136, 137, 139
Nelson, Richard, xiii
Nernst, Walther, 183, 185, 187, 189, 191, 192, 199, 200
Netherlands, 5, 27, 123, 124, 126, 132, 138–140, 143, 144, 214
Neutrality, 123, 131, 132, 134, 137, 138
Nicholas II, Tsar, 75, 85
Nicolson, Harold, 221
Nicolardot, Major, 209, 214–216, 219

Nitrate plant, artificial, 44
Nitrates, nitric acid, 2–4, 6–9, 10–12, 14, 16, 31, 34–36, 38, 48, 62, 83, 99, 107, 123, 129, 130, 134, 176, 177, 222, 223, 235, 236
Nitration, 126, 128–134, 138
Nitrobenzene, 124
Nitrocellulose see also Guncotton, 3, 33, 34, 38, 105, 133, 171, 172, 178, 223, 226, 231, 238, 242, 243, 249, 251
Nitro Chemical Company, West Virginia, 105, 175
Nitro-glycerine, 47, 51, 54, 55, 61, 123, 132
Nitrotoluene see Mononitrotoluenes
Nobel, Alfred, 33, 35, 37, 61, 206, 239, 250
Nobel Dynamite Trust (and its companies), 2, 3, 9, 37, 44, 239, 250, 251, 254
Nobel Prize for Chemistry, 68
Nollet, General Charles, 229, 231
Nylon, 177

Office des Produits Chimiques et Pharmaceutiques (OPCP) (Chemicals and Pharmaceuticals Office), 21, 23, 28, 152
Office National Industriel de l'Azote, 161
Oil, 125–128, 131, 136
Oldbury Electro Chemical Company, 35, 38, 115, 132
Oldenburg, Prince, 85, 88–92
Old Hickory, Tennessee, 176, 178, 253
Oleum, 3, 13, 129, 232
OPCP, see Office des Produits Chimiques et Pharmaceutiques
Oppau, Germany (see also BASF), 44
Ordnance Department, Prussian, see Feldzeugmeisterei
Organisation for the Prohibition of Chemical Weapons (OPCW), 240
Oudekerk a/d Amstel, 123, 124
Ozil, General, 151, 210

Painlevé, Paul, 68
Patart, Georges, 145, 155, 157, 161
Patents, 227, 235
Paternò, Emanuele, 67
Pelouze, Théophile-Jules, 61, 205
Petrochemicals, 124, 125
Petrol, 127–131, 133
Petroleum see Oil
Philippet, 209, 215
Phosgene, 85, 113, 116
Phosphorus, 23

Picric acid (trinitrophenol) *see also* Lyddite, 3, 4, 6, 7, 9, 11, 12, 14, 16, 31, 43, 62, 104, 105, 109–111
Pijzel, Daniel, 126, 130, 139
Planck, Max, 183
Plants, 'readiness', 2, 13, 14, 16, 222, 223
Plumer, Field Marshal Lord Herbert, 224
Poincaré, Henri, 206, 216, 217
Poirrier, A., 145, 156
Pont-de-Claix, 213
Pope, Sir William, 134, 135
Poudre B, 204–207, 217, 218, 250, 254
Poulenc Frères, 227
Powder factories (*see also* propellant factories), 3, 8, 10, 47, 48, 53, 148, 205, 206, 212, 213, 216, 226, 248
Prisoners of war, 54
Propellant factories, state owned, German, 2, 3, 8–10, 13, 15
Prussian War Ministry, 49–51, 54, 58
War Office (Kriegsamt), 14
Pyrite, 81, 82, 92, 93

Queensferry, North Wales, 38, 132
Quinan, Kenneth B., 38, 132, 133

Rathenau, Walther, 6, 9, 180, 194, 230, 235
Raw materials, 48, 50
Rayleigh, Lord, 189
Reich Industry Group (Reichsverband der Industrie), 231
Reisholz, near Düsseldorf, 127–130, 133, 134
Riedler, Alois, 188, 190, 196, 199, 201
Robertson, Sir Robert, 226
Royal Arsenal, Woolwich, 31, 33, 35, 44
Royal Artillery, 33, 248
Royal Dutch/Shell, 123–139, 253
Royal Gunpowder Factory, Waltham Abbey, Essex *see also* Waltham Abbey Royal Gunpowder Mills, 31, 33, 35, 42, 44–46, 248
Royal Naval Cordite Factory, Holton Heath, 33, 34, 43, 44
Royal Navy, 31
Rubens Heinrich, 183, 192
Russian Physical-Chemical Society, 87, 90
War-Chemical Committee, 76, 88, 94

SA des Matières Colorantes et des Produits Chimiques de Saint-Denis, 145, 148, 155, 156, 160, 208
Saint-Gobain, 28, 146, 148–150, 160, 166, 208
Salpetre (*see also* nitrates), 48
Salpeterkommission (Nitrate Commission), 180
Schlenk, Wilhem, 192
Schlieffen Plan, 2, 5, 6, 48, 58
Schmidt-Ott, Friedrich, 181–187, 191, 192, 195–199, 201, 202
Semet-Solvay Company, 105, 106
Service des Poudres, 25, 145, 148, 152, 154, 155, 160, 163–166, 204–208, 210, 212, 216–218, 234, 248
Sevran-Livry Powder Plant, 205, 206, 213
Shell *see* Royal Dutch/Shell
Shell Company, 38
Shell filling, 8, 11, 31, 34, 36, 37, 116, 233
Shell Oil (USA), 126
Shells scandal, 36
Shell Transport & Trading, 124, 126–128, 138
Sibert, William, General, 114
Single base propellants, 33
SIPE (Società Italiana Prodotti Esplodenti), 61, 65, 67
SIPS (Italian Society for the Advancement of Science), 65
Sluyterman van Loo, Willem, 127, 128, 131, 132, 135
Smokeless powder (nitrocellulose), 76, 105, 123, 133, 167, 169–175, 177, 178, 204, 248–250, 252
Società Bombrini Parodi-Delfino, 69
Société Chimique des Usines du Rhône, 150, 153
Société de Chimie Industrielle, 21, 23, 29, 164, 208, 210, 218
Société d'Eclairage Chauffage et Force Motrice, 150, 154, 157
Société des Crédits Industriels et de Dépots du Nord and Credit du Nord, 147
Société des Matières Colorantes (*see also* SA des Matières Colorantes et des Produits Chimiques de Saint-Denis), 208
Société du Traitement du Quinquina, 25, 27, 29, 30
Société Française des Glycérines, 214
Société Métallurgique de Pont-à-Vendin, 155
Société Norvégienne de l'Azote et des Force Motrices, 150, 157
Soda, 23, 28, 35, 147
Sodium nitrate (*see also* nitrates), 38
Somme offensive, 36
Sorbonne (The), 29, 152, 166, 210
Sorgues, 213, 217
Spandau Arsenal, 2, 8, 10, 17, 48, 52, 58, 59, 179, 180

Special Committee for Defense (Russia), 81–83, 87
Standard Oil, 124, 125, 168
Stettinius, Edward R. Sr., 106, 108
Stone, Norman, 75, 79
Strachan, Hew, xi
Sukhomlinov, V.V., 77
Sulphuric acid, 3, 23, 25, 31, 38, 81, 82, 92, 94, 95, 129, 130, 134, 147, 149, 150, 171, 176, 177, 222
Superphosphates, 23, 147, 150
Syndicat National des Matières Colorants *see also* Compagnie Nationale des Matières Colorantes, 145, 153
Synthetic ammonia, 38, 44, 125, 126

Tangiers Crisis, 206
Tear gas, 86, 100, 181, 209
Technische Hoogeschool, Delft, 126, 127, 133
Technology transfer, 110, 111, 117
Thermochemistry, 207, 249, 250
Thomas, Albert, 22, 29, 145, 154, 159, 203, 209
Thomson, J.J., Sir, 191
Thompson-Starrett Company, 175, 176
TNT (trinitrotoluene), 3, 6–12, 18, 34–36, 38, 45, 50, 53, 55, 61, 62, 77, 93, 104–106, 109–111, 119, 121, 124, 131–134, 143, 170, 173, 226, 232, 237
Toluene, 3–5, 7, 8, 18, 35, 38, 77, 79, 83, 84, 93, 104, 107, 119, 128, 129, 131, 136, 137, 165, 170
Trebilcock, Clive, 33
Troisdorf (Rheinisch-Westfälische Sprengstoff AG), 3, 6, 8
Turkish Petroleum Company, 161

Ufficio Invenzioni e Ricerche (Italy), 68, 69
UK Board of Trade, 224
UK Committee on High Explosives, 35
UK Department of Scientific and Industrial Research (DSIR), 39, 184, 197
UK Ministry of Munitions, 33, 36, 37, 39, 42, 43
Department of Explosives Supply, 42
Universities and industrial mobilization (Italy), 62, 67–70
Urbain, Georges, 210
US Army Ordnance Department, 103, 108–111, 115, 117, 118, 168, 174, 225
US Army Chemical Warfare Service, 114, 115, 117
US Bureau of Mines, 110, 112–115, 117
US Congress, 103, 104

US Navy, 107, 108, 118
US Ordnance Department, 103, 108–111, 115, 117, 118
US War Department, 107–109, 114, 118, 119, 122, 174, 241, 249
USA, 124, 128, 132, 135, 137
Usines Chimiques du Rhone, 23, 28, 151, 164, 227

Van Creveld, Martin, xi
Vereinigte Köln-Rottweiler Pulverfabriken, 48
Vereinigte Rheinisch-Westfälische Pulverfabriken, 37, 48
Versailles, Treaty of (and articles thereof), 16, 221, 223, 227, 228, 232–236, 239
Vieille, Paul, 205–207, 217, 248, 250, 254
Volterra, Vito, 68–70

Waffen und Munitionen Beschaffungsamt (WUMBA), 14
Waley Cohen, Robert, 127, 130, 134, 135, 137
Waltham Abbey Royal Gunpowder Mills *see also* Royal Gunpowder Factory, Waltham Abbey, Essex, 31, 33, 35, 42–46, 247
Warburg, Emil, 140, 183
War Chemical Committee (War Ministry) (Russia), 89, 90
War gases, 103, 112–118, 203, 209
War-Industry Committees (Russia), 76, 82–84, 87, 88
War Ministry, Austro-Hungarian, 3, 8, 9, 14
Wäser, Bruno, 183, 184, 188, 197
Watts, Hugh, 229, 230, 232, 234
Weizmann, Chaim, 33
Westfälisch-Anhaltinische Sprengstoff AG (WASAG), 3
Wild von Hohenborn, Adolf, 182, 186
Wilson, Woodrow, US President, 103, 106, 107, 115, 119, 122, 168, 175, 178
Women, 39, 40
Wüst, Friedrich, 188, 192, 199, 200

Xylenes, 135, 137

Ypérite, 70, 151, 208, 217

Zelinskii, N.D., 84, 88–90
Zelinskii-Kummant gas mask, 88–92
Zemgor (All-Russian Union of Zemstvos and Towns), 76, 88, 90
Zsigmondy, Richard, 188, 189

# *Archimedes*

## NEW STUDIES IN THE HISTORY AND PHILOSOPHY OF SCIENCE AND TECHNOLOGY

1. J.Z. Buchwald (ed.): *Scientific Credibility and Technical Standards in 19th and Early 20th Century Germany and Britain.* 1996      ISBN 0-7923-4241-0
2. K. Gavroglu (ed.): *The Sciences in the European Periphery During the Enlightenment.* 1999      ISBN 0-7923-5548-2; Pb 0-7923-6562-1
3. P. Galison and A. Roland (eds.): *Atmospheric Flight in the Twentieth Century,* 2000      ISBN 0-7923-6037-0; Pb 0-7923-6742-1
4. J.M. Steele: *Observations and Predictions of Eclipse Times by Early Astronomers.* 2000      ISBN 0-7923-6298-5
5. D-W. Kim: *Leadership and Creativity.* A History of the Cavendish Laboratory, 1871–1919. 2002      ISBN 1-4020-0475-3
6. M. Feingold: *The New Science and Jesuit Science: Seventeenth Century Perspective.* 2002      ISBN 1-4020-0848-1
7. F.L. Holmes, J. Renn, H-J. Rheinberger: *Reworking the Bench.* 2003      ISBN 1-4020-1039-7
8. J. Chabás, B.R. Goldstein: *The Alfonsine Tables of Toledo.* 2003      ISBN 1-4020-1572-0
9. F.J. Dijksterhuis: *Lenses and Waves.* Christiaan Huygens and the Mathematical Science of Optics in the Seventeenth Century. 2004      ISBN 1-4020-2697-8
10. L. Corry: *David Hilbert and the Axiomatization of Physics (1898–1918).* From Grundlagen der Geometrie to Grundlagen der Physik. 2004      ISBN 1-4020-2777-X
11. J.Z. Buchwald and A. Franklin (eds.): *Wrong for the Right Reasons.* 2005      ISBN 1-4020-3047-9
12. M. Feingold and V. Navarro-Brotons (eds.): *Universities and Science in the Early Modern Period.* 2006      ISBN 1-4020-3974-3
13. R.R. Hamerla: *An American Scientist on the Research Frontier.* Edward Morley, Community, and Radical Ideas in Nineteenth-Century Science. 2006      ISBN 1-4020-4088-1
14. J. Schickore and F. Steinle (eds.): *Revisiting Discovery and Justification.* Historical and Philosophical Perspectives on the context distinction. 2006      ISBN 1-4020-4250-7
15. K. Nickelsen: *Draughtsmen, Botanists and Nature.* The construction of Eighteenth-Century Botanical Illustrations. 2006      ISBN 1-4020-4819-X
16. R. MacLeod and J.A. Johnson (eds.): *Frontline and Factory.* Comparative Perspectives on the Chemical Industry at War, 1914–1924. 2006      ISBN 1-4020-5489-0

springer.com

Printed in the United States
84766LV00002B/96/A